# ADIPOSE TISSUE AND INFLAMMATION

## OXIDATIVE STRESS AND DISEASE



1. Oxidative Stress in Cancer, AIDS, and Neurodegenerative Diseases, *edited by Luc Montagnier, René Olivier, and Catherine Pasquier*
2. Understanding the Process of Aging: The Roles of Mitochondria, Free Radicals, and Antioxidants, *edited by Enrique Cadenas and Lester Packer*
3. Redox Regulation of Cell Signaling and Its Clinical Application, *edited by Lester Packer and Junji Yodoi*
4. Antioxidants in Diabetes Management, *edited by Lester Packer, Peter Rösen, Hans J. Tritschler, George L. King, and Angelo Azzi*
5. Free Radicals in Brain Pathophysiology, *edited by Giuseppe Poli, Enrique Cadenas, and Lester Packer*
6. Nutraceuticals in Health and Disease Prevention, *edited by Klaus Krämer, Peter-Paul Hoppe, and Lester Packer*
7. Environmental Stressors in Health and Disease, *edited by Jürgen Fuchs and Lester Packer*
8. Handbook of Antioxidants: Second Edition, Revised and Expanded, *edited by Enrique Cadenas and Lester Packer*
9. Flavonoids in Health and Disease: Second Edition, Revised and Expanded, *edited by Catherine A. Rice-Evans and Lester Packer*
10. Redox–Genome Interactions in Health and Disease, *edited by Jürgen Fuchs, Maurizio Podda, and Lester Packer*
11. Thiamine: Catalytic Mechanisms in Normal and Disease States, *edited by Frank Jordan and Mulchand S. Patel*
12. Phytochemicals in Health and Disease, *edited by Yongping Bao and Roger Fenwick*
13. Carotenoids in Health and Disease, *edited by Norman I. Krinsky, Susan T. Mayne, and Helmut Sies*
14. Herbal and Traditional Medicine: Molecular Aspects of Health, *edited by Lester Packer, Choon Nam Ong, and Barry Halliwell*

15. Nutrients and Cell Signaling, *edited by Janos Zempleni and Krishnamurti Dakshinamurti*
16. Mitochondria in Health and Disease, *edited by Carolyn D. Berdanier*
17. Nutrigenomics, *edited by Gerald Rimbach, Jürgen Fuchs, and Lester Packer*
18. Oxidative Stress, Inflammation, and Health, *edited by Young-Joon Surh and Lester Packer*
19. Nitric Oxide, Cell Signaling, and Gene Expression, *edited by Santiago Lamas and Enrique Cadenas*
20. Resveratrol in Health and Disease, *edited by Bharat B. Aggarwal and Shishir Shishodia*
21. Oxidative Stress and Age-Related Neurodegeneration, *edited by Yuan Luo and Lester Packer*
22. Molecular Interventions in Lifestyle-Related Diseases, *edited by Midori Hiramatsu, Toshikazu Yoshikawa, and Lester Packer*
23. Oxidative Stress and Inflammatory Mechanisms in Obesity, Diabetes, and the Metabolic Syndrome, *edited by Lester Packer and Helmut Sies*
24. Lipoic Acid: Energy Production, Antioxidant Activity and Health Effects, *edited by Mulchand S. Patel and Lester Packer*
25. Dietary Modulation of Cell Signaling Pathways, *edited by Young-Joon Surh, Zigang Dong, Enrique Cadenas, and Lester Packer*
26. Micronutrients and Brain Health, *edited by Lester Packer, Helmut Sies, Manfred Eggersdorfer, and Enrique Cadenas*
27. Adipose Tissue and Inflammation, *edited by Atif B. Awad and Peter G. Bradford*

# ADIPOSE TISSUE AND INFLAMMATION

Edited by

ATIF B. AWAD
PETER G. BRADFORD

CRC Press
Taylor & Francis Group
Boca Raton London New York

CRC Press is an imprint of the
Taylor & Francis Group, an **informa** business

CRC Press
Taylor & Francis Group
6000 Broken Sound Parkway NW, Suite 300
Boca Raton, FL 33487-2742



ISBN 13: 978-1-4200-9130-4 (hbk)



**Library of Congress Cataloging-in-Publication Data**

Adipose tissue and inflammation / editors, Atif B. Awad, Peter G. Bradford.
p. ; cm. -- (Oxidative stress and disease ; 28)
Includes bibliographical references and index.
ISBN 978-1-4200-9130-4 (hardcover : alk. paper)
1. Obesity--Complications. 2. Inflammation. 3. Adipose tissues--Pathophysiology. I. Awad, Atif B. II. Bradford, Peter G. III. Series: Oxidative stress and disease ; 28.
[DNLM: 1. Adipose Tissue--physiopathology. 2. Adipose Tissue--metabolism. 3. Inflammation--metabolism. 4. Inflammation--physiopathology. 5. Obesity--physiopathology. W1 OX626 v. 28 2010 / QS 532.5.A3 A23477 2010]

RC628.A335 2010
362.196'398--dc22 2009029624

**Visit the Taylor & Francis Web site at**
**http://www.taylorandfrancis.com**

**and the CRC Press Web site at**
**http://www.crcpress.com**

# Contents

# Series Preface

## OXYGEN BIOLOGY AND MEDICINE

Through evolution, oxygen—itself a free radical—was chosen as the terminal electron acceptor for respiration. The two unpaired electrons of oxygen spin in the same direction; thus, oxygen is a biradical. Other oxygen-derived free radicals such as superoxide anion or hydroxyl radicals formed during metabolism or by ionizing radiation are stronger *oxidants*, that is, endowed with higher chemical reactivities. Oxygen-derived free radicals are generated during metabolism and energy production in the body and are involved in regulation of signal transduction and gene expression, activation of receptors and nuclear transcription factors, oxidative damage to cell components, antimicrobial and cytotoxic actions of immune system cells, as well as in aging and age-related degenerative diseases. Conversely, cells conserve antioxidant mechanisms to counteract the effects of oxidants; these *antioxidants* may remove oxidants either in a highly specific manner (e.g., by superoxide dismutases) or in a less specific c manner (e.g., through small molecules such as vitamin E, vitamin C, and glutathione). *Oxidative stress* as classically defined is an *imbalance between oxidants and antioxidants.* Overwhelming evidence indicates that oxidative stress can lead to cell and tissue injury. However, the same free radicals that are generated during oxidative stress are produced during normal metabolism and, as a corollary, are involved in both human health and disease.

## UNDERSTANDING OXIDATIVE STRESS

In recent years, the research disciplines interested in oxidative stress have grown and enormously increased our knowledge of the importance of the cell redox status and the recognition of oxidative stress as a process with implications for many pathophysiological states. From this multi- and inter-disciplinary interest in oxidative stress emerges a concept that attests to the vast consequences of the complex and dynamic interplay of oxidants and antioxidants in cellular and tissue settings. Consequently, our view of oxidative stress is growing in scope and new future directions. Likewise, the term *reactive oxygen species*—adopted at some stage in order to highlight nonradical oxidants such as $H_2O_2$ and $O_2$—now fails to reflect the rich variety of other reactive species in free radical biology and medicine encompassing nitrogen-, sulfur-, oxygen-, and carbon-centered radicals. With the discovery of nitric oxide, nitrogen-centered radicals gathered momentum and have matured into an area of enormous importance in biology and medicine. Nitric oxide or nitrogen monoxide (NO), a free radical generated in a variety of cell types by nitric oxide synthases (NOSs), is involved in a wide array of physiological and pathophysiological phenomena such as vasodilation, neuronal signaling,and inflammation. Of great importance is the radical–radical reaction of nitric oxide with superoxide anion. This is among the most rapid nonenzymatic reactions in biology (well over the diffusion-controlled limits) and yields the potent nonradical oxidant, peroxynitrite. The involvement of this species in tissue injury through oxidation and nitration reactions is well documented.

Virtually all diseases thus far examined involve free radicals. In most cases, free radicals are secondary to the disease process, but in some instances causality is established by free radicals. Thus, there is a delicate balance between oxidants and antioxidants in health and disease. Their proper balance is essential for ensuring healthy aging.

Both reactive oxygen and nitrogen species are involved in the redox regulation of cell functions. Oxidative stress is increasingly viewed as a major upstream component in the signaling cascade involved in inflammatory responses, stimulation of cell adhesion molecules, and chemoattractant production and as an early component in age-related neurodegenerative disorders such as Alzheimer's, Parkinson's, and Huntington's diseases, and amyotrophic lateral sclerosis. Hydrogen peroxide is probably the most important redox signaling molecule that, among others, can activate NFκB, Nrf2, and other universal transcription factors. Increasing steady-state levels of hydrogen peroxide have been linked to a cell's redox status with clear involvement in adaptation, proliferation, differentiation, apoptosis, and necrosis.

The identification of oxidants in regulation of redox cell signaling and gene expression was a significant breakthrough in the field of oxidative stress: the classical definition of oxidative stress as an *imbalance between the production of oxidants and the occurrence of cell antioxidant defenses* proposed by Sies in 1985 now seems to provide a limited concept of oxidative stress, but it emphasizes the significance of cell redox status. Because individual signaling and control events occur through discrete redox pathways rather than through global balances, a new definition of oxidative stress was advanced by Dean P. Jones (*Antioxidants & Redox Signaling* [2006]) as a disruption of redox signaling and control that recognizes the occurrence of compartmentalized cellular redox circuits. Recognition of discrete thiol redox circuits led Jones to provide this new definition of oxidative stress. Measurements of GSH/GSSG, cysteine/cystine, or thioredox$^{inreduced}$/thioredoxin$^{oxidized}$ provide a quantitative definition of oxidative stress. Redox status is thus dependent on the degree to which tissue-specific cell components are in the oxidized state.

In general, the reducing environments inside cells help to prevent oxidative damage. In this reducing environment, disulfide bonds (S–S) do not spontaneously form because sulfhydryl groups are maintained in the reduced state (SH), thus preventing protein misfolding or aggregation. The reducing environment is maintained by metabolism and by the enzymes involved in maintenance of thiol/disulfide balance and substances such as glutathione, thioredoxin, vitamins E and C, and enzymes such as superoxide dismutases, catalase, and the selenium dependent glutathione reductase and glutathione and thioredoxin-dependent hydroperoxidases (periredoxins) that serve to remove reactive oxygen species (hydroperoxides). Also of importance is the existence of many tissue- and cell compartment-specific isoforms of antioxidant enzymes and proteins.

Compelling support for the involvement of free radicals in disease development originates from epidemiological studies showing that enhanced antioxidant status is associated with reduced risk of several diseases. Of great significance is the role that micronutrients play in modulation of redox cell signaling; this establishes a strong linking of diet and health and disease centered on the abilities of micronutrients to regulate redox cell signaling and modify gene expression.

These concepts are anticipated to serve as platforms for the development of tissue-specific therapeutics tailored to discrete, compartmentalized redox circuits. This, in essence, dictates principles of drug development-guided knowledge of mechanisms of oxidative stress. Hence, successful interventions will take advantage of new knowledge of compartmentalized redox control and free radical scavenging.

## OXIDATIVE STRESS IN HEALTH AND DISEASE

Oxidative stress is an underlying factor in health and disease. In this series of books, the importance of oxidative stress and diseases associated with organ systems of the body is highlighted by exploring the scientific evidence and clinical applications of this knowledge. This series is intended for researchers in the basic biomedical sciences and clinicians. The potential of such knowledge for healthy aging and disease prevention warrants further knowledge about how oxidants and antioxidants modulate cell and tissue function.

**Lester Packer**
**Enrique Cadenas**

# Preface

Obesity is a worldwide epidemic disorder that has become recognized in the 21st century as a principal health threat in most countries. Obesity is characterized by accumulation of excess body fat and is quantitatively defined as a body-mass index greater than 30. Several factors contribute to obesity and these can be broadly classified as genetic and environmental. Among the environmental influences, the combination of excess caloric intake and sedentary life contribute most significantly to the incidence of obesity. The American Obesity Association identifies obesity with more than 30 medical conditions. In particular, obesity is a risk factor for the development of common chronic diseases including hypertension, type 2 diabetes, metabolic syndrome, cardiovascular disease, several cancers, and a host of inflammatory disorders. Accumulating evidence implicates inflammation as an essential common thread among these chronic diseases as well as a key feature of obesity-associated morbidities. We must realize that this is not inflammation in the classic sense: obesity and its associated diseases manifest a low-grade, metabolically-associated inflammation; it is inflammation triggered by high caloric diets that involves many of the same mediators associated with classic inflammation.

Concurrent with this understanding of obesity as a chronic low-grade inflammatory disease, it is necessary to recognize adipose tissue as more than a storage site for fat. Adipose tissue is an essential endocrine organ that produces and secretes a host of hormones in response to varying physiologic and pathologic states. Obesity creates an identifiable and characteristic shift in the secreted profiles of these adipose-specific hormones, termed adipokines. These same adipokines promote low-grade systemic inflammation. For example, obesity and chronic inflammation are accompanied by suppression of adiponectin levels and elevation of resistin levels; the resultant effects on signal transduction converge to increase activation of nuclear factor kappa B (NF-κB) and accelerate production of tumor necrosis factor alpha (TNF-α). In turn, these events alter insulin signaling, decrease Akt activity, and impair translocation of the GLUT-4 glucose transporter to cell surfaces—all events that are characteristic of the insulin-resistant state common in obesity. In addition, the fat cells in obesity recruit macrophages into adipose tissue where they secrete their own host of inflammatory factors.

*Adipose Tissue and Inflammation* focuses on the contribution of adipose tissue to local and systemic inflammation and allows numerous themes to be drawn. At the start, epidemiologic time–trend analyses in populations worldwide indicate that obesity has increased sharply over the past 10 to 20 years and that in the United States the potential health consequences of this rise have been quantified such that obesity at age 40 is estimated to reduce life expectancy by at least 6 years. From investigative research of the endocrine nature of adipose tissue, we learn here that adipose tissue is better considered as an organ composed of both white and brown adipose tissue contained within two main subcutaneous depots and several specific visceral depots. Analysis of the endocrine nature of the adipose organ, detailed in this volume, reveals that about a quarter of the genes expressed in white adipose tissue encode secreted proteins and that the number of established and putative adipokines identified among these genes exceeds several dozen. The authors, all experts in their fields, report that essential among these adipokines, particularly in regard to their role as modulators of local and systemic inflammation, are leptin, adiponectin, TNF-α, numerous interleukins and prostaglandins, resistin, leukocyte chemoattractants (monocyte chemoattractant protein-1 and macrophage migration inhibitory factor-1), fibrinolytic proteins, and growth factor molecules. Detailed investigations of the inflammatory responses of the adipose organ reveal that classic inflammatory signal transducers such as NF-κB, JNK, PPAR, and iNOS are operative and that their continued regulation of adipose gene expression contributes to chronic inflammatory status in obesity.

We are fortunate to have the contributions from several leading edge experts in the area of obesity and inflammation and they report here their current findings obtained through basic, translational, and clinical research. Insulin is central in this research. Insulin affects adipose inflammation and we learn of the interdependent relationships among insulin resistance, central obesity, and inflammatory processes in adipose tissue. We learn of detailed examinations of the effects of insulin on the levels of key adipokines and the effects of inflammation on insulin sensitivity and other key regulators of glucose homeostasis, cell-cycle progression, and apoptosis in adipose tissue. Experts in their respective fields report on how obesity and adipose inflammation are modulated by systemic and local hormonal factors including growth hormone, glucocorticoids, and prostaglandins; by dietary factors including fatty acids, polyphenols, phytosterols, phytoestrogens, and antioxidants; by life-style changes involving diet, exercise and weight loss; and finally by new and investigative advances in pharmacotherapy.

*Adipose Tissue and Inflammation* features contributions from international experts in the fields of adiposity, inflammation, adipokines, and pharmaconutrition. We sincerely thank these contributors for sharing their expertise with the scientific community at large through their chapter authorships. In addition, we would like to thank the publisher, Taylor & Francis Group, for agreeing to publish this book. We thank the series editors, Dr. Lester Packer and Dr. Enrique Cadenas for their continued inspiration and our colleague Dr. Mulchard Patel for his enthusiastic encouragement. We also thank the publication staff, whose dedicated work to assist in production resulted in such a well constructed book. Last, but not least, we would like to thank the readers who are interested in learning about the most up-to-date advances in the area of adipose tissue and inflammation.

**Atif B. Awad, PhD**

**Peter G. Bradford, PhD**

# Editors

**Atif B. Awad** is an associate professor of nutrition at the State University of New York at Buffalo. He earned his PhD in nutrition from Rutgers University, New Brunswick, New Jersey, and completed postdoctoral training in biochemistry at the University of Iowa. He has supervised more than 80 MS and PhD students. He has published more than 80 papers and presented more than 70 abstracts at national and international scientific meetings. His research focus is in the area of dietary lipids and their effects on disease prevention, particularly the effects of phytochemicals on combating chronic diseases.

**Peter G. Bradford** is an associate professor of pharmacology and toxicology at the State University of New York at Buffalo. Dr. Bradford earned a BS *summa cum laude* in biology from the State University of New York at Albany and a PhD in biochemistry from the University of Rochester. He is author or coauthor of more than 50 research journal articles and book chapters and has served as an invited speaker for more than 50 university lectures, international scientific presentations, and various television and radio interviews. His work on hormone therapeutics has been cited by the *New York Times*, *Women's World Magazine*, and various pharmacology weblogs. He is a member of the ADA Council on Scientific Affairs and the ADA National Board Test Construction Committee. He is a long-standing member of the Endocrine Society.

# Contributors

**Ahmad Aljada**
Department of Biomedical Sciences
C.W. Post Long Island University
Brookville, New York, United States

**Atif B. Awad**
Department of Exercise and Nutrition Sciences
State University of New York at Buffalo
Buffalo, New York, United States

**Maria S. Baranowski**
St. Boniface Research Center
University of Manitoba
Winnipeg, Manitoba, Canada

**Peter G. Bradford**
Department of Pharmacology and Toxicology
State University of New York at Buffalo
Buffalo, New York, United States

**Saverio Cinti**
Department of Molecular Pathology and Innovative Therapies
University of Ancona
Ancona, Italy

**Vanessa C. DeClercq**
St. Boniface Research Center
University of Manitoba
Winnipeg, Manitoba, Canada

**Joseph Doria**
Department of Biomedical Sciences
C.W. Post Long Island University
Brookville, New York, United States

**Giamila Fantuzzi**
Department of Kinesiology and Nutrition
University of Illinois at Chicago
Chicago, Illinois, United States

**Kerry B. Goralski**
Department of Pharmacology
Dalhousie University
Halifax, Nova Scotia, Canada

**Melissa E. Gove**
Department of Kinesiology and Nutrition
University of Illinois at Chicago
Chicago, Illinois, United States

**Ryan W.T. Hunt**
St. Boniface Research Center
University of Manitoba
Winnipeg, Manitoba, Canada

**Michael J. LaMonte**
Department of Exercise and Nutritional Sciences
State University of New York at Buffalo
Buffalo, New York, United States

**Martha Lappas**
Department of Obstetrics and Gynaecology
University of Melbourne
Melbourne, Victoria, Australia

**Hong-Biao Liu**
Department of Medicine
State University of New York at Buffalo
Buffalo, New York, United States

**Mohammed H. Moghadasian**
Department of Human Nutritional Sciences
University of Manitoba
Winnipeg, Manitoba, Canada

**Nicholas M. Morton**
University of Edinburgh
Edinburgh, United Kingdom

**Alexander R. Moschen**
Department of Gastroenterology and Hepatology
Innsbruck Medical University
Innsbruck, Austria

**Rgia A. Othman**
Department of Human Nutritional Sciences
University of Manitoba
Winnipeg, Manitoba, Canada

**Fereidoon Shahidi**
Department of Biochemistry
Memorial University of Newfoundland
St. John's, Newfoundland, Canada

**Christopher J. Sinal**
Department of Pharmacology
Dalhousie University
Halifax, Nova Scotia, Canada

**Danielle M. Stringer**
St. Boniface Research Center
University of Manitoba
Winnipeg, Manitoba, Canada

**Thomas M. Stulnig**
Department of Medicine
Medical University of Vienna
Vienna, Austria

**Carla G. Taylor**
St. Boniface Research Center
University of Manitoba
Winnipeg, Manitoba, Canada

**Herbert Tilg**
Department of Gastroenterology and Hepatology
Innsbruck Medical University
Innsbruck, Austria

**Roberto Vettor**
Department of Medical and Surgical Sciences
Padua University
Padua, Italy

**Tongjian You**
Department of Social and Preventive Medicine
State University of New York at Buffalo
Buffalo, New York, United States

**Peter Zahradka**
St. Boniface Research Center
University of Manitoba
Winnipeg, Manitoba, Canada

**Maximilian Zeyda**
Department of Medicine
Medical University of Vienna
Vienna, Austria

**Ying Zhong**
Department of Biochemistry
Memorial University of Newfoundland
St. John's, Newfoundland, Canada

# 1 The Adipose Organ

*Saverio Cinti and Roberto Vettor*

## CONTENTS



## 1.1 ADIPOSE ORGAN CONCEPT

We recently developed a new concept: white and brown adipose tissues (WATs and BATs) are contained together into a dissectible "adipose organ" (Figure 1.1) [1–3]. It is composed of two main subcutaneous depots (anterior and posterior, forming about 60 to 70% of the organ) and several visceral depots (mediastinal, omental, mesenteric, perirenal, retroperitoneal, perigonadic, and perivesical). All depots are anatomically defined by a cleavage plane that allows a precise dissection of depots from surrounding structures.

The depot anatomy is preserved in different strains and at different ages. The white parts of the organ are made mainly by WAT. The brown parts are made mainly by BAT. White and brown adipocytes are often mixed and the colors of the mixed areas depend on the prevalence of one cell type. The relative amounts of white, brown, and mixed parts are genetically determined and depend on several factors such as age, sex,

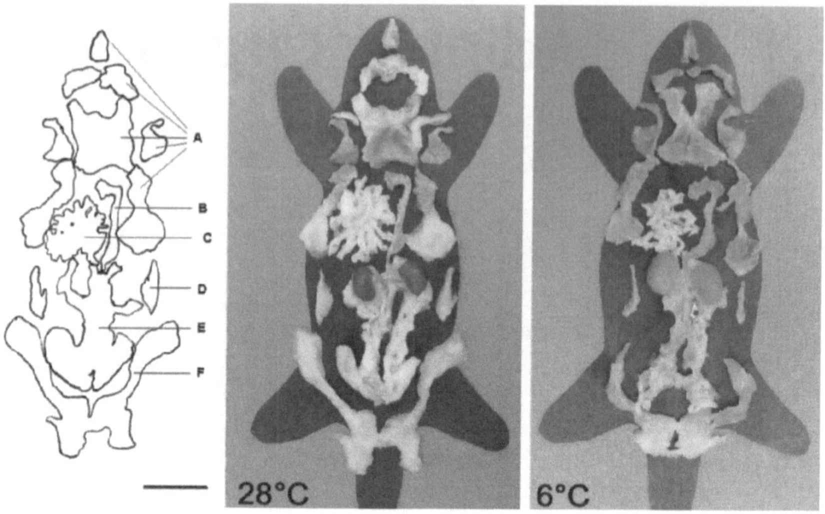


**FIGURE 1.1** Gross anatomy of adipose organs of adult female 129Sv mice. The subcutaneous and visceral depots were dissected and positioned on templates of the mice to show their locations within the animals. The mouse on the left was maintained at warm conditions (28°C for 10 days) and the one on the right at cold conditions (6°C for 10 days). Note the obvious transformation of the color of the organ due to increase of brown adipose tissue and decrease of white adipose tissue. The organ consists of two subcutaneous depots: A = anterior (deep cervical, superficial cervical, interscapular, subscapular, axillo-thoracic); F = posterior (dorso-lumbar, inguinal, gluteal); and several visceral depots: B = mediastinal, C = mesenteric, D = retroperitoneal and E = abdomino-pelvic (perirenal, periovarian, parametrial and perivesical). Bar = 1 cm. (*Source:* Murano I, Zingaretti CM, and Cinti S. (2005). *Adipocytes* 1, 121–130. With permission.)

environmental temperature, and nutritional status. In most small rodents, brown areas are visually evident in the interscapular, axillary, and cervical parts of the anterior subcutaneous depots and in the mediastinal and perirenal visceral depots.

In a recent paper we quantitatively described the anatomy of the adipose organs of Sv129 adult female mice. We calculated the total number of white and brown adipocytes contained in most depots (anterior subcutaneous, posterior subcutaneous, mediastinal, perirenal, perigonadic, perivesical, retroperitoneal, and mesenteric). Our data show that, in this strain, all subcutaneous and all visceral depots contain both white and brown adipocytes mixed together. In some depots, white adipocytes are more numerous (posterior subcutaneous, mesenteric, and retroperitoneal); in other depots, brown adipocytes are more numerous (anterior subcutaneous, mediastinal, and abdomino-pelvic, i.e., composed by a unique dissectible depot formed by perirenal, periovaric, parametrial, and perivesical parts) [4].

## 1.2 DISTINCT MORPHOLOGIES AND PHYSIOLOGIES OF WHITE AND BROWN ADIPOCYTES

White adipocytes are spherical cells (Figure 1.2) of variable sizes mainly dependent on the sizes of the stored lipid droplets. Differentiated white adipocytes can be very

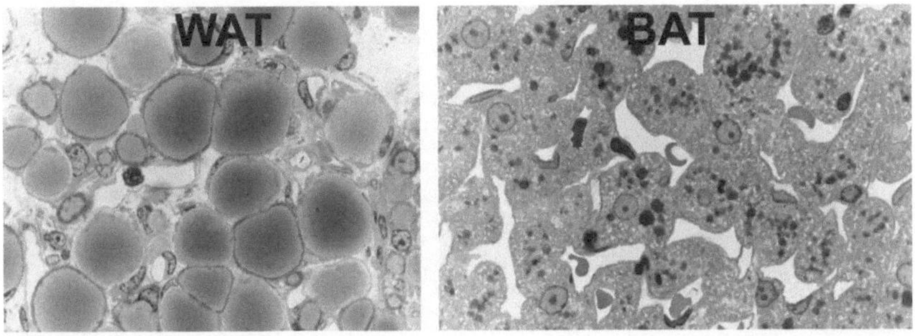


**FIGURE 1.2** Light microscopy of murine white (WAT, left) and brown (BAT, right) adipose tissues. White adipocytes are roundish with unilocular lipid droplets. Brown adipocytes are polyhedral with multilocular lipid droplets.

small (less than 10 μm diameter) in comparison with the average diameters of the adipocytes found in the different depots of adult mammals.

Brown adipocytes store triglycerides in form of numerous small vacuoles (multilocular cells) (Figure 1.2). The shape is mainly polygonal with a variable diameter which, in mice, is usually in the range of 15 to 50 μm. The most characteristic organelle of a brown adipocyte is the mitochondrion. It is spherical, big, and packed with laminar cristae (Figure 1.3). Usually, mitochondria are numerous in the cytoplasm of brown adipocytes and contain a characteristic protein known as uncoupling protein 1 (UCP1) and expressed only in this cell type [5]. In our examinations of tissue from adult animals, when an adipocyte appearing as a multilocular cell under a light microscopic is examined under an electron microscope, it always exhibits mitochondria with characteristic features of those found in brown adipocytes [3,6] independent of the presence of UCP1 in the cell. We believe that expression of UCP1 merely reflects the thermogenic capacity of brown adipocytes, and that these cells have a distinctive morphology (i.e. mainly a multilocular lipid content and characteristic mitochondria). We believe that the multilocular adipocytes found in the adipose organs of adult animals must be considered thermogenically hypo-functioning brown adipocytes when they are UCP1-negative and thermogenically active when they are UCP1-positive as determined by immunohistochemistry. In this chapter, we will use a nomenclature in accordance with this definition of brown adipocyte.

The adipose organ is diffuse within an organism and most of its depots receive vascular supplies by regional visceral or parietal nerve vascular bundles. The extension of the capillary network is quite different in the white and brown parts of the organ. In the brown areas, the density of the capillaries is much higher than in the white areas (Figure 1.4). The nerve supply to the adipose organ is different in both areas, with brown areas more innervated than white areas. In brown areas, numerous noradrenergic fibers are found in fat lobules, running along blood vessels and directly in contact with adipocytes [7]. Adrenergic receptors (α1, α2, β1, β2, and β3) are present in the adipose organ and β1 and β3 adrenoceptors are mainly present on adipocytes [8].

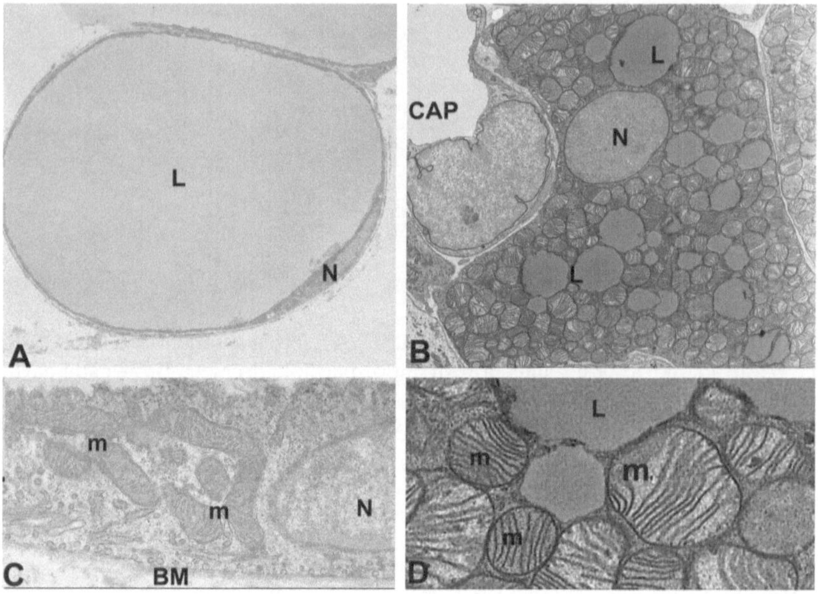


**FIGURE 1.3** Transmission electron microscopy. A: Mouse WAT. Most of the cell is represented by the unilocular lipid droplet. The cytoplasm forms a thin rim containing organelles. Mitochondria (m) are small, elongated with randomly oriented cristae (enlarged in C). N = nucleus; BM = basal membrane; L = lipid droplet. B: Mouse interscapular BAT. Brown adipocyte showing numerous mitochondria packed with transverse cristae (enlarged in D) in cytoplasm. Several small lipid droplets (L) are also visible. CAP = capillary lumen.

The density of parenchymal fibers varies according to the functional status of the organ. During cold exposure, the noradrenergic parenchymal fibers increase their density in the brown part of the organ [7,9]. During fasting, these fibers increase their density in the white part of the organ [10]. Vascular noradrenergic fibers are also immunoreactive for neuropeptide Y (NPY). The majority of these nerves also contain norepinephrine (NE) [10,11], suggesting that they belong to the sympathetic nerve supply to WAT blood vessels. Recently a parasympathetic innervation of WAT has been described, suggesting possible functional implications but the matter is still open to discussion [12–15].

The main functions of white adipocytes are storing and releasing highly energetic molecules, fatty acids (FAs), that supply fuel to the organism during intervals between meals. Brown adipocytes use FAs to produce heat (non-shivering thermogenesis). This function is due to the above mentioned mitochondrial UCP1 present exclusively in brown adipocytes [5,8,16–20]. The signal for brown adipocyte activation is a temperature below thermoneutrality—a temperature that induces activation of the sympathetic nervous system [20].

The presence and activity of BATs in the adipose organ seems to play a pivotal role for obesity prevention because genetic ablation of BAT and all beta adrenergic receptors induces obesity in mice [21,22], although mice lacking UCP1 are cold-sensitive

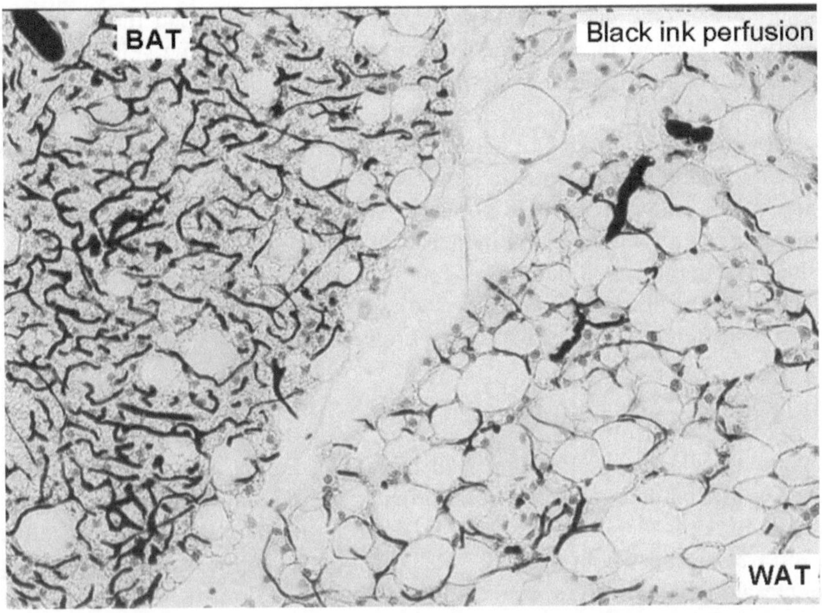


**FIGURE 1.4** Light microscopy showing capillary network in area of transition between WAT and BAT. Capillaries appear black because of black ink perfusion of mouse.

but not obese [23]. On the other hand, ectopic expression of UCP1 in WATs results in obesity resistance [24]. Accordingly, it has been recently shown that obesity-prone mice have less BAT or inducible BAT activity than obesity-resistant mice [25].

Another primary function of white adipocytes was discovered some years ago: production of leptin, a hormone able to influence animal behavior concerning food intake [26]. Leptin also induces energy dispersion (via BAT and locomotor activation) and has gonadotrophic properties. Brown adipocytes in their classic multilocular configuration (i.e., during thermogenic activity) are not immunoreactive for leptin [27,28]. A growing body of evidence suggests that the adipose organ produces several additional factors or adipokines, and that these control important functions such as glucose and lipid metabolism, blood coagulation, blood pressure, and steroid hormone modulation. The production of these adipokines by fat supports the concept of the adipose organ as an endocrine structure [29,30].

## 1.3 ABILITY OF ADIPOSE ORGAN TO MODIFY ANATOMY UNDER PHYSIOLOGIC STIMULI

### 1.3.1 Acclimatization to Different Temperatures

Pregnancy and lactation, obesity, fasting, and caloric restrictions are the most frequent physiologic and pathologic (e.g., when obesity induces co-morbidities such as diabetes) conditions in which the adipose organ shows plasticity. Here we will consider some of the cell biology aspects related to its plasticity.

The organs of cold acclimated mice are darker in color than those of warmth exposed mice, suggesting a change to a more brown phenotype (Figure 1.1). This reversible phenomenon is due to an increased number of brown adipocytes, capillaries, and nerves in the adipose organ [6,7,31–35]. The same phenomenon can be achieved by the administration of beta-3 adrenoceptor agonists [36–42] and is mostly suppressed in mice lacking beta-3 adrenoceptors [41,43], suggesting that noradrenergic fibers play a central role in adipocytes. Accordingly, after cold acclimatization, the density of noradrenergic fibers increases in all parts of the adipose organ [7,9,32]. Published and unpublished data produced by our laboratories and others favour the hypothesis that the newly formed brown adipocytes derive from a direct transformation (or transdifferentiation) of white into brown adipocytes [4,40,41].

### 1.3.2 Pregnancy and Lactation

The mammary glands comprise most of the adipose organ. They are composed of branched epithelial ducts infiltrating all subcutaneous adipose tissues and connected to nipples. In adult female mice, three bilateral nipples are connected to epithelial ducts infiltrating the whole anterior subcutaneous fat depot of the adipose organ. Two bilateral nipples are connected to epithelial ducts infiltrating the whole posterior subcutaneous fat depot of the adipose organ. Therefore virgin adult (post-pubertal) female mice are provided five bilateral incomplete mammary glands that are ready to become milk-secreting during pregnancy and lactation. The two subcutaneous depots containing the glands differ from those of male mice only by the presence of the above described branched epithelial ducts. The adipose component of these depots follows the general rules described above for the adipose organ: a mixed composition of white and brown adipocytes (with relative amounts depending mainly on age, strain, and environmental conditions). Of note, adipocytes of the mammary glands express the prolactin receptor [44].

During pregnancy and lactation, the mammary gland anatomy changes with a progressive reduction of adipocytes and the formation of milk-secreting lobulo-alveolar epithelial glands. This plastic phenomenon is reversible and at the end of lactation the milk-secreting components of the gland disappear to give room to the reappearing adipocytes and allow a complete reconstruction of the pre-gravidic anatomy of the gland. This phenomenon was previously viewed as due to "hiding" among the glands of the adipocytes that themselves undergo a de-lipidation process during pregnancy and a lipid re-filling process in the post-lactation period. Our recent morphological studies combined with the Cre-lox fate mapping technique suggested that adipocytes undergo a reversible adipo-epithelial transdifferentiation process in mammary glands during pregnancy and lactation [45].

### 1.3.3 Energy-Driven Changes

When the energy balance becomes positive, the adipose organ increases its white parts. White adipocytes undergo hypertrophy followed by hyperplasia. In fact, it has been proposed that adipocytes have a maximum volume and cannot be further expanded. This maximum volume, also referred to as critical cell size, is genetically determined and specific for each fat depot [46]. Adipocytes of critical cell size trigger an increase in cell numbers [47,48]. Not all depots have the same tendency to hypertrophy and hyperplasia; the former seems more characteristic of epididymal and mesenteric depots, the latter of inguinal and perirenal depots [46].

Adipose tissue expresses numerous factors that may be implicated in modulation of adipogenesis: IGF-1, TGF-β, TNF-α, macrophage colony-stimulating factor (MCSF), angiotensin-2, autotaxin-lysophosphatidic acid (ATX-LPA), leptin, resistin, and others [49]. Interestingly, it has been shown in mice that obesity induced by high fat diet is hypertrophic, while obesity induced by hypothalamic lesions due to administration of monosodium glutamate is hyperplastic [50]. It has been recently suggested that adipocyte precursors can derive from bone marrow [51], but our data and those from other authors favor a vascular in-site origin [52,53]. Additionally, WATs of obese mice and humans are infiltrated by macrophages and the level of infiltration correlates with body-mass index (BMI) and mean sizes of adipocytes [54–56]. This infiltration seems to be an important cause for the insulin resistance associated with obesity.

We recently observed that macrophages are mainly located at the level of dead adipocytes in white adipose tissues of obese mice, obese humans, and in transgenic mice that are lean but have hypertrophic adipocytes (HSL knock-out mice) [57]. The common notion that obese people with visceral fat accumulation are more prone to diabetes than obese persons with subcutaneous fat predominance may arise from the fact that visceral adipocytes seem to be more susceptible to cell death than subcutaneous adipocytes. In other words, hypertrophic visceral adipocytes reach critical size and this triggers programmed adipocyte death, whereas hypertrophic subcutaneous adipocytes are less susceptible to this apoptosis and their longevity allows or causes macrophage infiltration and insulin resistance in visceral fat before they occur in subcutaneous fat [58].

The brown part of the adipose organ is modified under conditions of positive energy balance. In obese mice, the rate of apoptosis of brown adipocytes increases and this is strongly attenuated in mice lacking TNF-α receptors [59]. In obese animals, the morphology of brown adipocytes gradually changes into one more similar to that of white adipocytes, including transformation of the multilocular lipid depot into a unilocular one. This is accompanied by activation of the leptin gene and these cells become immunoreactive for leptin [27,28] thus providing further evidence for a reversible transdifferentiation between the two type of adipocytes.

The morphology of the adipose organ during fasting is quite characteristic. Under fasting, a variable number of slimmed cells are present in the white part of the organ. The slimmed cells are barely visible under light microscopy but are easily recognized by electron microscopy, i.e. they have a specific ultrastructural morphology: cytoplasmic irregularities with thin projections and numerous invaginations rich in pinocytotic vesicles. In acute fasting, completely de-lipidized adipocytes can be found near apparently unaffected unilocular cells. Vasculogenesis and neurogenesis are observed in white adipose tissues of fasted animals. Capillaries are often surrounded by the thin cytoplasmic projections of slimmed adipocytes. Neurogenesis is mainly supported by an increase of noradrenergic fibers [10]. Under chronic caloric restriction, the reduction in size of adipocytes is homogeneously distributed [60].

## 1.4 ADIPOSE ORGANS OF HUMANS

Although the morphology of human adipose tissues is very similar to that of murine adipose tissues, several anatomical peculiarities must be highlighted. In humans, anatomical dissection reveals a clear distinction of the two important compartments of the adipose organ without giving adequate details of other components not discernible by dissection. The new highly sensitive imaging techniques make it possible to detect and measure depots other than subcutaneous and visceral adipose tissues that collectively contribute to total-body adipose tissue [61]. Computed tomography (CT) and magnetic resonance imaging (MRI) allow better definition of the precise margins of these compartments and ready quantification of total as well as regional (perirenal, mesenteric, etc.) amounts of adipose tissue [62]. With high resolution MRI, it is possible to quantify the adipose tissue from bone marrow.

Adipose tissue is also present within many organs and tissues, in particular within skeletal muscle where it is normally not detectable by CT and MRI under physiological conditions. However, it is possible to calculate the lipid content of intermuscular adipose tissue (IMAT) by subtracting intramyocellular lipid content measured by magnetic resonance spectroscopy from total tissue lipid content measured by chemical shift imaging [63]. Results from analyses using these new imaging techniques show us that the real volume of adipose tissue determined by dissection and subsequent histological analysis underestimates whole-body adipose tissue. Moreover, the techniques allow the grouping of adipose tissue compartments according to structure–function relationships and particularly endocrine and metabolic activities.

### 1.4.1 Subcutaneous Depots

Subcutaneous adipose tissue (SAT) is certainly the best defined compartment and exhibits clear anatomic demarcations (deposits found between the dermis and the aponeuroses and fasciae of muscles including mammary adipose tissue). We can also distinguish with the new imaging techniques [64] superficial and a deep subcutaneous adipose tissues (fat depots found between the skin and fascial plane in the lower trunk and gluteal–thigh area and depots found between this fascia plane and the muscle fascia, respectively) (Figure 1.5). This distinction has been recently emphasized because of both morphological and metabolic differences between the two adipose tissue depots; the deeper depot behaves more like visceral adipose tissue [65].

### 1.4.2 Internal Depots

Internal adipose tissue can be divided into visceral adipose tissue (VAT) and non-strictly visceral adipose tissue. The reliability and accuracy of visceral adipose tissue estimates by imaging methods have been extensively studied. The coefficient of variation for VAT measurements seems to be greater with MRI (~ 9 to 18%) than

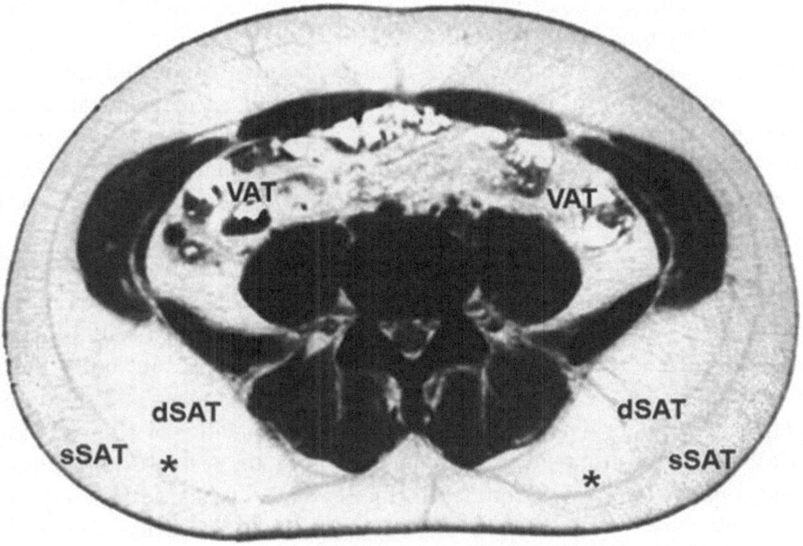


**FIGURE 1.5** Computed tomography scan at abdominal level of obese subject. Visceral adipose tissue (VAT) and subcutaneous abdominal adipose tissue (SAT) with subtle fascial plane (asterisks) separating into superficial (sSAT) and deep (dSAT) portions. (*Source:* Modified from Iacobellis G, Corradi D, and Sharma AM. (2005). *Nat Clin Pract Cardiovasc Med* 2, 536–543.)

with CT (~2% ) [64]. Non-strictly visceral internal adipose tissue (internal adipose tissue minus visceral adipose tissue) includes the intramuscular (between fascicles), perimuscular, intermuscular, and paraosseal adipose tissue [64].

An absolute or relative VAT expansion has been associated with increased risk of morbidity and mortality for cardiovascular disease (CVD) and metabolic diseases [66,67], but important differences in the metabolic and functional properties among the different depots within the VAT compartment have been recognized. Therefore, a more detailed classification of visceral adipose tissue has been proposed. The first partition of VAT must be among the intrathoracic (ITAT), intra-abdominal (IAAT), and intrapelvic (IPAT) adipose tissues that may be roughly separated using the new imaging techniques because of the irregular margins between VAT and other tissues and organs [64]. The ITAT is mainly distributed around the heart and its physiological role is in the early stages of investigation.

### 1.4.3 Intrathoracic Adipose Tissue (ITAT): Epicardial Adipose Tissue

Much interest has focused on the importance of IAAT because we know that it confers significantly higher risks for type 2 diabetes and cardiovascular disease. However, ITAT including mediastinal and epicardial fat has also been studied [68], and shown to be related to CVD and myocardial function through unidentified mechanisms. It is well known that variable amounts of fat cover the epicardial surface of the human heart but this finding is not a constant anatomical characteristic in all species. In fact, in laboratory rodents, this fraction of the visceral fat is minimal [69], perhaps explaining why epicardial adipose tissue has been poorly investigated until recently. Its function is far from fully explained and the concept of mechanical protection appears to be inconsistent based on the lack of such fat pads in several animal species. Its origin is also unknown although some evidence suggests that epicardial fat may evolve from brown adipose tissue during embryogenesis [69]. This hypothesis has not been confirmed.

Iacobellis et al. [70] published a comprehensive description of the morphological and functional aspects of epicardial fat. In recent years, the attention on epicardial fat biology and its clinical impact has grown exponentially. In the adult heart, WAT is present along the atrioventricular and interventricular grooves including the apex. Minor amounts of fat are also located in the subepicardial parts of atria walls and appendages. It is important to underline that a small quantity of adipose tissue of the epicardial surface also penetrates the myocardium, thus establishing a tight anatomic (no apparent separation from fascia and the same coronary blood supply) and functional relationship with the muscular components of the heart.

In some pathological conditions, epicardial fat progressively develops, sometimes covering the whole epicardial surface. The amount of epicardial fat, far too great to be simply related to overall adiposity, seems more closely related to VAT [71,72], age [73], and heart hypertrophy, but not ischemia [71]. The positive relationship of the amount of epicardial fat and ventricular myocardial mass was also noted in an echocardiographic study [74]. Cardiac adipose tissue may supply energy for the adjacent myocardium and serve as a buffer against toxic levels of fatty acids (FAs) [69]. By contrast, the high lipolytic activity of epicardial fat suggests that this tissue may also serve as a ready source of FAs to meet increased myocardial energy demands. Since this tissue is not depleted during starvation, it seems simplistic to consider epicardial adipose tissue as a compartment devoted to the storage of excess calories to be released not in physiological conditions, but rather in emergency situations like ischemia. Marchington et al. [69] found that epicardial adipose tissue has a greater capacity for FA release than adipose tissue elsewhere in the body. The high lipolysis observed in epicardial adipose tissue may be due to several factors included reduced antilipolytic effect of insulin and the increased activity of β3 adrenergic receptors.

Epicardial AT expresses an inflammatory profile of proteins, but the mechanisms responsible are yet to be elucidated [75]. A higher expression of MCP-1, IL-1, IL-6, and TNF-α along with more disseminate inflammatory cell infiltrates were seen in epicardial adipose stores than in SAT. Despite these observations, no clear relationship between epicardial AT metabolic and inflammatory pattern and obesity, type 2 diabetes, or atherogenic dyslipidemia has been seen. Lower expression of adiponectin along with higher expression of resistin has also been observed in human epicardial fat. Inconsistent data link these alteration to the inflammatory response of the adipose tissue. The release of bioactive molecules from the pericoronary tissues may alter vascular [76] and myocardial function and exert potentially worsening effects on coronary artery disease progression. Conversely, this inflammatory reaction may lead to a more pronounced angiogenic response that may be beneficial for the development of collateral circulation in patients with ischemic heart disease [75].

### 1.4.4 Intrathoracic Adipose Tissue (ITAT): Thymus

The thymus is a lymphoid organ that selects T cells for release to the peripheral immune system. Unfortunately, thymopoiesis is highly susceptible to damage by physiologic stressors and may contribute to immune deficiencies that occur in a variety of clinical settings. The thymus is critical for establishing the immune system during childhood, but begins to shrink just before puberty. In adults, thymic tissue is replaced by adipose tissue (involution); however, thymocytes are continually produced into old age. In adults, it is difficult to distinguish between the cortex and medulla by the concentration of thymocytes although Hassall's corpuscles can still be identified in the medulla.

It is interesting to note that leptin-deficient (ob/ob) mice exhibit severe thymic atrophy, suggesting that this hormone is required for normal thymopoiesis. In a recent study, leptin showed a selective thymostimulatory role in settings of leptin deficiency and endotoxin administration-induced thymic atrophy. Thus this major adipokine and perhaps others may be useful for protecting the thymus from damage by augmenting T cell reconstitution in these clinical states [77]. The role of adipose tissue surrounding the thymus may be seen not simply as the result of thymic involution but may function as an active neighboring organ.

### 1.4.5 Intra-Abdominal Adipose Tissue (IAAT)

IAAT has been also subdivided into intraperitoneal and retroperitoneal adipose types, taking into consideration the parietal peritoneum [78] or as an alternative, the straight line across the anterior border of L4–L5 and the psoas muscles, continuing on a tangent just before the posterior limits of the ascending and descending colon, and extending to the abdominal wall. However, the lack of precise limits of the intraperitoneal and retroperitoneal spaces makes it very difficult to obtain a detailed quantitative estimation of the amount of fat stored in this area. Abdominal VAT is synonymous with IAAT [79]. IPAT is usually quantified with IAAT but the two deposits clearly differ functionally and morphologically; IPAT is entirely extraperitoneal [80].

Abate et al. proposed that metabolic differences exist between intraperitoneal (drained by the portal vein) and retroperitoneal adipose tissue which flows into the inferior vena cava [81]. It is clear that the direct exposure of liver cells through the portal circulation to high concentrations of FA and/or other metabolites derived from intraperitoneal adipose tissue is responsible for the increased frequency of dyslipidemia, hyperinsulinemia, and other metabolic complications associated with abdominal obesity [66,67,82]. Ideally, the study of VAT should include all adipose tissues in the thoracic, abdominal, and pelvic cavities. The use of CT and MRI with a smaller field of view, higher resolution, and thinner slices may allow separation of all adipose tissue depots from one another. Retroperitoneal components such as pararenal adipose tissue are clearly visible on some conventional MRI scans.

### 1.4.6 Non-Strictly Visceral Internal Adipose Tissue: Muscular Adipose Tissue

Particular attention must be devoted to the adipose tissue components present within or near the muscles. We can distinguish intermuscular adipose tissue (IMAT), paraosseal adipose tissue, and perimuscular adipose tissue which is not so easily distinguishable from the adjacent adipose tissue compartments even with the current imaging techniques. The very recent approach of using microdissection [83] in human cadaver and animal studies has provided a means of accurately estimating the small volumes of perimuscular and intramuscular adipose tissue depots.

Attention has focused on the content, localization, and composition of fat within skeletal muscle as determinants of insulin resistance, but less information is available on the impacts of their intracellular or interfibrillar localization on insulin action [84]. In morbidly obese patients, weight loss induced by biliary–pancreatic-diversion provoked a significant amelioration in insulin resistance with a parallel change in intramyocellular—but not perivascular or interfibrillar—lipid accumulation [85]. It is now clear that intramyocellular fat is mostly due to the accumulation of triglycerides within muscle cells whereas the perivascular or interfibrillar lipids correspond to adipose cells along the blood vessels and in the intermuscular spaces [86]. It is possible with MRI to define and localize IMAT which can be found between muscle bundles and is clearly separated from SAT by a well defined fascia. There is a strong direct linear correlation between total adipose tissue and IMAT in men and women of different ethnic groups [84]. IMAT increases during aging [87], but also in sedentary young subjects, under different pathological conditions such as partial lipodystrophy, and in both men and women who have metabolic syndrome [65].

What is the origin of fat cells surrounding muscle bundles? They may derive from different progenitors normally present in adult skeletal muscles: mesenchymal stem cells, muscle-derived stem cells, and satellite cells. After birth, muscle regeneration is mostly mediated by satellite cells, a unique population of committed stem cells located adjacent to the plasma membranes of myofibers. It has been demonstrated that mouse satellite cells behave as multipotent stem cells.

We studied the differentiation capacities of human satellite cells and in particular their adipogenic conversion. We proved by morphological analysis, mRNA expression, and immunohistochemistry that human satellite cells possess a clear adipogenic potential that may explain the presence of mature adipocytes within skeletal muscles under pathological conditions such as obesity, type 2 diabetes, and age-related sarcopenia [88]. Moreover, we recently reported that primary stem cell cultures derived from skeletal muscle differentiate into adipocytes when cultured in high glucose. High glucose induces reactive oxygen species (ROS) production and protein kinase C beta (PKCβ) activation. These two events appear crucial in this differentiation process that can be directly induced by oxidizing agents and inhibited by PKCβ siRNA silencing. The differentiated adipocytes, when implanted in vivo, form viable and vascularized adipose tissue. Overall, the data highlight a previously uncharacterized differentiation route triggered by high glucose that drives resident stem cells present in muscles to form adipose depots. This process may represent a feed-forward cycle between the regional increase in adiposity and insulin resistance that plays a key role in the pathogenesis of diabetes mellitus [89].

### 1.4.7 Other Internal Adipose Tissue Components: Perivascular Adipose Tissue

In addition to the VAT and SAT, adipose tissue is also found in the close vicinity of blood vessels and it is known as perivascular white adipose tissue (pAT). pAT secretes cytokines such as IL-1, MCP-1, TNF-α, pro-atherogenic chemokines, and pro-angiogenic peptides. These factors appear to contribute directly to alterations of the function and structure of the vascular wall, including chronic inflammation, infiltration of leukocytes at the interface between human pAT and the adventitia of atherosclerotic aortas, alterations of vascular tone, proliferation of smooth muscle cells, neo-angiogenesis, and the development of obesity-associated atherosclerosis and cardiovascular complications. The effects of other obesity-related risk factors such as dyslipidemia, hypertension, and insulin resistance on pAT remain unexplored, but it is conceivable that these factors modulate the adipogenesis and functionality of pAT depots, e.g., through local shear stress [90].

As obesity develops, hypertrophy and hyperplasia of perivascular adipocytes result in altered patterns of secretion of the adipokines including TNF-α, TGF-β, IL-6, and chemokines like IP-10 and MCP-1 that promote chemotaxis of leukocytes to the vascular endothelium and their migration into the vascular walls. The consequence is a local inflammatory burst with production of additional cytokines and chemokines by chemo-attracted leukocytes and adipocytes.

The local production of matrix metalloproteinases and various angiogenic factors by AT increases neo-vascularization that in turn supplies the oxygen and nutrients necessary for the development and maintenance of local inflammation and plaque formation. Some pAT-derived chemokines such as MCP-1 and IP-10 stimulate the migration and proliferation of smooth muscle cells. Finally, angiotensin II and TNF-α, along with the local overproduction of ROS, contribute to a lower bioavailability of NO, putatively leading to vasoconstriction and impaired insulin-mediated vasodilatation, ultimately enhancing insulin resistance. Taken together, pAT exhibits all the characteristics of a local promoter of atherosclerosis. However, this appealing concept requires confirmation by further mechanistic research and intervention trials. Although the direct causal role of pAT in the process of atherogenesis will be challenging to establish, it may potentially provide a novel target for the prevention and treatment of obesity-associated cardiovascular complications [91]. pAT, on the other hand, has been reported to lower vascular tone through the release of a transferable, thermosensitive, non-lipid factor that stimulates the generation of NO by endothelium and through an endothelium-independent mechanism involving hydrogen peroxide ($H_2O_2$) and subsequent activation of soluble guanylyl cyclase (sGC) [92].

### 1.4.8 Brown Adipose Tissue

Like the murine adipose organ, the human adipose organ contains BAT. It is easy to understand that thermodispersion in humans is much lower than in rodents due to the relationship between surface and volume (S/V) of the human body that alone justifies a reduced need for brown adipose tissue in adult humans. Newborns have a different S/V relationship and considerable brown adipose tissue is present at that stage. Nevertheless, brown adipocytes dispersed among white adipocytes have been described in several histological studies (including studies showing the presence of UCP1) [93,94]. BAT in human newborns has been described in most sites described for rodents and UCP1 gene expression was found in biopsies from VATs of lean and obese adult patients. The authors calculated the presence of one brown adipocyte for every 100 to 200 white adipocytes in VATs of lean adult humans [95].

BAT was reported to be increased in outdoor workers in northern Europe [96] and in patients with pheochromocytoma (a noradrenaline-secreting tumor). Furthermore, rare hibernomas, BAT tumors occurring in several anatomical sites including SAT and VAT, have been described (about 100 cases are discussed in the literature and we recently observed a case in which brown adipocytes expressed UCP1 and exhibited the classic electron microscopic profile with typical mitochondria).

Positron emission tomography using fluorodeoxyglucose (FDG PET) has been applied to brown fat. Extensive use of this technique identified amounts of brown adipose tissue in adult humans [97,98]. The anatomical sites described as normal for human BAT are the root of the neck, the roots of the upper limbs, and the intercostal spaces near the vertebral column [99]. Of note, the density of human BAT increases after cold exposure, especially during winter, as revealed by PET [100]. In biopsies of the perithyroid areas of the necks of human adults (corresponding to one of the PET-positive areas in other patients of the same age and BMI levels), we found UCP1-positive brown adipocytes by immunohistochemistry analysis.

The physiological role of BAT in humans is debated, but the possibility of increasing it artificially to treat obesity and related disorders cannot be excluded. It is interesting to note that human adults with reduced brown phenotypes of abdominal subcutaneous adipose tissue exhibit reduced insulin sensitivity [101] and that human white adipocyte precursors may be induced in vitro to express UCP1 by administration of drugs [102].

### 1.4.9 Development and Turnover

In addition to specific differences in the distribution of adipose tissue, humans differ from other primates and subprimates, by the presence of a significant amount of body fat in utero and at birth. It has been hypothesized that these sites may serve as supplemental energy stores for maintaining adequate feeding to the enlarged human brain when the flow of energy fuels from the mother is sharply decreased [103]. However, the in utero and perinatal periods are crucial for the regulation of whole-body energy balance and the development of childhood obesity [104]. Two other critical periods are the adiposity rebound between ages 4 and 6 during which BMI, after a rise in infancy and subsequent decline, begins to increase again, and adolescence, when important changes in the quantity and location of body fat occur [105].

In girls, body fat changes from ~17 to ~24% of body mass throughout adolescence. On the other hand, body fat in boys decreases over this same period. In contrast to girls, boys lose body fat, but the central deposition of body fat increases almost five-fold, whereas this increase in females is only approximately three-fold [106].

In lean adults, the human adipose organ constitutes about 8 to 18% of the body weight in males and 14 to 28% in females (about 5% in monkeys) [107]. Gender greatly affects the morphological aspects and endocrine–metabolic functions of the adipose organ. Women have a higher percentage of body fat than men and tend to store adipose tissue preferentially in the lower body (gluteal and femoral) regions contrasting with the male pattern of fat distribution in the upper body (abdominal visceral and thorax) depots [108]. The increased gluteal–femoral adiposity in women is associated with increased lipid turnover due to increases in both stimulated lipolysis and triglyceride synthesis, resulting in a larger fat cell size in these depots [109]. In contrast, increases in abdominal adipose tissue in men are accompanied by increased lipoprotein lipase (LPL) activity and decreased stimulated lipolysis in these depots [109,110]. These gender differences disappear after menopause and may justify the changes in fat distribution [111]. These differences in adipocyte metabolism, including both basal and stimulated lipolysis, may be determined by sex steroids, particularly estrogen which increases LPL activity in the gluteal–femoral region, leading to typical female adipose tissue distribution [112]. The reduced lipolytic activity in the gluteal–femoral region seems to be due to a relative preponderance of antilipolytic activity of $\alpha 2$ adrenoceptors over the lipolytic $\beta$ adrenoceptors [111].

Several other aspects of adipose organ functioning clearly show a sexual dimorphism through the expression, synthesis, and release of different adipokines [113]. The development of the human adipose organ ends at puberty, mainly due to a proliferative process [114]. In massively obese humans, the adipose organ can increase four times and reach 60 to 70% of body weight [102]. The factors determining fat mass in adult humans are not fully understood, but increased lipid storage in mature adipocytes is considered to be a major determinant. In an insightful report, Spalding, et al. showed that adipocyte number is a crucial factor for determining fat mass in adults [115]. However, the number of fat cells remains very constant in adulthood in lean and obese individuals, even after marked weight loss, indicating that the number of adipocytes is set during childhood and adolescence.

To establish the dynamics within a stable population of adipocytes in adults, adipocyte turnover was measured by analyzing the integration of $^{14}C$ derived from nuclear bomb tests in genomic DNA. Using this creative experimental approach, it was estimated that approximately 10% of fat cells are renewed annually at all adult ages and levels of BMI. Moreover, neither adipocyte death nor generation rate seemed to be altered in early onset obesity, suggesting a tight regulation of fat cell numbers in obesity during adulthood [115]. In case of fasting or caloric restriction, the adipose organ reduces its volume and adipocytes reduce their size. The reduction in adipocyte size is important because their size correlates with insulin sensitivity [116]. Completely de-lipidized adipocytes can be found in the adipose tissues of subjects with negative energy balance. The morphology of de-lipidized adipocytes is similar to that described above for the equivalent cells of mice and rats. The fate of these de-lipidized adipocytes is still debated although some authors suggest that they undergo apoptosis [117].

## REFERENCES


1. Cinti S. (1999). *The Adipose Organ*. Kurtis, Milan.
2. Cinti S. (2001). The adipose organ: morphological perspectives of adipose tissues. *Proc Nutr Soc* 60, 319–328.
3. Cinti S. (2002). Adipocyte differentiation and transdifferentiation: plasticity of the adipose organ. *J Endocrinol Invest* 25, 823–835.
4. Murano I, Zingaretti CM, and Cinti S. (2005). The adipose organ of Sv129 mice contains a prevalence of brown adipocytes and shows plasticity after cold exposure. *Adipocytes* 1, 121–130.
5. Frontini A et al. (2007). Thymus uncoupling protein 1 is exclusive to typical brown adipocytes and is not found in thymocytes. *J Histochem Cytochem* 55, 183–189.
6. Cousin B et al. (1992). Occurrence of brown adipocytes in rat white adipose tissue: molecular and morphological characterization. *J Cell Sci* 103, 931–942.
7. Murano I, Barbatelli G, Giordano A, and Cinti S. (2009). Noradrenergic parenchymal nerve fiber branching after cold acclimatisation correlates with brown adipocyte density in mouse adipose organ. *J Anat* 214, 171–178.
8. Cannon B and Nedergaard J. (2004). Brown adipose tissue: function and physiological significance. *Physiol Rev* 84, 277–359.
9. De Matteis R, Ricquier D, and Cinti S. (1998). TH-, NPY-, SP-, and CGRP-immunoreactive nerves in interscapular brown adipose tissue of adult rats acclimated at different temperatures: an immunohistochemical study. *J Neurocytol* 27, 877–886.
10. Giordano A et al. (2005). Regional-dependent increase of sympathetic innervation in rat white adipose tissue during prolonged fasting. *J Histochem Cytochem* 53, 679–687.
11. Cannon B and Nedergaard J. (1986). Brown adipose tissue thermogenesis in neonatal and cold–adapted animals. *Biochem Soc Trans* 14, 233–236.
12. Giordano A et al. (2007). Reply to Kreier and Buijs: no sympathy for the claim of parasympathetic innervation of white adipose tissue. *Am J Physiol Regul Integr Comp Physiol* 293, R550–R552.
13. Giordano A et al. (2006). White adipose tissue lacks significant vagal innervation and immunohistochemical evidence of parasympathetic innervation. *Am J Physiol Regul Integr Comp Physiol* 206, R1243–R1255.
14. Kreier F and Buijs RM (2007). Evidence for parasympathetic innervation of white adipose tissue, clearing up some vagaries. *Am J Physiol Regul Integr Comp Physiol* 293, R548–R549.
15. Kreier F et al. (2002). Selective parasympathetic innervation of subcutaneous and intra–abdominal fat: functional implications. *J Clin Invest* 110, 1243–1250.
16. Cannon B, Hedin A, and Nedergaard J. (1982). Exclusive occurrence of thermogenin antigen in brown adipose tissue. *FEBS Lett* 150, 129–132.
17. Cinti S et al. (1989). Immunoelectron microscopical identification of the uncoupling protein in brown adipose tissue mitochondria. *Biol Cell* 67, 359–362.
18. Klaus S. (2001). Adipose Tissues. Eureka Software, Austin, TX.
19. Ricquier D, Casteilla L, and Bouillaud F. (1991). Molecular studies of the uncoupling protein. *FASEB J* 5, 2237–2242.
20. Trayhurn P and Nicholls D. (1986). *Brown Adipose Tissue*. Edward Arnold, London.

21. Bachman ES et al. (2002). β-AR signaling required for diet-induced thermogenesis and obesity resistance. *Science* 297, 843–845.
22. Lowell BB et al. (1993). Development of obesity in transgenic mice after genetic ablation of brown adipose tissue. *Nature* 366, 740–742.
23. Enerback S et al. (1997). Mice lacking mitochondrial uncoupling protein are cold–sensitive but not obese. *Nature* 387, 90–94.
24. Kopecky J et al. (1996). Reduction of dietary obesity in aP2-Ucp transgenic mice: physiology and adipose tissue distribution. *Am J Physiol* 270, E768–E775.
25. Almind K et al. (2007). Ectopic brown adipose tissue in muscle provides a mechanism for differences in risk of metabolic syndrome in mice. *Proc Natl Acad Sci USA* 104, 2366–2371.
26. Zhang Y et al. (1994). Positional cloning of the mouse obese gene and its human homologue. *Nature* 372, 425–432.
27. Cancello R et al. (1998). Leptin and UCP1 genes are reciprocally regulated in brown adipose tissue. *Endocrinology* 139, 4747–4750.
28. Cinti S et al. (1997). Immunohistochemical localization of leptin and uncoupling protein in white and brown adipose tissue. *Endocrinology* 138, 797–804.
29. Kershaw EE and Flier JS. (2004). Adipose tissue as an endocrine organ. *J Clin Endocrinol Metab* 89, 2548–2556.
30. Trayhurn P and Wood IS. (2004). Adipokines: inflammation and the pleiotropic role of white adipose tissue. *Br J Nutr* 92, 347–355.
31. Giordano A et al. (1998). Sensory nerves affect the recruitment and differentiation of rat periovarian brown adipocytes during cold acclimation. *J Cell Sci* 111, 2587–2594.
32. Giordano A et al. (1996). Tyrosine hydroxylase, neuropeptide Y, substance P, calcitonin gene-related peptide and vasoactive intestinal peptide in nerves of rat periovarian adipose tissue: an immunohistochemical and ultrastructural investigation. *J Neurocytol* 25, 125–136.
33. Guerra C et al. (1998). Emergence of brown adipocytes in white fat in mice is under genetic control: effects on body weight and adiposity. *J Clin Invest* 102, 412–420.
34. Loncar D. (1991). Convertible adipose tissue in mice. *Cell Tissue Res* 266, 149–161.
35. Young P, Arch JR, and Ashwell M. (1984). Brown adipose tissue in the parametrial fat pad of the mouse. *FEBS Lett* 167, 10–14.
36. Champigny O et al. (1991). β3 adrenergic receptor stimulation restores message and expression of brown fat mitochondrial uncoupling protein in adult dogs. *Proc Natl Acad Sci USA* 88, 10774–10777.
37. Collins S et al. (1997). Strain-specific response to β3 adrenergic receptor agonist treatment of diet-induced obesity in mice. *Endocrinology* 138, 405–413.
38. Ghorbani M, Claus TH, and Himms-Hagen J. (1997). Hypertrophy of brown adipocytes in brown and white adipose tissues and reversal of diet-induced obesity in rats treated with a β3 adrenoceptor agonist. *Biochem Pharmacol* 54, 121–131.
39. Ghorbani M and Himms-Hagen J. (1997). Appearance of brown adipocytes in white adipose tissue during CL 316,243-induced reversal of obesity and diabetes in Zucker fa/fa rats. *Int J Obes Relat Metab Disord* 21, 465–475.
40. Granneman JG et al. (2005). Metabolic and cellular plasticity in white adipose tissue I: effects of β3 adrenergic receptor activation. *Am J Physiol Endocrinol Metab* 289, E608–E616.
41. Himms-Hagen J et al. (2000). Multilocular fat cells in WAT of CL 316,243-treated rats derive directly from white adipocytes. *Am J Physiol Cell Physiol* 279, C670–C681.
42. Toseland CD et al. (2001). Comparison of adipose tissue changes following administration of rosiglitazone in the dog and rat. *Diabetes Obes Metab* 3, 163–170.

43. Jimenez M et al. (2003). β3 adrenoceptor knockout in C57BL/6J mice depresses the occurrence of brown adipocytes in white fat. *Eur J Biochem* 270, 699–705.
44. Ling C et al. (2000). Prolactin (PRL) receptor gene expression in mouse adipose tissue increases during lactation and in PRL-transgenic mice. *Endocrinology* 141, 3564–3572.
45. Morroni M et al. (2004). Reversible transdifferentiation of secretory epithelial cells into adipocytes in the mammary gland. *Proc Natl Acad Sci USA* 101, 16801–16806.
46. DiGirolamo M et al. (1998). Qualitative regional differences in adipose tissue growth and cellularity in male Wistar rats fed ad libitum. *Am J Physiol* 274, R1460–R1467.
47. Bjorntorp P. (1991). Visceral fat accumulation: the missing link between psychosocial factors and cardiovascular disease? *J Intern Med* 230, 195–201.
48. Miller WH Jr and Faust IM. (1982). Alterations in rat adipose tissue morphology induced by a low-temperature environment. *Am J Physiol* 242, E93–E96.
49. Ahima RS and Flier JS. (2000). Adipose tissue as an endocrine organ. *Trends Endocrinol Metab* 11, 327–332.
50. Imai T et al. (2001). Selective ablation of retinoid X receptor alpha in hepatocytes impairs their lifespan and regenerative capacity. *Proc Natl Acad Sci USA* 98, 4581–4586.
51. Crossno JT Jr et al. (2006). Rosiglitazone promotes development of a novel adipocyte population from bone marrow-derived circulating progenitor cells. *J Clin Invest* 116, 3220–3228.
52. Cinti S et al. (1984). A morphological study of the adipocyte precursor. *J Submicrosc Cytol* 16, 243–251.
53. Tang W et al. (2008). White fat progenitor cells reside in the adipose vasculature. *Science* 322, 583–586.
54. Cancello R et al. (2005). Reduction of macrophage infiltration and chemo-attractant gene expression changes in white adipose tissue of morbidly obese subjects after surgery-induced weight loss. *Diabetes* 54, 2277–2286.
55. Weisberg SP et al. (2003). Obesity is associated with macrophage accumulation in adipose tissue. *J Clin Invest* 112, 1796–1808.
56. Xu H et al. (2003). Chronic inflammation in fat plays a crucial role in the development of obesity-related insulin resistance. *J Clin Invest* 112, 1821–1830.
57. Cinti S et al. (2005). Adipocyte death defines macrophage localization and function in adipose tissue of obese mice and humans. *J Lipid Res* 46, 2347–2355.
58. Murano I et al. (2008). Dead adipocytes, detected as crown-like structures, are prevalent in visceral fat depots of genetically obese mice. *J Lipid Res* 49, 1562–1568.
59. Nisoli E et al. (2000). Tumor necrosis factor-α mediates apoptosis of brown adipocytes and defective brown adipocyte function in obesity. *Proc Natl Acad Sci USA* 97, 8033–8038.
60. Napolitano L and Gagne HT. (1963). Lipid-depleted white adipose cells: an electron microscope study. *Anat Rec* 147, 273–293.
61. Thomas EL et al. (1998). Magnetic resonance imaging of total body fat. *J Appl Physiol* 85, 1778–1785.
62. Barnard ML et al. (1996). Development of a rapid and efficient magnetic resonance imaging technique for analysis of body fat distribution. *NMR Biomed* 9, 156–164.
63. Lunati E et al. (2001). In vivo quantitative hydrolipidic map of perirenal adipose tissue by chemical shift imaging at 4.7 Tesla. *Int J Obes Relat Metab Disord* 25, 457–461.
64. Shen W et al. (2003). Adipose tissue quantification by imaging methods: a proposed classification. *Obes Res* 11, 5–16.
65. Kelley DE et al. (2000). Subdivisions of subcutaneous abdominal adipose tissue and insulin resistance. *Am J Physiol Endocrinol Metab* 278, E941–E948.
66. Bjorntorp P. (1992). Metabolic abnormalities in visceral obesity. *Ann Med* 24, 3–5.

67. Pischon T et al. (2008). General and abdominal adiposity and risk of death in Europe. *New Engl J Med* 359, 2105–2120.
68. Sharma AM. (2004). Mediastinal fat, insulin resistance, and hypertension. *Hypertension* 44, 117–118.
69. Marchington JM, Mattacks CA, and Pond CM. (1989). Adipose tissue in the mammalian heart and pericardium: structure, foetal development and biochemical properties. *Comp Biochem Physiol B* 94, 225–232.
70. Iacobellis G, Corradi D, and Sharma AM. (2005). Epicardial adipose tissue: anatomic, biomolecular and clinical relationships with the heart. *Nat Clin Pract Cardiovasc Med* 2, 536–543.
71. Corradi D et al. (2004). The ventricular epicardial fat is related to the myocardial mass in normal, ischemic and hypertrophic hearts. *Cardiovasc Pathol* 13, 313–316.
72. Iacobellis G et al. (2003). Echocardiographic epicardial adipose tissue is related to anthropometric and clinical parameters of metabolic syndrome: a new indicator of cardiovascular risk. *J Clin Endocrinol Metab* 88, 5163–5168.
73. Schejbal V. (1989). Epicardial fatty tissue of the right ventricle: morphology, morphometry and functional significance. *Pneumologie* 43, 490–499.
74. Iacobellis G et al. (2004). Relation between epicardial adipose tissue and left ventricular mass. *Am J Cardiol* 94, 1084–1087.
75. Mazurek T et al. (2003). Human epicardial adipose tissue is a source of inflammatory mediators. *Circulation* 108, 2460–2466.
76. Lohn M et al. (2002). Periadventitial fat releases a vascular relaxing factor. *FASEB J* 16, 1057–1063.
77. Hick RW et al. (2006). Leptin selectively augments thymopoiesis in leptin deficiency and lipopolysaccharide-induced thymic atrophy. *J Immunol* 177, 169–176.
78. van der Kooy K and Seidell JC. (1993). Techniques for the measurement of visceral fat: a practical guide. *Int J Obes Relat Metab Disord* 17, 187–196.
79. Busetto L et al. (2000). Visceral fat loss evaluated by total body magnetic resonance imaging in obese women operated with laparascopic adjustable silicone gastric banding. *Int J Obes Relat Metab Disord* 24, 60–69.
80. Rendell M et al. (2001). Relationship between abdominal fat compartments and glucose and lipid metabolism in early postmenopausal women. *J Clin Endocrinol Metab* 86, 744–749.
81. Abate N et al. (1994). Estimation of adipose tissue mass by magnetic resonance imaging: validation against dissection in human cadavers. *J Lipid Res* 35, 1490–1496.
82. Macor C et al. (1997). Visceral adipose tissue impairs insulin secretion and insulin sensitivity but not energy expenditure in obesity. *Metabolism* 46, 123–129.
83. Guo Z, Mishra P, and Macura S. (2001). Sampling the intramyocellular triglycerides from skeletal muscle. *J Lipid Res* 42, 1041–1048.
84. Gallagher D et al. (2005). Adipose tissue in muscle: a novel depot similar in size to visceral adipose tissue. *Am J Clin Nutr* 81, 903–910.
85. Greco AV et al. (2002). Insulin resistance in morbid obesity: reversal with intramyocellular fat depletion. *Diabetes* 51, 144–151.
86. Kim J et al. (2004). Intermuscular adipose tissue-free skeletal muscle mass: estimation by dual energy x-ray absorptiometry in adults. *J Appl Physiol* 97, 655–660.
87. Song MY et al. (2004). Sarcopenia and increased adipose tissue infiltration of muscle in elderly African American women. *Am J Clin Nutr* 79, 874–880.
88. De Coppi P et al. (2006). Rosiglitazone modifies the adipogenic potential of human muscle satellite cells. *Diabetologia* 49, 1962–1973.
89. Aguiari P et al. (2008). High glucose induces adipogenic differentiation of muscle-derived stem cells. *Proc Natl Acad Sci USA* 105, 1226–1231.

90. Henrichot E et al. (2005). Production of chemokines by perivascular adipose tissue: a role in the pathogenesis of atherosclerosis? *Arterioscler Thromb Vasc Biol* 25, 2594–2599.
91. Thalmann S and Meier CA. (2007). Local adipose tissue depots as cardiovascular risk factors. *Cardiovasc Res* 75, 690–701.
92. Coppola A et al. (2007). A central thermogenic-like mechanism in feeding regulation: an interplay between arcuate nucleus T3 and UCP2. *Cell Metab* 5, 21–33.
93. Garruti G and Ricquier D. (1992). Analysis of uncoupling protein and its mRNA in adipose tissue deposits of adult humans. *Int J Obes Relat Metab Disord* 16, 383–390.
94. Kortelainen ML et al. (1993). Immunohistochemical detection of human brown adipose tissue uncoupling protein in an autopsy series. *J Histochem Cytochem* 41, 759–764.
95. Oberkofler H et al. (1997). Uncoupling protein gene: quantification of expression levels in adipose tissues of obese and non-obese humans. *J Lipid Res* 38, 2125–2133.
96. Huttunen P, Hirvonen J, and Kinnula V. (1981). The occurrence of brown adipose tissue in outdoor workers. *Eur J Appl Physiol Occup Physiol* 46, 339–345.
97. Gelfand MJ et al. (2005). Pre-medication to block [$^{18}$F] FDG uptake in the brown adipose tissue of pediatric and adolescent patients. *Pediatr Radiol* 35, 984–990.
98. Hany TF et al. (2002). Brown adipose tissue: a factor to consider in symmetrical tracer uptake in the neck and upper chest region. *Eur J Nucl Med Mol Imaging* 29, 1393–1398.
99. Nedergaard TB, Bengtsson T, and Cannon B (2007). Unexpected evidence for active brown adipose tissue in adult humans. *Am J Physiol Endocrinol Metab* 293, E444–E452.
100. Masayuki SYO et al. (2007). Human brown adipose tissue evaluated by FDG PET: activation by cold exposure. *Int J Obes* 31, S32–S36.
101. Yang X, Enerback S, and Smith U. (2003). Reduced expression of FOXC2 and brown adipogenic genes in human subjects with insulin resistance. *Obes Res* 11, 1182–1191.
102. Prins JB and O'Rahilly S. (1997). Regulation of adipose cell number in man. *Clin Sci (Lond).* 92, 3–11.
103. Kuzawa CW. (1998). Adipose tissue in human infancy and childhood: an evolutionary perspective. *Am J Phys Anthropol* 27, 177–209.
104. Dietz WH. (1994). Critical periods in childhood for the development of obesity. *Am J Clin Nutr* 59, 955–959.
105. Mueller WH. (1982). The changes with age of the anatomical distribution of fat. *Soc Sci Med* 16, 191–196.
106. Goran MI, Kaskoun M, and Shuman WP. (1995). Intra–abdominal adipose tissue in young children. *Int J Obes Relat Metab Disord* 19, 279–283.
107. Pond CM and Mattacks CA. (1987). The anatomy of adipose tissue in captive Macaca monkeys and its implications for human biology. *Folia Primatol (Basel).* 48, 164–185.
108. Blaak E. (2001). Gender differences in fat metabolism. *Curr Opin Clin Nutr Metab Care* 4, 499–502.
109. Fried SK et al. (1993). Lipolysis in intra-abdominal adipose tissues of obese women and men. *Obes Res* 1, 443–448.
110. Edens NK et al. (1993). In vitro lipid synthesis in human adipose tissue from three abdominal sites. *Am J Physiol* 265, E374–E379.
111. Rebuffe–Scrive M et al. (1986). Metabolism of mammary, abdominal, and femoral adipocytes in women before and after menopause. *Metabolism* 35, 792–797.
112. D'Eon TM et al. (2005). Estrogen regulation of adiposity and fuel partitioning: evidence of genomic and non-genomic regulation of lipogenic and oxidative pathways. *J Biol Chem* 280, 35983–35991.
113. Trujillo ME and Scherer PE. (2006). Adipose tissue-derived factors: impact on health and disease. *Endocr Rev* 27, 762–778.

114. Chumlea WC et al. (1981). Size and number of adipocytes and measures of body fat in boys and girls 10 to 18 years of age. *Am J Clin Nutr* 34, 1791–1797.
115. Spalding KL et al. (2008). Dynamics of fat cell turnover in humans. *Nature* 453, 783–787.
116. Stern JS et al. (1972). Adipose cell size and immunoreactive insulin levels in obese and normal weight adults. *Lancet* 2, 948–951.
117. Nisoli E et al. (2006). White adipocytes are less prone to apoptotic stimuli than brown adipocytes in rodent. *Cell Death Differ* 13, 2154–2156.

# 2 Adipose Tissue as Endocrine Organ

*Kerry B. Goralski and Christopher J. Sinal*

## CONTENTS



## 2.1 OVERVIEW

The global prevalence of obesity has increased at an alarming rate, particularly in the highly developed countries of North America and Europe. This excess adiposity is associated with various negative psychosocial impacts and is a primary cause of disability, reduced economic productivity, and shortened life spans [1,2]. Obese individuals are also at increased risk for a number of serious comorbidities such as hypertension, type 2 diabetes, cardiovascular disease, dyslipidemia, gall bladder disease, sleep apnea, and some cancers [3–7].

Most troubling is the increased prevalence of childhood obesity, which indicates that obesity and related comorbidities will remain serious health concerns for many years to come. Despite intensive research, our understanding of the pathogenic relationship between obesity and obesity-associated metabolic disorders remains incomplete. In addition to serving an important metabolic role, adipose tissue is an active endocrine organ that secretes a variety of chemical signals collectively termed adipokines. Currently, the total number of established and putative adipokines exceeds 50. This chapter is intended to introduce the concept of adipose tissue as an endocrine organ and present a select group of representative adipokines that play established roles in energy homeostasis and inflammation.

## 2.2 ADIPOSE TISSUE AS ENDOCRINE ORGAN

Adipocytes, the major constituent cell types of adipose tissues, possess the metabolic machinery to synthesize fatty acids (lipogenesis) and store them in the form of triglycerides during periods of abundant energy supply. In mammals, adipose tissue exists in various depots throughout the body (primarily subcutaneous and visceral) and in two distinct forms, brown adipose tissue and white adipose tissue [8]. In humans, brown adipose tissue is generally found only in infants and is specialized for heat production through non-shivering thermogenesis [9].

The lipid stored in brown adipocytes is primarily used as a fuel for this function. In contrast, lipid stored in white adipocytes serves as a long-term energy reserve that can be mobilized to meet the general energy requirements of the organism in times of caloric deficit [10]. In humans, the vast majority of adipose tissue is white and the expansion of this mass occurs when chronic energy intake that is primarily responsible for obesity exceeds the energy expenditure. Historically, white adipose tissue was considered primarily with respect to energy storage and mobilization. However, in recent years, our understanding of the physiological and pathophysiological roles of white adipose tissue has undergone a major revolution driven by the identification of a large and diverse group of signaling molecules that are synthesized and secreted by this tissue.

It has been estimated that 20 to 30% of genes expressed in white adipose tissue encode secreted proteins [11,12]. While most cells in white adipose tissue are adipocytes, non-adipocyte cell types are also present including those that comprise the adipose tissue matrix (endothelial, smooth muscle, and fibroblast cells) and the stromal vascular components (monocytes, macrophages, and pre-adipocytes) [13]. By strict definition, the *adipokine* term was devised in reference to cytokine molecules (adipocytokines) secreted by adipocytes. However, in recent years this term is more commonly used to cover a broad range of biologically active molecules secreted by white adipose tissue.

Adipokines include pro-inflammatory cytokines and cytokine-related proteins, complement and complement-related proteins, fibrinolytic proteins, proteins of the renin–angiotensin system, and a variety of other biologically active proteins exerting hormone-like actions (Figure 2.1). Some adipokines such as leptin are synthesized and secreted almost exclusively by adipocytes, while others such as adiponectin are produced and secreted by both adipocytes and non-adipocyte cells (Figure 2.2) [13,14]. Other adipokines such as tumor necrosis factor-alpha (TNF-$\alpha$) and interleukins-6 and -8 (IL-6, IL-8) originate largely from non-adipocyte cells (Figure 2.2) [13]. Many adipokines perform local autocrine or paracrine actions that affect adiposity, adipocyte metabolism, and inflammatory responses in white adipose tissue. Adipokines also play important roles in the regulation of systemic energy metabolism through endocrine and systemic actions in the brain, liver, and muscle. The serum levels of many adipokines are profoundly affected by degree of adiposity [15–23], indicating that the synthesis and secretion of these signaling molecules are dynamic and modifiable. This led to the hypothesis that dysregulation of adipokine secretion, particularly of those that influence systemic insulin sensitivity and/or inflammation, underlies increased risks for type 2 diabetes and cardiovascular disease in the obese.

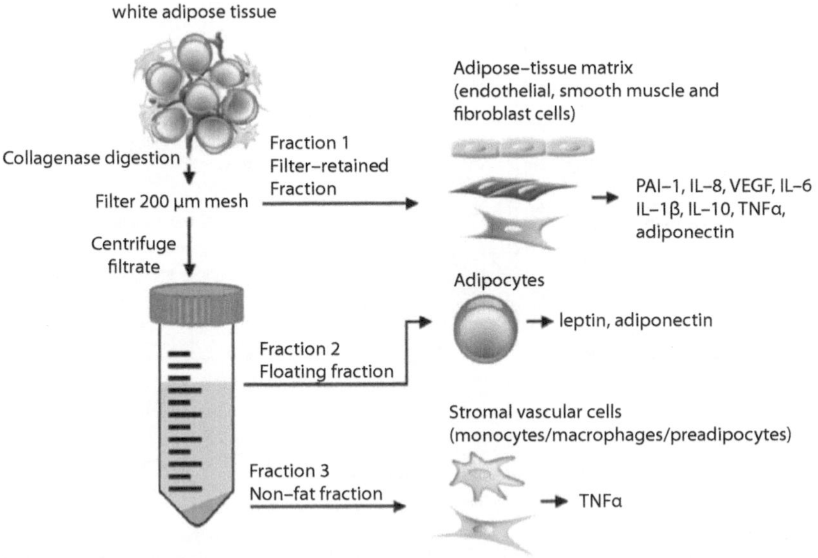


**FIGURE 2.1** Cellular localization of adipokine secretion. Using a simple collagenase digestion, filtration, and centrifugation procedure, human white adipose tissue can be resolved into three cellular fractions. Fraction 1 is non-fat and is retained on the filter. It represents undigested vascular endothelial and smooth muscle cells and connective tissue fibroblasts. Following centrifugation of the filtrates, two additional fractions are obtained. Fraction 2 consists of lipid-filled adipocytes and floats following centrifugation. Fraction 3 is a non-fat (stromal–vascular) fraction that pellets during centrifugation and contains monocytes/macrophages and preadipocytes. Using this procedure Fain and colleagues characterized the relative distribution of adipokine secretion from each of the cellular compartments [13,158,186,188,190,191].

Our current recognition of white adipose tissue as an endocrine organ and the linkage between dysregulation of this function and metabolic disorders derives in large measure from the early efforts of research groups to elucidate the function of leptin, the cardinal adipokine.

## 2.3 LEPTIN

The critical role of the hypothalamus in energy homeostasis was realized in the second half of the 20th century. The earliest evidence emerged from studies in rats where it was observed that lesions of the ventromedial hypothalamus resulted in hyperphagia and obesity [24] while lesions of the lateral hypothalamus caused reduced food intake and a lean morphology [25]. These findings led Kennedy [26] to formulate the "lipostat" hypothesis that predicted the existence of circulating humoral (endocrine) factors released peripherally in proportion to fat mass and function to regulate energy balance in the brain.

The first identification of one of these factors was aided tremendously by the characterization of two strains of obese mutant mice by the researchers at the Jackson

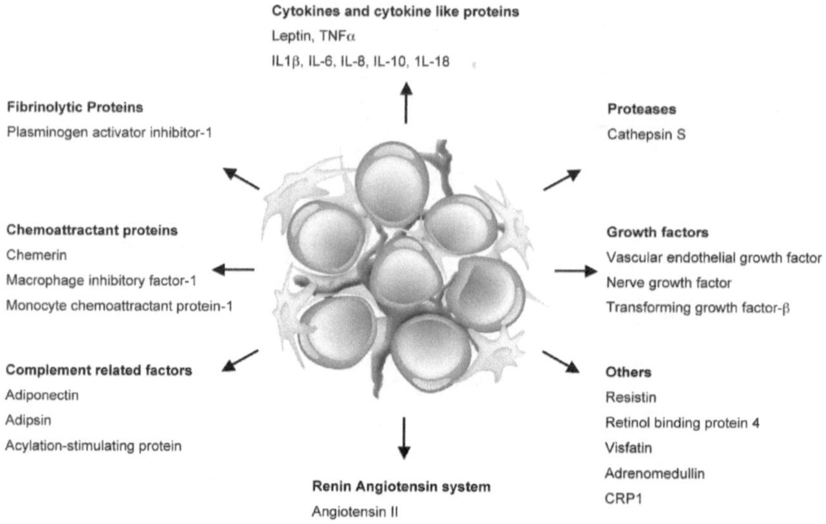


**FIGURE 2.2** Representative summary of adipose-derived signalling molecules. The growing list of these signaling molecules, collectively termed adipokines, consists of several families of biologically active proteins including pro-inflammatory cytokines and cytokine-related proteins, complement and complement-related proteins, fibrinolytic proteins, proteins of the renin–angiotensin system, chemo-attractant proteins, growth factors and a variety of other biologically active proteins with hormone-like actions. Adipokines such as leptin, ASP, TNF-α, IL-6, and chemerin have local autocrine and paracrine actions that regulate adipocyte metabolism, preadipocyte differentiation into adipocytes, and recruitment of immune cells (macrophages) and inflammation in white adipose tissue. Through endocrine actions, adipokines including leptin, adiponectin, resistin, TNF-α, and IL-6 have important roles in the regulation of inflammation and metabolic and vascular homeostasis. It is now widely accepted that the dysregulation of adipokine secretion in obesity is linked to the development of to chronic low-grade inflammation and insulin resistance that are central components of vascular and metabolic diseases. (Key citations for adipokines: leptin [32], TNF [122], resistin [152], chemerin [179], adiponectin [73], visfatin [180], RBP4 [181], PAI-1 [182], adipsin [183], ASP [184], angiotensin II [185], TGF-β [186], MCP-1 [187], IL-1β, IL-6, IL-8, IL-10, and VEGF [13], IL-18, cathepsin S, macrophage inhibitory factor, and nerve growth factor [188], CRP [189].)

Laboratory. The first mouse strain was described in 1950 [27] as exhibiting rapid weight gain beginning at approximately 4 to 6 weeks of age; by the age of 10 months the mutant mice weighed approximately four times the weights of normal littermates. While these obese mice were sterile, heterozygote matings indicated a recessive gene mutation designated *ob* (obese). A second mutant strain with an obese and frankly diabetic phenotype was subsequently identified in 1966 [28]. The recessive mutation in this case was designated *db* (diabetes). Further characterization of both the homozygous *ob/ob* and *db/db* mouse strains revealed many common phenotypic traits including obesity, hyperphagia, hyperglycemia, hyperinsulinemia, insulin resistance, and impaired thermogenesis [29].

Parabiosis studies utilize a surgical procedure that allows the conjoining of animals with different physiologic or genetic characteristics to share blood supplies and

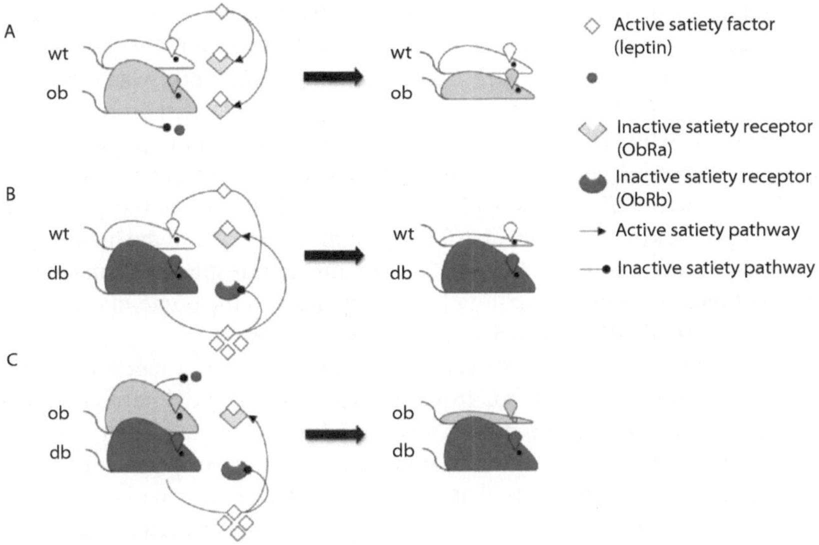


**FIGURE 2.3** Summary of parabiosis experiments that contributed to the discovery of leptin and the leptin receptor. When a wild-type mouse was paired with an *ob/ob* parabiot, functional leptin was provided by the wild-type partner and this restored food intake and weight gain to near normal levels in the *ob/ob* mouse (A). In contrast, *db/db* mice were refractory to the leptin produced by the wild type (B) and *db/db* (C) parabiot due to a lack of functional leptin receptor. The adverse hypophagic effect in the wild-type or *ob/ob* parabiot when paired with a *db/db* mouse was explained by overproduction of leptin by the *db/db* parabiot.

subsequent circulating endocrine factors. The classical parabiosis studies of *ob/ob* and *db/db* mice conducted by Coleman and Hummel [30,31] were instrumental in defining the natures of *ob* and *db* mutations and linking them to the same metabolic pathway (Figure 2.3). In experiments in which a normal wild type mouse was parabiosed with an *ob/ob* mouse, the hyperphagia and rate of weight gain in the obese mouse were reduced substantially (Figure 2.3A).

The absence of any effects in the normal wild-type parabiot indicated that the *ob/ob* mouse was deficient in some endocrine satiety factor rather than producing an altered form of that factor with adverse effects in the normal parabiot. In contrast, when a normal mouse was parabiosed with a *db/db* mouse, the obese mouse was largely unaffected but the normal mouse rapidly lost weight and died of starvation within 2 months of the surgical procedure (Figure 2.3B). This suggested that the *db/db* mouse was resistant to the endocrine satiety factor produced by the normal mouse and also produced a satiety factor that adversely affected the normal parabiot. When an *ob/ob* mouse was parabiosed with a *db/db* mouse, the *ob/ob* mouse lost weight, developed hypoglycemia and hypoinsulinemia, and eventually succumbed to starvation (Figure 2.3C). In comparison, the obese and diabetic phenotype of the *db/db* mouse was unaffected. Collectively, these results indicated that the *ob/ob* mice were deficient in an endocrine satiety signal but possessed an intact satiety center. In contrast, the *db/db* mice appeared to produce this signal but had defective satiety centers.

The identity of this proposed satiety factor remained unknown for a further 20 years until Jeffrey Friedman and colleagues defined the product of the *ob* gene as leptin [32]. Derived from the Greek *leptos* meaning thin, leptin was identified by positional cloning as a highly conserved 16-kDa cytokine-related protein expressed predominantly in white adipose tissue. *Ob/ob* mice were found to possess nonsense mutations in their leptin genes that resulted in the generation of a non-functional protein product. Soon the *db* gene was identified to encode the leptin receptor (ObR), a member of the cytokine receptor superfamily highly expressed in the hypothalamus, white adipose tissue, and testes [33,34]. The *db* mutation results in a truncated ObR that does not mediate leptin signalling [35]. In contrast to the *ob/ob* mice, the *db/db* strain expressed markedly elevated levels of functional leptin.

Thus, the emergence of modern genetic analysis provided a mechanistic explanation for the observations derived from classical physiological parabiosis studies (Figure 2.3). The discovery of leptin as a white adipose-secreted satiety factor profoundly modernized our view of white adipose tissue from an organ solely responsible for energy storage and mobilization to reveal that this tissue is an endocrine organ that regulates metabolism and body weight through the release of a diverse spectrum of adipokines.

In humans, adiposity and gender are major determinants of circulating leptin levels [36–39]. Regardless of fat mass, fasting plasma leptin levels are generally higher in women than men. This is attributed in part to larger subcutaneous white adipose tissue depots in women and the effects of male and female reproductive hormones on leptin production in white adipose tissue [40]. Plasma leptin levels follow both diurnal and pulsatile patterns in humans. Peak plasma leptin levels occur at night and nadir levels occur in the morning hours [38]. The major pulsatory secretion appears to occur as a delayed post-prandial response, 2 to 3 hours following meals and subsequent to a rise in insulin level [38]. Plasma leptin levels decrease with short-term fasting and increase a few hours after re-feeding [39,41–43]. The short-term reductions in plasma leptin that occur with fasting directly correlate with serum insulin and glucose levels, are increased within hours after initiating a glucose infusion, and can be prevented by maintenance of euglycemia [39,41]. A number of other factors including estrogenic hormones, glucocorticoids, and acute exposure to TNF-$\alpha$ are known to stimulate leptin release from white adipose tissue [40]. In comparison, androgens, thiazolinediones, sympathetic nervous system stimulation, and chronic exposure to TNF-$\alpha$ inhibit leptin secretion [40].

Early observations that leptin administration to leptin-deficient obese mice reversed obesity and corrected metabolic abnormalities generated considerable excitement with respect to the potential therapeutic use of this adipokine as an anti-obesity or anti-diabetic agent [44–46]. However, disappointing results from human clinical trials demonstrating minimal efficacy in treating obesity did much to dissipate this fervor [47–49]. As most overweight and obese persons already have elevated leptin levels, the current general consensus is that responsiveness to leptin is decreased, leading to a state of leptin resistance. Reduced blood–brain barrier transport of leptin in obesity, impaired diurnal and pulsatile secretion of leptin, and leptinergic receptor blockade in white adipose tissue are postulated to contribute to a reduction in leptin sensitivity [38,50–52].

Nonetheless, given that leptin administration can blunt increased hunger associated with weight reduction [53], the potential remains to utilize leptin as an adjuvant to prevent the regain of weight that usually occurs after dieting. Similar to obesity, an absence of white adipose tissue (lipodystrophy) also produces insulin resistance, diabetes, hypertriglyceridemia, and other undesirable physiologic changes in rodents and humans [54–58]. Leptin replacement therapy has shown promise as an approach for reversing the metabolic complications of generalized lipodystrophy and lipodystrophy in HIV-infected individuals treated with highly active antiretroviral therapy [59,60]. This approach may also prove advantageous by reducing the need for other antidiabetic agents in patients with lipodystrophy [59].

It is well known that leptin activates anorexigenic pathways (cocaine- and amphetamine-related transcripts and propiomelanocortin) that decrease feeding and inhibit orexigenic (neuropeptide Y and agouti-related protein) pathways that enhance food intake in the arcuate nucleus of the hypothalamus [61]. When energy expenditure exceeds energy intake, body fat stores decrease, leading to decreased leptin secretion and a corresponding increase in food intake. When energy intake exceeds energy expenditure, body fat stores increase, leading to increased leptin levels and a subsequent reduction in feeding. Through these mechanisms in the non-obese individual, the action of leptin within the hypothalamus accurately balances food intake with energy expenditure maintaining body fat stores within narrowly defined limits [53,61,62]. However, alterations in body weight "set point" may arise due to leptin resistance in the hypothalamus and thus contribute to the development of obesity [62]. This idea is supported by observations that mice and humans that lack functional leptin or the ObR are hyperphagic, have lower metabolic rates, and develop massive obesity [32,37,63].

While the central effects of leptin to regulate energy intake and expenditure are well established, pleiotropic effects in peripheral tissues also occur. For instance, leptin produces overall catabolic actions in white adipose tissue by decreasing glucose uptake into adipocytes, promoting lipolysis of triglycerides by hormone-sensitive lipase and inhibiting lipoprotein lipase [52,64]. In skeletal muscle, leptin induces AMP-activated protein kinase (AMPK) phosphorylation and activity leading to downstream stimulation of fatty acid oxidation [65].

High concentrations of leptin have also been shown to enhance glucose-stimulated insulin secretion from pancreatic islet cells [66]. Increasing evidence indicates that leptin has pro-inflammatory and immunomodulatory functions [67,68]. Mice homozygous for *ob* or *db* gene mutations show impaired immune responses and are more susceptible to infections attributed to altered T cell-mediated immune responses [69]. Macrophage infiltration into white adipose tissue and corresponding low-grade inflammation has been implicated as an early event in the metabolic complications of obesity [70,71]. Experimental evidence indicates that elevated levels of leptin contribute to the development and persistence of this localized inflammatory response in obesity [72].

## 2.4 ADIPONECTIN

In the mid 1990s, several research groups independently described adiponectin as a novel 30-kDa protein secreted primarily by white adipose tissue [73–75]. Since those initial reports, adiponectin has been the most intensively studied adipokine next to leptin. This abundant serum protein belonging to the complement factor C1q family has a pleiotropic role in regulating inflammation, energy metabolism, and vascular function. In lean individuals, adiponectin is secreted to a similar degree from subcutaneous and visceral white adipose tissues [73,74,76]. However, in obese individuals, adiponectin secretion from visceral (but not subcutaneous) white adipose tissue is significantly reduced [13].

While expression of adiponectin is restricted to adipocytes in rodents [74], it is released in similar amounts from adipocytes and non-adipocyte cells of the tissue matrices of human white adipose tissue [13]—the major source of circulating adiponectin. However, modifiable adiponectin expression and/or secretion were reported for skeletal muscle, cardiac myocytes, hepatic endothelial cells, and osteoblasts, suggesting potential local physiological roles at these sites [77–80]. Full-length adiponectin is synthesized as a monomer that contains collagenous and globular domains. In the blood, full-length adiponectin exists as low, medium, and high molecular weight aggregates (LMW, MMW and HMW, respectively). The LMW form is a homotrimer consisting of three full-length adiponectin monomers. The MMW form is a hexamer composed of two disulfide-linked homotrimers. The HMW forms are multimeric adiponectin complexes formed by further assembly of the hexamers and contain 12 to 18 adiponectin monomers [76,81–83].

These various forms of adiponectin interact with two subtypes of adiponectin receptors (AdipoRs). AdipoR1 is expressed in brain, heart, kidney, liver, lung, spleen, and testes with the highest expression occurring in skeletal muscle. AdipoR2 is most highly expressed in the liver [84]. Using bacterial-generated recombinant proteins, Yamauchi and colleagues demonstrated that Adipo1 has a higher affinity for the globular subdomain of adiponectin and lower affinity for the naturally occurring full-length adiponectin [84,85]. Conversely, adipoR2 has intermediate affinity for globular and full-length adiponectins. While full-length adiponectin can undergo proteolytic cleavage to release globular adiponectin, the natural presence of the globular fragment in the circulation remains questionable [86–88]. It is important to note that recombinant full-length adiponectin generated in mammalian cells compared to the full-length protein generated in bacteria is more efficacious with respect to a number of measured metabolic endpoints [89,90]. Thus, it is likely that a lack of post-translational processing and oligomerization into HMW forms of bacterially derived full-length adiponectin prevents optimal interaction of adiponectin with AdipoR1 and 2 [87].

More recently, T-cadherin, a cell-surface receptor found on endothelial and smooth muscle cells, has been demonstrated to serve as a third putative receptor for MMW and HMW adiponectin [91]. The multitude of tissues that express adiponectin receptors, combined with different forms of circulating adiponectin likely contribute to the diverse anti-inflammatory, anti-atherogenic, anti-proliferative, and anti-diabetic metabolic effects of this adipokine.

At variance with the other adipokines described in this chapter, circulating adiponectin concentrations are decreased in obese and type 2 diabetic rodents and humans with insulin resistance [12,92–95]. Conversely, elevated levels of adiponectin in elderly women with peripheral adiposity are associated with increased insulin sensitivity and hypothesized to provide protection against the pro-atherogenic effects of inflammatory cytokines [96]. In addition to overall fat mass, visceral fat mass, an important risk factor for insulin resistance, is an independent negative predictor of adiponectin levels [92].

In agreement with these observations, adiponectin secretion from omental white adipose explants prepared from obese and diabetic patients was reduced to 67 and 34% of controls, respectively [14]. The mechanisms leading to a reduction in adiponectin with obesity are not clearly understood but may involve elevated TNF-α and IL-6 levels in white adipose tissue, both of which inhibit adiponectin expression and secretion from adipocytes and the vascular matrix. In support of this idea, blockage of TNF-α with etanercept enhanced the corresponding release of adiponectin from human white adipose tissue explants by 30% [14].

Increasing evidence indicates that the reduced circulating adiponectin levels characteristic of obesity are causative factors in the inflammatory, metabolic, and vascular complications of this disorder [97]. Collectively, experiments with mouse models of obesity and adiponectin- and adipoR-deficient mouse models show that adiponectin mediates a number of beneficial anti-diabetic metabolic effects including reduction of plasma triglycerides and free fatty acids along with reversal of hyperglycemia and insulin resistance [12,95,98]. For example, overexpression of adiponectin in *ob/ob* mice resulted in preferential distribution of lipid to white adipose tissue and prevention of ectopic fat distribution to muscle and liver, resulting in the maintenance of a normal metabolic profile in the face of increased white adipose tissue mass [99].

The beneficial effects of adiponectin on circulating triglycerides and insulin resistance are mediated through activation of AdipoR1 and R2 signalling in the liver where this adipokine activates PPARα and AMPK signaling pathways, ultimately increasing the β-oxidation of fatty acids and inhibiting hepatic gluconeogenesis and insulin-stimulated hepatic glucose output [84]. AdipoR1 and R2 are also expressed in the hypothalamus and administration of adiponectin into the lateral ventricle decreases body weight, blood glucose, blood insulin, and serum triglycerides in *ob/ob* mice, indicating that this adipokine may regulate metabolism and insulin sensitivity in the CNS [100].

With regard to insulin sensitivity, the overall distribution of adiponectin between LMW and HMW forms is more physiologically important than absolute levels of adiponectin [101]. This is supported by the observation that insulin sensitivity directly correlated with the ratio of HMW adiponectin to total adiponectin but not the absolute amount of circulating HMW or total adiponectin [101]. Further support for this idea arises from the observations that treatment of humans and mice with insulin-sensitizing thiazolidinediones (TZDs) selectively increased HMW adiponectin, whereas total adiponectin was only marginally increased. Moreover, the ability of TZDs to promote HMW adiponectin secretion from adipocytes and increase plasma levels of HMW adiponectin presents a novel mechanism that may contribute to the insulin-sensitizing effects of these agents [101–104].

The observation that humans with various point mutations in the *adiponectin* gene are deficient in HMW adiponectin, insulin-resistant, and diabetic confirms that anti-diabetic activity resides with the multimeric form of adiponectin [87]. Of interesting note, females have higher circulating levels of HMW adiponectin compared to males [83,87,96]. This prompted speculation that the sexual dimorphism in adiponectin distribution may contribute in part to the lower prevalence of insulin resistance and atherosclerosis in females [83,105,106].

Adiponectin exerts an anti-inflammatory effect through inhibition of the NF-κB transcription factor in adipocytes and endothelial cells and blockade of the production of the pro-inflammatory IL-6 and TNF-α cytokines [107,108]. Thus, the loss of adiponectin secretion in obesity has been suggested to contribute to inflammatory responses and endothelial dysfunction that leads to atherosclerotic vascular changes [109,110]. This link between reduced adiponectin and vascular disease is supported by a number of findings in animals. Adiponectin inhibits the development of atherosclerosis in ApoE knockout mice [111]. As a second example, adiponectin knockout mice develop hypertension on a high salt diet that can be ameliorated by adiponectin administration [112]. A similar role for adiponectin in protection against vascular disease appears to extend to humans. For example, high plasma adiponectin concentrations are associated with a lower risk of myocardial infarction in men [113] and decreased adiponectin levels are observed in hypertensive men compared to healthy controls [114].

## 2.5 TUMOR NECROSIS FACTOR-ALPHA (TNF-α)

TNF-α was originally characterized as cachectin, a cytokine factor produced from activated macrophages [115]. It is now well established that TNF-α is an adipokine with multiple biological functions including cell proliferation and death, metabolism, inflammation, and immune function. Most white adipose-secreted TNF-α originates from stromal vascular cells [13,116]. TNF-α acts in a paracrine fashion to inhibit adipogenesis and lipogenesis and activate lipolysis in adipocytes [115,117–119]. These combined actions reduce white adipose tissue mass and may contribute to the cachexia and hyperlipidemia that occur with certain infections and malignancies.

Increased expression and secretion of TNF-α from white adipose tissue has been reported in obese and insulin-resistant humans and rodents. [116,120–122]. In contrast, body weight reduction in obese humans is associated with lower white adipose TNF-α mRNA expression and improved insulin sensitivity [120]; in obese rodents, neutralizing antibodies for TNF-α can reverse insulin resistance [122]. The results of these studies support a causative role for TNF-α in mediating the pathogenic effects of obesity including insulin resistance, diabetes, and cardiovascular diseases [23,123].

At the molecular level, TNF-α contributes to insulin resistance through the inhibition of insulin-stimulated glucose uptake and lipoprotein lipase activity in white adipose tissue and insulin-stimulated glucose uptake and fatty acid metabolism in muscle [124–126]. White adipose tissue secretion of TNF-α (but not plasma TNF-α) was a predictor of the insulin-resistant state, suggesting that local effects of TNF-α in white adipose tissue may indirectly lead to insulin resistance in other organs [116].

One proposed mechanism whereby TNF-α may promote systemic insulin resistance indirectly is through a reduction in secretion of adiponectin, an anti-inflammatory adipokine that antagonizes many of the actions of TNF-α [76]. In recent years, anti-inflammatory therapies targeting TNF-α activity have emerged as experimental strategies for the treatment of obesity-related diseases. In double-blind placebo-controlled clinical studies, treatment of patients with metabolic syndrome with the etanercept TNF-α blocker reduced the levels of the CRP, IL-6, fibrinogen, and resistin inflammatory markers [127,128]. However, any potential beneficial effects of interrupting the inflammatory cascade occurring with abdominal obesity appear to be offset by increased muscle adiposity and a decrease in the ratio of HMW adiponectin to total adiponectin—a strong predictor of insulin sensitivity [128].

## 2.6 INTERLEUKIN-6 (IL-6)

IL-6 is a multi-functional adipokine that regulates immune responses and metabolism as well as the growth and differentiation of a variety of cell types. White adipose tissue secretion accounts for approximately 30% of the total amount of circulating IL-6, with the majority derived from non-adipocyte matrix cell constituents [13, 129, 130]. The biological effects of IL- 6 are mediated by the IL-6 receptor expressed by a variety of cell types including adipocytes, monocytes, hepatocytes, neurons, and glial cells [131–134]. Additionally, IL-6 may mediate inflammatory responses via binding to a soluble form of the IL-6 receptor in biological fluids and subsequent interactions with membrane-bound gp130 signal-transducing subunits on various cell types [135,136]. IL-6 acts in a paracrine fashion within white adipose tissue to reduce adiponectin secretion from adipocytes and reduce the production and activity of the lipase lipogenic enzyme lipoprotein [137,138].

Consistent with this, IL-6 has been linked to white adipose wasting that occurs in cancer cachexia [139]. IL-6 systematically induces hypertriglyceridemia by stimulating triglyceride secretion from hepatocytes [140]. In obese humans, elevated plasma blood levels of IL-6 reflect increased expression and secretion from white adipose tissue and are believed to be predictive of type 2 diabetes [116,130,141,142]. Furthermore, IL-6 alters the expression and/or functions of key proteins involved in insulin signalling and glucose transport, indicating a causative link between IL-6 and insulin resistance in skeletal muscle, liver, and adipocytes [141,143–145]. Interestingly, mice deficient in IL-6 were reported to develop mature onset obesity and decreased glucose tolerance that could be reversed partially by delivery of IL-6 into the lateral ventricle of the brain, but not when administered peripherally [146]. However, a more recent study failed to detect mature onset obesity, abnormal lipid metabolism, and hyperglycemia in IL-6 deficient mice [147]. These findings may reflect a more complex role for IL-6 in the regulation of metabolism that involves differential effects in the CNS versus the periphery.

## 2.7 RESISTIN

Resistin, a member of the resistin-like molecule (RELM) family of proteins, is a relatively recently discovered adipokine with dual roles in inflammation and metabolism [148]. Similar to adiponectin, resistin circulates in multiple forms including a trimer, a hexamer, and higher molecular weight oligomers [149,150]. Resistin was originally identified as a novel insulin resistance factor induced during 3T3-L1 adipocyte differentiation but down-regulated in response to treatment of mature adipocytes with TZDs [151,152]. Steppan and colleagues went on to demonstrate that the recombinant resistin protein impairs glucose tolerance and insulin action in mice [152]. In comparison, neutralization of resistin improves blood glucose and insulin sensitivity in mice with dietary obesity and enhances glucose uptake by adipocytes [152].

Thus, it was postulated that modification of resistin levels may serve as a causative link between obesity and insulin resistance and as a mechanism whereby TZDs exert anti-diabetic effects. Support for an attenuating effect of resistin on insulin action was provided by the subsequent observations that resistin knockout mice displayed decreased hepatic gluconeogenesis whereas replacement of resistin to these mice increased hepatic glucose output [153].

In comparing the trimer and hexamer isoforms of resistin, the former was more biologically active than the latter with respect to impairing hepatic insulin action in vivo [150]. In rat skeletal muscle, resistin decreased insulin-stimulated glucose uptake though inhibition of IRS-1 signaling and translocation of the GLUT4 facilitated glucose transporter to cell membranes [154]. Thus, apart from the liver, skeletal muscle may act as a secondary target for the insulin-attenuating effects of this adipokine.

Rodents exhibited dissociation between *resistin* expression in white adipose tissue, which decreases with obesity, and serum resistin levels that increase with obesity [152,155]. In humans, conflicting experimental data has generated some controversy regarding the localization of white adipose tissue expression and secretion of resistin. For example, Nagaev et al. reported that *resistin* expression was undetectable or detectable only at very low levels in a small proportion of human adipocytes and white adipose tissue samples [156]. Similarly Yang et al. reported that *resistin* expression in human white adipose tissue was less than 1% of that in mouse white adipose tissue. In contrast, McTernan et al. detected higher levels of resistin protein in human subcutaneous and omental white adipose tissue compared to rat white adipose tissue and demonstrated greater resistin secretion from pre-adipocytes compared to adipocytes [157].

Fain and coworkers demonstrated highly variable levels of resistin secretion from human subcutaneous and visceral white adipose tissue explants attributable to non-fat cells rather than adipocytes [158]. Other experimental evidence indicates that macrophages and pre-adipocytes are the primary sites of resistin expression in humans and that macrophages are quantitatively the major sources of resistin secretion in human white adipose tissue [156,159,160].

Thus, in obesity, infiltration of macrophages into the white adipose tissue may be responsible for increased resistin secretion from that compartment. Physiologically, the locally elevated levels of resistin may be important for altered adipocyte function as human resistin has been shown to stimulate pre-adipocyte proliferation and lipolysis in mature adipocytes [161]. Furthermore, elevated white adipose tissue resistin may contribute to localized inflammatory responses in obesity because human resistin stimulates the secretion of other pro-inflammatory cytokines including TNF-α and IL-12 [162].

In addition to the differences in white adipose tissue expression patterns, human and mouse resistins are only 59% identical at the protein level, leading to questions about a conserved role of resistin in the pathogenesis of human obesity [163]. The functional role of resistin is further obscured by widely varying results from human studies. In support of resistin involvement in inflammation and insulin resistance, McTernan and coworkers found a 20% increase in serum resistin levels in type 2 diabetics compared to controls [164]. Consistent with a potential pro-inflammatory function, the levels of resistin correlated moderately with the CRP inflammatory marker. However, they found no association with adiposity or fasting insulin levels. In keeping with the effects of TZDs to lower resistin levels in mice, rosiglitazone blocked insulin-dependent secretion of resistin from isolated human adipocytes [164]. Similarly, in non-diabetic humans with metabolic syndrome and in HIV-infected men with insulin resistance and increased adiposity, TZD treatment produced small reductions in resistin levels [165–167].

Some evidence indicates that single nucleotide polymorphisms of the resistin gene may be associated with insulin sensitivity in humans [168–171]. However, a number of other studies failed to identify a link of resistin and insulin resistance and type 2 diabetes in humans. For example, no relationship was found between adipocyte *resistin* expression and body weight, insulin sensitivity, or other metabolic parameters [159]. Nor did serum resistin levels correlate with various markers of adiposity (BMI, waist-to-hip ratio, fat mass), degree of insulin resistance, serum lipid profile, or serum leptin levels [172]. There was no difference in serum resistin levels of lean healthy and obese insulin-resistant nondiabetic and type 2 diabetic adolescents [172,173].

A second pediatric study found that resistin levels were higher in female compared to male children but found no difference in resistin levels between obese and lean children and no correlation between resistin levels and markers of insulin resistance in their subjects [149]. In adult patients treated with highly active antiretroviral therapy, no association between resistin levels and insulin resistance, fat redistribution or fat wasting, and metabolic abnormalities was detected [174]. While the jury is still out regarding the metabolic functions of resistin in humans, it is worth noting that human studies to date have been largely correlative. Studies in which resistin is directly administered or inhibited may shed more light on the role of this protein in human insulin resistance and diabetes.

## 2.8 CONCLUSIONS

Until relatively recently, white adipose tissue was regarded primarily as an organ of energy storage and mobilization. The discovery of leptin ushered in a new era of adipose biology that established the important regulatory role of this tissue in systemic energy homeostasis. Since then research of adipokines and the endocrine function of white adipose tissue has exploded and the known biological roles of adipokines have expanded to include pleiotropic regulatory effects on energy metabolism, vascular function, blood pressure regulation, inflammation, and immunomodulation.

It is now well established that changes in fat mass affect the secretion of numerous adipokines as well as the long-term risk for the development of prevalent diseases such as type 2 diabetes and cardiovascular disease. This suggests the intriguing possibility that manipulation of adipokine secretion and/or activity may be used as a novel therapeutic approach for the treatment of obesity and these obesity-associated comorbidities. Despite the largely disappointing results of trials utilizing exogenous leptin therapy as an anti-obesity agent, therapies that target or utilize adipokines still hold promise. For example, TZDs such as rosiglitazone and pioglitazone are among the most effective insulin-sensitizing drugs currently in use. A major effect of these drugs is to normalize white adipose tissue function and morphology, decrease serum leptin and TNF-α levels and increase serum adiponectin levels [103,104,175–178]. Thus, reversals of white adipose tissue dysfunction and aberrant adipokine secretion are likely significant aspects of the therapeutic efficacy of TZDs. Further research to fully elucidate the biological functions of adipokines and identify mechanisms that regulate the synthesis and secretion of these critical signaling molecules will contribute to the development of novel therapeutic approaches targeting adipokines.

## REFERENCES

1. Gorstein, J. and R.N. Grosse, The indirect costs of obesity to society. *Pharmacoeconomics*, 1994. 5: 58–61.
2. Pender, J.R. and W.J. Pories, Epidemiology of obesity in the United States. *Gastroenterol Clin N Am*, 2005. 34: 1–7.
3. McTiernan, A., Obesity and cancer: the risks, science, and potential management strategies. *Oncology*, 2005. 19: 871–881.
4. Muoio, D.M. and C.B. Newgard, Obesity-related derangements in metabolic regulation. *Annu Rev Biochem*, 2006. 75: 367–401.
5. Poirier, P. et al., Obesity and cardiovascular disease: pathophysiology, evaluation, and effect of weight loss: an update of the 1997 American Heart Association Scientific Statement on Obesity and Heart Disease. *Circulation*, 2006. 113: 898–918.
6. Poirier, P. et al., Obesity and cardiovascular disease: pathophysiology, evaluation, and effect of weight loss. *Arterioscler Thromb Vasc Biol*, 2006. 26: 968–976.
7. Wolk, R. and V.K. Somers, Obesity-related cardiovascular disease: implications of obstructive sleep apnea. *Diabetes Obes Metab*, 2006. 8: 250–260.
8. Cinti, S. et al., Adipocyte death defines macrophage localization and function in adipose tissue of obese mice and humans. *J Lipid Res*, 2005. 46: 2347–2355.
9. Cannon, B. and J. Nedergaard, Brown adipose tissue: function and physiological significance. *Physiol Rev*. 2004. 84: 277–359.
10. Cinti, S., The adipose organ: prostaglandins, *Leukot Essent Fatty Acids*, 2005. 73: 9–15.
11. Maeda, K. et al., Analysis of an expression profile of genes in the human adipose tissue. *Gene*, 1997. 190: 227–235.
12. Maeda, N. et al., Diet-induced insulin resistance in mice lacking adiponectin/ACRP30. *Nat Med*, 2002. 8: 731–737.

13. Fain, J.N. et al., Comparison of the release of adipokines by adipose tissue, adipose tissue matrix, and adipocytes from visceral and subcutaneous abdominal adipose tissues of obese humans. *Endocrinology*, 2004. 145: 2273–2282.
14. Fain, J.N. et al., Regulation of adiponectin release and demonstration of adiponectin mRNA as well as release by the non-fat cells of human omental adipose tissue. *Int J Obes*, 2008. 32: 429–435.
15. Itoh, K. et al., Relationship between changes in serum leptin levels and blood pressure after weight loss. *Hypertens Res*, 2002. 25: 881–886.
16. Folsom, A.R. et al., Impact of weight loss on plasminogen activator inhibitor (PAI-1), factor VII, and other hemostatic factors in moderately overweight adults. *Arterioscler Thromb*, 1993. 13: 162–169.
17. Primrose, J.N. et al., Reduction in factor VII, fibrinogen and plasminogen activator inhibitor-1 activity after surgical treatment of morbid obesity. *Thromb Haemost*, 1992. 68: 396–399.
18. Ziccardi et al., Reduction of inflammatory cytokine concentrations and improvement of endothelial functions in obese women after weight loss over one year. *Circulation*, 2002. 105: 804–809.
19. Dandona, P. et al., Tumor necrosis factor-α- in sera of obese patients: fall with weight loss. *J Clin Endocrinol Metab*, 1998. 83: 2907–2910.
20. Zhang, B. et al., Down-regulation of the expression of the obese gene by an antidiabetic thiazolidinedione in Zucker diabetic fatty rats and db/db mice. *J Biol Chem*, 1996. 271: 9455–9459.
21. Samad, F. et al., Elevated expression of transforming growth factor-β in adipose tissue from obese mice. *Mol Med,* 1997. 3: 37–48.
22. Samad, F., M. Pandey, and D.J. Loskutoff, Tissue factor gene expression in the adipose tissues of obese mice. *Proc Natl Acad Sci USA*, 1998. 95: 7591–7596.
23. Yudkin, J.S. et al., C-reactive protein in healthy subjects: associations with obesity, insulin resistance, and endothelial dysfunction: a potential role for cytokines originating from adipose tissue? *Arterioscler Thromb Vasc Biol*, 1999. 19: 972–978.
24. Hetherington, A. and S.W. Ranson, Hypothalamic lesions and adiposity in the rat. *Anat Rec*, 1940. 78: 149–172.
25. Anand, B.K. and J.R. Brobeck, Localization of a "feeding center" in the hypothalamus of the rat. *Proc Soc Exp Biol Med*, 1951. 77: 323–324.
26. Kennedy, G.C., The role of depot fat in the hypothalamic control of food intake in the rat. *Proc R Soc Lond B Biol Sci*, 1953. 140: 578–596.
27. Ingalls, A.M., M.M. Dickie, and G.D. Snell, Obese: a new mutation in the house mouse. *J Hered*, 1950. 41: 317–318.
28. Hummel, K.P., M.M. Dickie, and D.L. Coleman, Diabetes: a new mutation in the mouse. *Science*, 1966. 153: 1127–1128.
29. Coleman, D.L., Obese and diabetes: two mutant genes causing diabetes–obesity syndromes in mice. *Diabetologia*, 1978. 14: 141–148.
30. Coleman, D.L., Effects of parabiosis of obese with diabetes and normal mice. *Diabetologia*, 1973. 9: 294–298.
31. Coleman, D.L. and K.P. Hummel, Effects of parabiosis of normal with genetically diabetic mice. *Am J Physiol*, 1969. 217: 1298–1304.
32. Zhang, Y. et al., Positional cloning of the mouse obese gene and its human homologue. *Nature,* 1994. 372: 425–432.
33. Chen, H. et al., Evidence that the diabetes gene encodes the leptin receptor: identification of a mutation in the leptin receptor gene in db/db mice. *Cell*, 1996. 84: 491–495.
34. Tartaglia, L.A. et al., Identification and expression cloning of a leptin receptor, OB-R. *Cell*, 1995. 83: 1263–1271.

35. Lee, G.H. et al., Abnormal splicing of the leptin receptor in diabetic mice. *Nature*, 1996. 379: 632–635.
36. Grinspoon, S. et al., Serum leptin levels in women with anorexia nervosa. *J Clin Endocrinol Metab*, 1996. 81: 3861–3863.
37. Montague, C.T. et al., Congenital leptin deficiency is associated with severe early-onset obesity in humans. *Nature*, 1997. 387: 903–908.
38. Saad, M.F. et al., Diurnal and ultradian rhythmicity of plasma leptin: effects of gender and adiposity. *J Clin Endocrinol Metab*, 1998. 83: 453–459.
39. Boden, G. et al., Effect of fasting on serum leptin in normal human subjects. *J Clin Endocrinol Metab*, 1996. 81: 3419–3423.
40. Considine, R.V., Regulation of leptin production. *Rev Endocr Metab Disord*, 2001. 2: 357–363.
41. Grinspoon, S.K. et al., Effects of fasting and glucose infusion on basal and overnight leptin concentrations in normal-weight women. *Am J Clin Nutr*, 1997. 66: 1352–1356.
42. MacDougald, O.A. et al., Regulated expression of the obese gene product (leptin) in white adipose tissue and 3T3-L1 adipocytes. *Proc Natl Acad Sci USA*, 1995. 92: 9034–9037.
43. Kolaczynski, J.W. et al., Responses of leptin to short-term fasting and refeeding in humans: a link with ketogenesis but not ketones. *Diabetes*, 1996. 45: 1511–1515.
44. Halaas, J.L. et al., Weight-reducing effects of the plasma protein encoded by the obese gene. *Science*, 1995. 269: 543–546.
45. Pelleymounter, M.A. et al., Effects of the obese gene product on body weight regulation in ob/ob mice. *Science,* 1995. 269: 540–543.
46. Campfield, L.A. et al., Recombinant mouse OB protein: evidence for a peripheral signal linking adiposity and central neural networks. *Science*, 1995. 269: 546–549.
47. Heymsfield, S.B. et al., Recombinant leptin for weight loss in obese and lean adults: a randomized, controlled, dose-escalation trial. *JAMA*, 1999. 282: 1568–1575.
48. Hukshorn, C.J. et al., The effect of pegylated recombinant human leptin (PEG-OB) on weight loss and inflammatory status in obese subjects. *Int J Obes Relat Metab Disord*, 2002. 26: 504–509.
49. Zelissen, M. et al., Effect of three treatment schedules of recombinant methionyl human leptin on body weight in obese adults: a randomized, placebo-controlled trial. *Diabetes Obes Metab*, 2005. 7: 755–761.
50. Caro, J.F. et al., Decreased cerebrospinal fluid/serum leptin ratio in obesity: a possible mechanism for leptin resistance. *Lancet*, 1996. 348: 159–161.
51. Van Heek, M. et al., Diet-induced obese mice develop peripheral, but not central, resistance to leptin. *J Clin Invest*, 1997. 99: 385–390.
52. Wang, M.Y. et al., Fat storage in adipocytes requires inactivation of leptin's paracrine activity: implications for treatment of human obesity. *Proc Natl Acad Sci USA*, 2005. 102: 18011–18016.
53. Rosenbaum, M. et al., Leptin reverses weight loss-induced changes in regional neural activity responses to visual food stimuli. *J Clin Invest*, 2008. 118: 2583–2591.
54. Asilmaz, E. et al., Site and mechanism of leptin action in a rodent form of congenital lipodystrophy. *J Clin Invest*, 2004. 113: 414–424.
55. Seip, M. and O. Trygstad, Generalized lipodystrophy, congenital and acquired (lipoatrophy). *Acta Paediatr Suppl*, 1996. 413: 2–28.
56. Oral, E.A. et al., Leptin replacement therapy modulates circulating lymphocyte subsets and cytokine responsiveness in severe lipodystrophy. *J Clin Endocrinol Metab*, 2006. 91: 621–628.
57. Kim, J.K. et al., Mechanism of insulin resistance in A-ZIP/F-1 fatless mice. *J Biol Chem*, 2000. 275: 8456–8460.

58. Pajvani, U.B. et al., Fat apoptosis through targeted activation of caspase 8: a new mouse model of inducible and *reversible* lipoatrophy. *Nat Med*, 2005. 11: 797–803.
59. Oral, E.A. et al., Leptin replacement therapy for lipodystrophy. *New Engl J Med*, 2002. 346: 570–578.
60. Lee, J.H. et al., r-metHuLeptin therapy in replacement doses improves insulin resistance and metabolic profile in patients with lipoatrophy and metabolic syndrome induced by the highly active antiretroviral therapy (HAART). *J Clin Endocrinol Metab*, 2006.
61. Spiegelman, B.M. and J.S. Flier, Obesity and the regulation of energy balance. *Cell*, 2001. 104: 531–543.
62. Friedman, J.M., Obesity in the new millennium. *Nature,* 2000. 404: 632–634.
63. Trayhurn, J., The development of obesity in animals: the role of genetic susceptibility. *Clin Endocrinol Metab*, 1984. 13: 451–474.
64. Muller, G. et al., Leptin impairs metabolic actions of insulin in isolated rat adipocytes. *J Biol Chem*, 1997. 272: 10585–10593.
65. Minokoshi, Y. et al., Leptin stimulates fatty acid oxidation by activating AMP-activated protein kinase. *Nature*, 2002. 415: 339–343.
66. Cawthorne, M.A. et al., Peripheral metabolic actions of leptin. *Proc Nutr Soc*, 1998. 57: 449–453.
67. Otero, M. et al., Leptin: a metabolic hormone that functions like a pro-inflammatory adipokine. *Drug News Perspect*, 2006. 19: 21–26.
68. Otero, M. et al., Toward a pro-inflammatory and immunomodulatory emerging role of leptin. *Rheumatology* (Oxford), 2006.
69. Lord, G.M. et al., Leptin modulates the T cell immune response and reverses starvation-induced immunosuppression. *Nature*, 1998. 394: 897–901.
70. Weisberg, S.P. et al., Obesity is associated with macrophage accumulation in adipose tissue. *J Clin Invest*, 2003. 112: 1796–1808.
71. Xu, H. et al., Chronic inflammation in fat plays a crucial role in the development of obesity-related insulin resistance. *J Clin Invest*, 2003. 112: 1821–1830.
72. Curat, C.A. et al., From blood monocytes to adipose tissue-resident macrophages: induction of diapedesis by human mature adipocytes. *Diabetes*, 2004. 53: 1285–1292.
73. Scherer, E. et al., A novel serum protein similar to C1q, produced exclusively in adipocytes. *J Biol Chem*, 1995. 270: 26746–26749.
74. Hu, E., Liang, and B.M. Spiegelman, AdipoQ is a novel adipose-specific gene dysregulated in obesity. *J Biol Chem*, 1996. 271: 10697–10703.
75. Maeda, K. et al., cDNA cloning and expression of a novel adipose specific collagen-like factor, apM1 (AdiPose most abundant hene transcript 1). *Biochem Biophys Res Commun*, 1996. 221: 286–289.
76. Whitehead, J.P. et al., Adiponectin: a key adipokine in the metabolic syndrome. *Diabetes Obes Metab*, 2006. 8: 264–280.
77. Berner, H.S. et al., Adiponectin and its receptors are expressed in bone-forming cells. *Bone*, 2004. 35: 842–849.
78. Delaigle, A.M. et al., Induction of adiponectin in skeletal muscle of type 2 diabetic mice: in vivo and in vitro studies. *Diabetologia,* 2006. 49: 1311–1323.
79. Ding, G. et al., Adiponectin and its receptors are expressed in adult ventricular cardiomyocytes and upregulated by activation of peroxisome proliferator-activated receptor-$\gamma$. *J Mol Cell Cardiol*, 2007. 43: 73–84.
80. Wolf, A.M. et al., Upregulation of the anti-inflammatory adipokine adiponectin in acute liver failure in mice. *J Hepatol*, 2006. 44: 537–543.
81. Fruhbeck, G., *Vasoactive Factors and Inflammatory Mediators Produced in Adipose Tissue, in Adipose Tissue and Adipokines in Health and Disease*, Fantuzzi, G. and T. Mazzone, Eds., 2007, Humana Press, Totowa, NJ, pp. 47–59.

82. Combs, T.P. et al., Sexual differentiation, pregnancy, calorie restriction, and aging affect the adipocyte-specific secretory protein adiponectin. *Diabetes*, 2003. 52: 268–276.
83. Xu, A. et al., Testosterone selectively reduces the high molecular weight form of adiponectin by inhibiting its secretion from adipocytes. *J Biol Chem*, 2005. 280: 18073–18080.
84. Yamauchi, T. et al., Cloning of adiponectin receptors that mediate antidiabetic metabolic effects. *Nature*, 2003. 423: 762–769.
85. Yamauchi, T. et al., Adiponectin stimulates glucose utilization and fatty acid oxidation by activating AMP-activated protein kinase. *Nat Med,* 2002. 8: 1288–1295.
86. Fruebis, J. et al., Proteolytic cleavage product of 30-kDa adipocyte complement-related protein increases fatty acid oxidation in muscle and causes weight loss in mice. *Proc Natl Acad Sci USA*, 2001. 98: 2005–2010.
87. Waki, H. et al., Impaired multimerization of human adiponectin mutants associated with diabetes: molecular structure and multimer formation of adiponectin. *J Biol Chem*, 2003. 278: 40352–40363.
88. Waki, H. et al., Generation of globular fragment of adiponectin by leukocyte elastase secreted by monocytic cell line THP-1. *Endocrinology*, 2005. 146: 790–796.
89. Berg, A.H. et al., The adipocyte-secreted protein Acrp30 enhances hepatic insulin action. *Nat Med*, 2001. 7: 947–953.
90. Combs, T.P. et al., Endogenous glucose production is inhibited by the adipose-derived protein Acrp30. *J Clin Invest*, 2001. 108: 1875–1881.
91. Hug, C. et al., T-cadherin is a receptor for hexameric and high molecular weight forms of Acrp30/adiponectin. *Proc Natl Acad Sci USA*, 2004. 101: 10308–10313.
92. Gavrila, A. et al., Serum adiponectin levels are inversely associated with overall and central fat distribution but are not directly regulated by acute fasting or leptin administration in humans: cross-sectional and interventional studies. *J Clin Endocrinol Metab*, 2003. 88: 4823–48231.
93. Chen, M.P. et al., Elevated plasma level of visfatin/pre-B cell colony-enhancing factor in patients with type 2 diabetes mellitus. *J Clin Endocrinol Metab*, 2006. 91: 295–299.
94. Arita, Y. et al., Paradoxical decrease of an adipose-specific protein, adiponectin, in obesity. *Biochem Biophys Res Commun*, 1999. 257: 79–83.
95. Yamauchi, T. et al., The fat-derived hormone adiponectin reverses insulin resistance associated with both lipoatrophy and obesity. *Nat Med*, 2001. 7: 941–946.
96. Tanko, L.B. et al., Novel associations between bioavailable estradiol and adipokines in elderly women with different phenotypes of obesity: implications for atherogenesis. *Circulation*, 2004. 110: 2246–2252.
97. Tilg, H. and A.R. Moschen, Adipocytokines: mediators linking adipose tissue, inflammation and immunity. *Nat Rev Immunol*, 2006. 6: 772–783.
98. Yamauchi, T. et al., Targeted disruption of AdipoR1 and AdipoR2 causes abrogation of adiponectin binding and metabolic actions. *Nat Med*, 2007. 13: 332–339.
99. Kim, J.Y. et al., Obesity-associated improvements in metabolic profile through expansion of adipose tissue. *J Clin Invest*, 2007. 117: 2621–2637.
100. Qi, Y. et al., Adiponectin acts in the brain to decrease body weight. *Nat Med*, 2004. 10: 524–529.
101. Pajvani, U.B. et al., Complex distribution, not absolute amount of adiponectin, correlates with thiazolidinedione-mediated improvement in insulin sensitivity. *J Biol Chem*, 2004. 279: 12152–12162.
102. Choi, K.C. et al., Effect of PPAR-$\alpha$- and -$\gamma$ agonists on the expression of visfatin, adiponectin, and TNF–$\alpha$ in visceral fat of OLETF rats. *Biochem Biophys Res Commun*, 2005. 336: 747–753.
103. Yu, J.G. et al., The effect of thiazolidinediones on plasma adiponectin levels in normal, obese, and type 2 diabetic subjects. *Diabetes*, 2002. 51: 2968–2974.

104. Maeda, N. et al., PPAR-γ ligands increase expression and plasma concentrations of adiponectin, an adipose-derived protein. *Diabetes,* 2001. 50: 2094–2099.
105. Czubryt, M.P. et al., The role of sex in cardiac function and disease. *Can J Physiol Pharmacol*, 2006. 84: 93–109.
106. Quyyumi, A.A., Women and ischemic heart disease: pathophysiologic implications from the Women's Ischemia Syndrome Evaluation (WISE) Study and future research steps. *J Am Coll Cardiol*, 2006. 47): S66–S71.
107. Ouchi, N. et al., Adiponectin, an adipocyte-derived plasma protein, inhibits endothelial NF-κB signaling through a cAMP-dependent pathway. *Circulation*, 2000. 102: 1296–1301.
108. Ajuwon, K.M. and M.E. Spurlock, Adiponectin inhibits LPS-induced NF-κB activation and IL-6 production and increases PPAR γ2 expression in adipocytes. *Am J Physiol Regul Integr Comp Physiol*, 2005. 288: R1220–R1225.
109. Goldstein, B.J. and R. Scalia, Adiponectin: a novel adipokine linking adipocytes and vascular function. *J Clin Endocrinol Metab*, 2004. 89: 2563–2568.
110. Shimabukuro, M. et al., Hypoadiponectinemia is closely linked to endothelial dysfunction in man. *J Clin Endocrinol Metab*, 2003. 88: 3236–3240.
111. Yamauchi, T. et al., Globular adiponectin protected ob/ob mice from diabetes and ApoE-deficient mice from atherosclerosis. *J Biol Chem*, 2003. 278: 2461–2468.
112. Ohashi, K. et al., Adiponectin replenishment ameliorates obesity-related hypertension. *Hypertension*, 2006. 47: 1108–1116.
113. Pischon, T. et al., Plasma adiponectin levels and risk of myocardial infarction in men. *JAMA*, 2004. 291: 1730–1737.
114. Dzielinska, Z. et al., Decreased plasma concentration of a novel anti-inflammatory protein—adiponectin—in hypertensive men with coronary artery disease. *Thromb Res*, 2003. 110: 365–369.
115. Beutler, B. et al., Identity of tumour necrosis factor and the macrophage-secreted factor cachectin. *Nature*, 1985. 316: 552–554.
116. Kern, A. et al., Adipose tissue tumor necrosis factor and interleukin-6 expression in human obesity and insulin resistance. *Am J Physiol Endocrinol Metab*, 2001. 280: E745–E751.
117. Torti, F.M. et al., A macrophage factor inhibits adipocyte gene expression: an in vitro model of cachexia. *Science*, 1985. 229: 867–869.
118. Cornelius et al., Regulation of lipoprotein lipase mRNA content in 3T3-L1 cells by tumour necrosis factor. *Biochem J*, 1988. 249: 765–769.
119. Beutler, B.A., I.W. Milsark, and A. Cerami, Cachectin/tumor necrosis factor: production, distribution, and metabolic fate in vivo. *J Immunol,* 1985. 135: 3972–3977.
120. Kern, A. et al., The expression of tumor necrosis factor in human adipose tissue: regulation by obesity, weight loss, and relationship to lipoprotein lipase. *J Clin Invest*, 1995. 95: 2111–2119.
121. Hotamisligil, G.S. et al., Increased adipose tissue expression of tumor necrosis factor-α in human obesity and insulin resistance. *J Clin Invest*, 1995. 95: 2409–2415.
122. Hotamisligil, G.S., N.S. Shargill, and B.M. Spiegelman, Adipose expression of tumor necrosis factor-α: direct role in obesity-linked insulin resistance. *Science*, 1993. 259: 87–91.
123. Katsuki, A. et al., Serum levels of tumor necrosis factor-α are increased in obese patients with noninsulin-dependent diabetes mellitus. *J Clin Endocrinol Metab*, 1998. 83: 859–862.
124. Bullo, M. et al., TNF-α expression of subcutaneous adipose tissue in obese and morbid obese females: relationship to adipocyte LPL activity and leptin synthesis. *Int J Obes Relat Metab Disord*, 2002. 26: 652–658.

125. Hotamisligil, G.S. et al., IRS-1-mediated inhibition of insulin receptor tyrosine kinase activity in TNF-α- and obesity-induced insulin resistance. *Science,* 1996. 271: 665–668.
126. Mingrone, G. et al., Skeletal muscle triglycerides lowering is associated with net improvement of insulin sensitivity, TNF-α reduction and GLUT4 expression enhancement. *Int J Obes Relat Metab Disord*, 2002. 26: 1165–1172.
127. Bernstein, L.E. et al., Effects of etanercept in patients with the metabolic syndrome. *Arch Intern Med*, 2006. 166: 902–908.
128. Lo, J. et al., Effects of TNF-α neutralization on adipocytokines and skeletal muscle adiposity in the metabolic syndrome. *Am J Physiol Endocrinol Metab*, 2007. 293: E102–E109.
129. Fried, S.K., D.A. Bunkin, and A.S. Greenberg, Omental and subcutaneous adipose tissues of obese subjects release interleukin-6: depot difference and regulation by glucocorticoid. *J Clin Endocrinol Metab*, 1998. 83: 847–850.
130. Mohamed-Ali, V. et al., Subcutaneous adipose tissue releases interleukin-6, but not tumor necrosis factor-α, in vivo. *J Clin Endocrinol Metab*, 1997. 82: 4196–4200.
131. Heinrich, C. et al., Principles of interleukin (IL)-6-type cytokine signalling and its regulation. *Biochem J*, 2003. 374: 1–20.
132. Schobitz, B., D.A. Voorhuis, and E.R. De Kloet, Localization of interleukin-6 mRNA and interleukin-6 receptor mRNA in rat brain. *Neurosci Lett*, 1992. 136: 189–192.
133. Bauer, J. et al., Regulation of interleukin-6 receptor expression in human monocytes and hepatocytes. *FEBS Lett*, 1989. 249: 27–30.
134. Path, G. et al., Human breast adipocytes express interleukin-6 (IL-6) and its receptor system: increased IL-6 production by β-adrenergic activation and effects of IL-6 on adipocyte function. *J Clin Endocrinol Metab*, 2001. 86: 2281–2288.
135. Rose-John, S. et al., The IL-6/sIL-6R complex as a novel target for therapeutic approaches. *Expert Opin Ther Targets*, 2007. 11: 613–624.
136. Peters, M., K.H. Meyer zum Buschenfelde, and S. Rose-John, The function of the soluble IL-6 receptor in vivo. *Immunol Lett*, 1996. 54: 177–184.
137. Greenberg, A.S. et al., Interleukin-6 reduces lipoprotein lipase activity in adipose tissue of mice in vivo and in 3T3-L1 adipocytes: a possible role for interleukin-6 in cancer cachexia. *Cancer Res*, 1992. 52: 4113–4116.
138. Fasshauer, M. et al., Adiponectin gene expression and secretion is inhibited by interleukin-6 in 3T3-L1 adipocytes. *Biochem Biophys Res Commun*, 2003. 301: 1045–1050.
139. Strassmann, G. et al., Evidence for the involvement of interleukin-6 in experimental cancer cachexia. *J Clin Invest*, 1992. 89: 1681–1684.
140. Nonogaki, K. et al., Interleukin-6 stimulates hepatic triglyceride secretion in rats. *Endocrinology*, 1995. 136: 2143–2149.
141. Rotter, V., I. Nagaev, and U. Smith, Interleukin-6 (IL-6) induces insulin resistance in 3T3-L1 adipocytes and is, like IL-8 and tumor necrosis factor-α, overexpressed in human fat cells from insulin-resistant subjects. *J Biol Chem*, 2003. 278: 45777–45784.
142. Roytblat, L. et al., Raised interleukin-6 levels in obese patients. *Obes Res*, 2000. 8: 673–675.
143. Kim, H.J. et al., Differential effects of interleukin-6 and -10 on skeletal muscle and liver insulin action in vivo. *Diabetes*, 2004. 53: 1060–1067.
144. Klover, J., A.H. Clementi, and R.A. Mooney, Interleukin-6 depletion selectively improves hepatic insulin action in obesity. *Endocrinology*, 2005. 146: 3417–3427.
145. Klover, J. et al., Chronic exposure to interleukin-6 causes hepatic insulin resistance in mice. *Diabetes*, 2003. 52: 2784–2789.
146. Wallenius, V. et al., Interleukin-6-deficient mice develop mature-onset obesity. *Nat Med*, 2002. 8: 75–79.

147. Di Gregorio, G.B. et al., Lipid and carbohydrate metabolism in mice with a targeted mutation in the IL-6 gene: absence of development of age-related obesity. *Am J Physiol Endocrinol Metab*, 2004. 287: E182–E187.
148. Steppan, C.M. et al., A family of tissue-specific resistin-like molecules. *Proc Natl Acad Sci USA*, 2001. 98: 502–506.
149. Gerber, M. et al., Serum resistin levels of obese and lean children and adolescents: biochemical analysis and clinical relevance. J *Clin Endocrinol Metab*, 2005. 90 4503–4509.
150. Patel, S.D. et al., Disulfide-dependent multimeric assembly of resistin family hormones. *Science*, 2004. 304: 1154–1158.
151. Steppan, C.M. and M.A. Lazar, Resistin and obesity-associated insulin resistance. *Trends Endocrinol Metab*, 2002. 13: 18–23.
152. Steppan, C.M. et al., The hormone resistin links obesity to diabetes. *Nature*, 2001. 409: 307–312.
153. Banerjee, R.R. et al., Regulation of fasted blood glucose by resistin. *Science*, 2004. 303: 1195–1198.
154. Palanivel, R. et al., Regulation of insulin signalling, glucose uptake and metabolism in rat skeletal muscle cells upon prolonged exposure to resistin. *Diabetologia*, 2006. 49: 183–190.
155. Rajala, M.W. et al., Regulation of resistin expression and circulating levels in obesity, diabetes, and fasting. *Diabetes*, 2004. 53: 1671–1679.
156. Nagaev, I. and U. Smith, Insulin resistance and type 2 diabetes are not related to resistin expression in human fat cells or skeletal muscle. *Biochem Biophys Res Commun*, 2001. 285: 561–564.
157. McTernan, G. et al., Increased resistin gene and protein expression in human abdominal adipose tissue. *J Clin Endocrinol Metab*, 2002. 87: 2407–2412.
158. Fain, J.N. et al., Resistin release by human adipose tissue explants in primary culture. *Biochem Biophys Res Commun*, 2003. 300: 674–678.
159. Janke, J. et al., Resistin gene expression in human adipocytes is not related to insulin resistance. *Obes Res*, 2002. 10: 1–5.
160. Curat, C.A. et al., Macrophages in human visceral adipose tissue: increased accumulation in obesity and a source of resistin and visfatin. *Diabetologia*, 2006. 49: 744–747.
161. Ort, T. et al., Recombinant human FIZZ3/resistin stimulates lipolysis in cultured human adipocytes, mouse adipose explants, and normal mice. *Endocrinology*, 2005. 146: 2200–2209.
162. Silswal, N. et al., Human resistin stimulates the pro-inflammatory cytokines TNF-α and IL-12 in macrophages by NF-κB-dependent pathway. *Biochem Biophys Res Commun*, 2005. 334: 1092–1101.
163. Ghosh, S. et al., The genomic organization of mouse resistin reveals major differences from the human resistin: functional implications. *Gene*, 2003. 305: 27–34.
164. McTernan, G. et al., Resistin and type 2 diabetes: regulation of resistin expression by insulin and rosiglitazone and the effects of recombinant resistin on lipid and glucose metabolism in human differentiated adipocytes. *J Clin Endocrinol Metab*, 2003. 88: 6098–6106.
165. Kamin, D. et al., Resistin levels in human immunodeficiency virus-infected patients with lipoatrophy decrease in response to rosiglitazone. *J Clin Endocrinol Metab*, 2005. 90: 3423–3426.
166. Samaha, F.F. et al., Effects of rosiglitazone on lipids, adipokines, and inflammatory markers in nondiabetic patients with low high-density lipoprotein cholesterol and metabolic syndrome. *Arterioscler Thromb Vasc Biol*, 2006. 26: 624–630.

167. Szapary,O. et al., Effects of pioglitazone on lipoproteins, inflammatory markers, and adipokines in nondiabetic patients with metabolic syndrome. *Arterioscler Thromb Vasc Biol*, 2006. 26: 182–188.
168. Bouchard, L. et al., Human resistin gene polymorphism is associated with visceral obesity and fasting and oral glucose stimulated C peptide in the Quebec Family Study. *J Endocrinol Invest*, 2004. 27: 1003–1009.
169. Smith, S.R. et al., A promoter genotype and oxidative stress potentially link resistin to human insulin resistance. *Diabetes*, 2003. 52: 1611–1618.
170. Wang, H. et al., Human resistin gene: molecular scanning and evaluation of association with insulin sensitivity and type 2 diabetes in Caucasians. *J Clin Endocrinol Metab*, 2002. 87: 2520–2524.
171. Pizzuti, A. et al., An ATG repeat in the 3' untranslated region of the human resistin gene is associated with a decreased risk of insulin resistance. *J Clin Endocrinol Metab*, 2002. 87: 4403–4406.
172. Lee, J.H. et al., Circulating resistin levels are not associated with obesity or insulin resistance in humans and are not regulated by fasting or leptin administration: cross-sectional and interventional studies in normal, insulin-resistant, and diabetic subjects. *J Clin Endocrinol Metab*, 2003. 88: 4848–4856.
173. Pfutzner, A. et al., Evaluation of human resistin assays with serum from patients with type 2 diabetes and different degrees of insulin resistance. *Clin Lab*, 2003. 49: 571–576.
174. Barb, D. et al., Circulating resistin levels are not associated with fat redistribution, insulin resistance, or metabolic profile in patients with the highly active antiretroviral therapy-induced metabolic syndrome. *J Clin Endocrinol Metab*, 2005. 90: 5324–5328.
175. He, W. et al., Adipose-specific peroxisome proliferator-activated receptor γ knockout causes insulin resistance in fat and liver but not in muscle. *Proc Natl Acad Sci USA*, 2003. 100: 15712–15717.
176. Yin, W.H. et al., Differential effects of peroxisome proliferator-activated receptor ligands and sulfonylurea plus statin treatment on plasma concentrations of adipokines in type 2 diabetes with dyslipidemia. *Diabetes Metab*, 2006. 32: 229–235.
177. Katsuki, A. et al., Troglitazone reduces plasma levels of tumour necrosis factor-α in obese patients with type 2 diabetes. *Diabetes Obes Metab*, 2000. 2: 189–191.
178. Kato, K. et al., Improvement by the insulin-sensitizing agent, troglitazone, of abnormal fibrinolysis in type 2 diabetes mellitus. *Metabolism*, 2000. 49: 662–665.
179. Goralski, K.B. et al., Chemerin, a novel adipokine that regulates adipogenesis and adipocyte metabolism. *J Biol Chem*, 2007. 282: 28175–28188.
180. Fukuhara, A. et al., Visfatin: a protein secreted by visceral fat that mimics the effects of insulin. *Science*, 2005. 307: 426–430.
181. Yang, Q. et al., Serum retinol binding protein 4 contributes to insulin resistance in obesity and type 2 diabetes. *Nature*, 2005. 436: 356–362.
182. Eriksson, et al., Adipose tissue secretion of plasminogen activator inhibitor-1 in non-obese and obese individuals. *Diabetologia*, 1998. 41: 65–71.
183. Choy, L.N., B.S. Rosen, and B.M. Spiegelman, Adipsin and an endogenous pathway of complement from adipose cells. *J Biol Chem*, 1992. 267: 12736–12741.
184. Cianflone, K. and M. Maslowska, Differentiation-induced production of ASP in human adipocytes. *Eur J Clin Invest*, 1995. 25: 817–825.
185. Karlsson, C. et al., Human adipose tissue expresses angiotensinogen and enzymes required for its conversion to angiotensin II. *J Clin Endocrinol Metab*, 1998. 83: 3925–3929.
186. Fain, J.N., D.S. Tichansky, and A.K. Madan, Transforming growth factor β1 release by human adipose tissue is enhanced in obesity. *Metabolism*, 2005. 54: 1546–1551.

187. Christiansen, T., B. Richelsen, and J.M. Bruun, Monocyte chemoattractant protein-1 is produced in isolated adipocytes, associated with adiposity and reduced after weight loss in morbid obese subjects. *Int J Obes*, 2005. 29: 146–150.
188. Fain, J.N., D.S. Tichansky, and A.K. Madan, Most of the interleukin 1 receptor antagonist, cathepsin S, macrophage migration inhibitory factor, nerve growth factor, and interleukin 18 release by explants of human adipose tissue is by the non-fat cells, not by the adipocytes. *Metabolism*, 2006. 55: 1113–1121.
189. Anty, R. et al., The inflammatory C–reactive protein is increased in both liver and adipose tissue in severely obese patients independently from metabolic syndrome, Type 2 diabetes, and NASH. *Am J Gastroenterol*, 2006. 101: 1824–1833.
190. Fain, J.N. and A.K. Madan, Regulation of monocyte chemoattractant protein 1 (MCP-1) release by explants of human visceral adipose tissue. *Int J Obes*, 2005. 29: 1299–1307.
191. Fain, J.N. et al., Release in vitro of adipsin, vascular cell adhesion molecule 1, angiotensin 1-converting enzyme, and soluble tumor necrosis factor receptor 2 by human omental adipose tissue as well as by the nonfat cells and adipocytes. *Metabolism*, 2007. 56: 1583–1590.

# 3 Epidemiology of Obesity

*Michael J. LaMonte*

## CONTENTS



## 3.1 INTRODUCTION

Obesity is a serious threat to public health worldwide [1–3]. In developed countries, obesity prevalence historically has been greatest among those of lower socioeconomic status, whereas in developing populations obesity was a condition of affluence. However, the prevalence of obesity (body mass index [BMI] ≥30 kg/m$^2$) has risen steadily during recent decades among all the major population subgroups (e.g., gender, age, social class), leading obesity to replace more traditional public health concerns such as infectious diseases and malnutrition as a principal cause of illness [2]. Approximately 250 million individuals are clinically obese and it is estimated that by 2025 this number will increase to more than 300 million [2]. Among U.S. adults who at age 40 are non-obese, the estimated residual lifetime risk of becoming obese is close to 50% in both women and men [4].

Diabetes, hypertension, and coronary heart disease (CHD) are major consequences of obesity [5,6]. Obesity at age 40 has been estimated to reduce life expectancy by at least 6 years [7]. Based on the rapid increase in obesity among adolescents during the past 15 years [2], the negative effects of obesity on health and longevity may considerably worsen in the forthcoming quarter century. The economic burdens of managing obesity-related risk factors, treating obesity-related comorbidities, and covering lost wages and productivity arising from obesity-related disabilities are staggering. The direct costs associated with obesity account for ≈6% ($99 billion) of total healthcare expenditures in the U.S. [8] and ≈3% in Europe [2]. Although obesity is an avoidable risk factor for premature morbidity and mortality, the current obesity epidemic is a sobering international problem that carries a heavy societal toll.

## 3.2 ASSESSMENT AND CLASSIFICATION OF OBESITY

In order to characterize and compare the distribution and determinants of obesity among defined populations, a standardized case definition and assessment method must be used to classify obese and non-obese phenotypes. Differences among studies in the reported prevalence of obesity and in the strength and pattern of association for obesity exposures and health outcomes depend, in part, on the methods of exposure assessment. Synonymous and inappropriate use of the *overweight* and *obesity* terms leads to further discrepancies regarding the true population distributions of both conditions and their associated health risks [2,9,10].

*Overweight* refers to body mass in excess of a standard, often a percentage of "ideal" weight or a weight-for-height criterion derived from actuarial tables of life insurance companies [9]. Some individuals can be overweight without being obese, for example, a heavily muscled athlete [11]. *Obesity* refers to excessive body fat or *adiposity* [9]. Body fat is diffuse and essentially inaccessible for direct quantification. Laboratory procedures for estimating body fat mass include hydrodensitometry, isotope dilution, dual x-ray absorptiometry (DXA), magnetic resonance imaging (MRI), and computed tomography (CT) scans. Although these methods provide highly accurate assessments of body composition (fat mass and lean mass) [9], their use in clinical practice or population research is limited by feasibility issues and by lack of standard criteria to define high risk levels of fat mass or percent body fat.

The most common method of assessing and classifying obesity status is body mass index (BMI) computed as weight in kilograms divided by the square of height in meters [9,10,12]. By standardizing body mass to the square of stature, this measure accounts for differences in body mass expected among individuals of different statures—a weight of 100 kg has a completely different meaning for an individual who stands 150 cm tall compared to one who is 190 cm tall.

An important assumption when using BMI as a surrogate measure of adiposity is that most variations in weight among individuals of the same height are due to differences in fat mass. This assumption tends to be true at the population level but may not be for a given individual. Body composition varies considerably with sex, age, race–ethnicity, nutritional status, and physical conditioning; thus a criterion-referenced BMI scale may not correspond to the same levels of body fat within or between populations [12–15].

Women tend to have 10% higher body fat than men at a given BMI, and the relationship between BMI and fat mass weakens with advanced age. One of the best examples of an inappropriate use of BMI as a measure of adiposity is a report showing that National Football League players are nearly all overweight, if not obese [16]. Despite its inability to discriminate between the various aspects of body composition, BMI is a simple, inexpensive measure that is mostly independent of stature, is highly correlated with total body fat ($r = 0.70$ to 0.89) in women and men representing a wide range of ages and BMIs [13–15,17], and is characterized by extensive international reference data on its distribution and association with health outcomes in a variety of populations.

Another important distinction when assessing and classifying obesity is the pattern or distribution of body fat [1,2,9]. Considerable variation in fat distribution exists among individuals of similar BMI or percent body fat [9,18]. Of particular concern is android fat distribution, otherwise known as abdominal obesity, that appears to be associated with a variety of neuroendocrine and metabolic disturbances that increase risk of diabetes, hypertension, and CHD, independent of overall adiposity [9,18,19]. Several methods have been used to quantify fat distribution including more costly and burdensome laboratory measures such as CT and MRI scans and simple anthropometric measures such as site-specific circumferences and related ratios, site-specific skinfolds and related ratios, and sagittal abdominal diameter [9,18,19]. The simplest approach to assessing and classifying abdominal obesity in clinical and research settings is the waist circumference measurement obtained using an anthropometric tape measure level with the iliac crest at the end of a normal expiration. None of the other anthropometric measures of fat distribution appears to provide significant additional information on health risk beyond that provided by waist circumference measurement [18,19].

Correlations among BMI, waist circumference, total body fat mass, and abdominal fat mass (from CT scan) are high [17–19]. For example, correlations with BMI were $r = 0.93$ for waist, $r = 0.72$ for abdominal fat mass, and $r = 0.94$ for total body fat mass. Correlations with waist circumference were $r = 0.77$ and $r = 0.92$ for abdominal and total body fat mass, respectively [17]. Given the large amount of shared variation among BMI, waist circumference, and measures of total and abdominal fat mass, some authorities argue that BMI is a more than adequate primary means of assessing and classifying obesity in clinical and population settings [17,20]. The use of waist circumference, however, or another anthropometric measure of fat distribution, may improve risk assessment at a given BMI.

Defining the level at which excessive body weight or excessive body fat characterizes overweight and obesity, respectively, is difficult and remains a matter of debate. One proposed classification system considers men and women to be obese at percent body fat levels of >25% and >33%, respectively [21]. Ideally, a health-oriented approach would be used to develop classifications of overweight and obesity based on criterion levels above which population health risks increase. Very few prospective studies have included measures of body fat, and thus it is difficult to determine percent body fat criteria that define obesity on the basis of increased risk for various health outcomes.

An extensive evidence base exists, however, on the associations of BMI and anthropometric measures of fat distribution with health outcomes in a variety of populations. Historically, a variety of BMI-defined criteria have been used to classify overweight and obesity [10]. Comparisons of reported population trends and study findings have been difficult and led to considerable debate on defining healthy body weight and composition. The World Health Organization [2] and National Institutes of Health [1] recently adopted similar BMI cut-points for use in clinical and research settings (Table 3.1). Individuals are classified as underweight, normal weight, overweight, or obese based on recommended BMI cut-points that are

**TABLE 3.1**
**Classification of Weight Status according to BMI and Health Risk Based Jointly on BMI and Waist Circumference**

| | | Health Risk[a] Compared with Normal BMI and Waist Circumference | |
|---|---|---|---|
| BMI, Kg/m$^2$ | Classification | Waist: Men <102 cm; Women <88 cm | Waist: Men ≥102 cm; Women ≥88 cm |
| <18.5 | Underweight | Increased | – |
| 18.5 to 24.9 | Normal weight | Average | Increased |
| 25.0 to 29.9 | Overweight | Increased | High |
| 30.0 to 34.9 | Class I obesity | High | Very high |
| 35.0 to 39.9 | Class II (severe) obesity | Very high | Very high |
| ≥40 | Class III (very severe) obesity | Extremely high | Extremely high |

*Source:* National Heart, Lung, and Blood Institute. *Clinical Guidelines on the Identification, Evaluation, and Treatment of Overweight and Obesity in Adults: The Evidence Report.* Rockville, MD: National Institutes of Health, 1998.

[a] Risk of heart disease, hypertension, and diabetes mellitus.

assumed to segregate health risk similarly in women and men and across the adult age range. Further, both WHO and NIH provide criteria for waist circumference as a measure of abdominal obesity for use as an adjuvant to characterizing health risk at a given level of BMI.

Investigators from the U.S. Health and Nutrition Examination Survey (HANES) recently reported the following threshold levels of percent body fat (from bioelectrical impedance) among apparently healthy adults for BMI cut-points shown in Table 3.1: BMI = 25 (overweight), ≈29% fat in men, ≈31% fat in women; BMI = 30 (obesity), ≈29% fat in men, ≈37% fat in women [22]. Because available data on the health risks associated with body size and composition are particularly variable among populations within the Asia-Pacific region, different BMI criteria for overweight and obesity have been developed for Asians (overweight, BMI 23.0 to 25.0; obesity, BMI ≥ 25.0) and Pacific Islanders (overweight, BMI 26.0 to 32.0; obesity, BMI ≥ 32.0) [2].

## 3.3 PREVALENCE OF OBESITY

Time trend analyses of several populations worldwide indicate that the average body mass and the prevalence of obesity have sharply increased in the past 10 to 20 years [2]. Among European centers in the WHO MONICA study, obesity prevalence ranged from 4 to 22% in men and 8 to 43% in women from 1989 to 1996 [23]. Among women in 38 developing countries, the prevalence of obesity during the 1990s was 0.1% in South Asia, 2.5% in Sub-Saharan Africa, 5.6% in Latin America and the Caribbean, 15.5% in the Central Eastern Europe Commonwealth of Independent States, and 17.2% in Middle East and North Africa; comparatively, obesity prevalence during the same period was

20.7% among U.S. women [24]. Between 1993 and 2003, obesity prevalence in Great Britain rose from 13.4 to 22.7% in men, and from 15.8 to 22.4% in women [25].

Among U.S. adults aged 20 to 74 years, the average height has increased by about 1 inch and the average weight by 24 pounds between 1960 and 2002 [26]. The modest average weight gain of 0.5 pound/year among U.S. adults has been paralleled by a sharp rise in obesity prevalence during the same 42-year interval (Figure 3.1). A notable increase in obesity prevalence is seen in 1988 through 1994 followed by a steady progressive rise thereafter through 2003 and 2004. The secular increase in obesity prevalence has been consistent in major subgroups of the U.S. population (Table 3.2). The pattern of increase in overall obesity is similar in women and men, although prevalence is higher in women at each time interval. Among individuals born in the U.S. between 1887 and 1975, the 50th, 75th, and 90th percentiles of BMI at age 50 were estimated to have increased by 0.69, 1.00, and 1.37 kg/m$^2$, respectively, within successive 10-year increments of birth date [3]. These data suggest that increases in BMI-defined overweight and obesity may be accelerated in an aging U.S. population. The prevalence of severe obesity (BMI ≥40 kg/m$^2$) showed a marked increase between 1960 to 1962 and 2003 to 2004 in U.S. women (1.4% to 6.9%) and men (0.3% to 2.8%) [27,28], a condition for which the economic and individual burden is staggering.

Abdominal obesity also is on the rise in U.S. and European populations [25,29,30]. In 2005, the global prevalence of abdominal obesity was estimated to be 20% in men and 48% in women. Between 1993 and 2003 in Great Britain, average (± standard deviation [SD]) waist circumference (cm) increased from 92.9 ± 10.7 to 96.5 ± 11.6 in men and from 80.9 ± 11.3 to 85.2 ± 12.6 in women [25]. Among U.S. adults, the secular trends for increased prevalence of abdominal obesity parallel those seen for overall obesity during the 1960 to 1962 and 2003 to 2004 interval (Table 3.2). Notably,

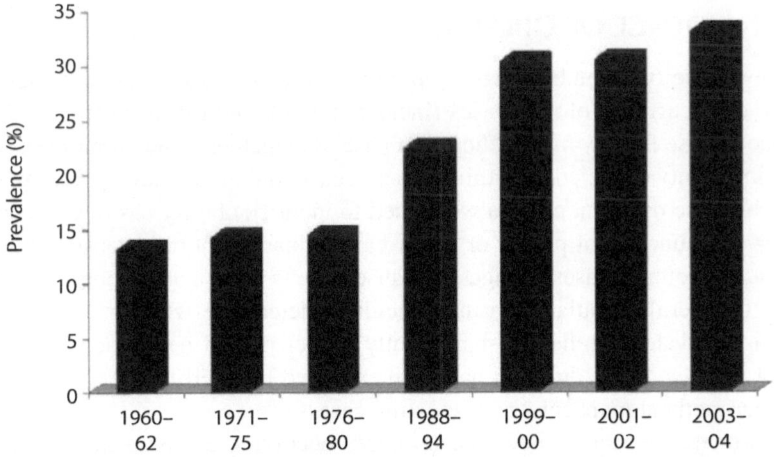


**FIGURE 3.1** Secular trends in prevalence of obesity (BMI ≥30 kg/m$^2$) among U.S. adults ≥20 years of age (HANES 1960–1962 to 2003–2004) [27,28].

**TABLE 3.2**
**Trends in Prevalence (%) of Abdominal and Overall Obesity among U.S. Adults (≥20 Years of Age) by Sex, Age, and Race–Ethnicity (HANES, 1960–1962 and 2003–2004)**

| | Obesity[a] | | | Abdominal Obesity[a] | | |
|---|---|---|---|---|---|---|
| **Population** | **1960–62** | **1988–94** | **2003–04** | **1960–62** | **1988–94** | **2003–04** |
| All | 13.4 | 22.3 | 32.2 | 16.2 | 37.3 | 52.4 |
| **Sex** | | | | | | |
| Male | 10.6 | 19.6 | 31.1 | 12.7 | 27.7 | 42.3 |
| Female | 16.2 | 24.9 | 33.2 | 19.4 | 46.3 | 62.0 |
| **Age (Years)** | | | | | | |
| 20–39 | 9.4 | 17.6 | 28.5 | 19.0 | 22.3 | 38.0 |
| 40–59 | 15.7 | 28.8 | 36.8 | 21.3 | 45.8 | 58.3 |
| ≥60 | 16.9 | 21.7 | 31.0 | 28.1 | 53.2 | 64.2 |
| **Race–Ethnicity** | | | | | | |
| White | — | 20.0 | 30.6 | 15.1 | 37.2 | 51.5 |
| Black | — | 21.3 | 45.0 | 20.5 | 42.6 | 55.2 |
| Hispanic | — | 23.1 | 36.8 | — | 46.4 | 55.7 |

[a] Obesity = BMI ≥30 kg/m$^2$. Abdominal obesity = waist circumference ≥102 cm in men and ≥88 cm in women. Prevalence estimates (%) unadjusted; shown as reported or approximated from reported data [27–29,74]. Dashes lines indicate data not reported.

half of U.S. adults have abdominal obesity, the prevalence of which is nearly two in three for women and two in three for adults ≥60 years.

## 3.4 INCIDENCE OF OBESITY

The prevalence of a condition at any point in time depends on the case definition, the rate of occurrence of new cases (incidence rate), and the average case duration (balance of case fatality rates, efficacy of case management, and cure rates). Secular trends in obesity prevalence within or between populations may appear to change simply because different criteria were used to identify obesity cases. Case duration changes as a function of poorer or better case management or because the underlying incidence of new cases changes. Because little is known about obesity incidence within the overall population, it is difficult to determine whether recent changes in obesity prevalence reflect a significantly higher rate of new case occurrence or whether the mean case duration has been extended by decreased case fatality rates and better medical management of existing cases.

To estimate the rate of new obesity cases occurring during a defined observation period, a cohort of initially non-obese individuals would need to have their BMIs characterized and undergo regular periodic re-assessments during the observation period. This would allow for computation of the number of new obesity cases standardized to the amount of time each non-obese (at risk) individual was under observation until the exact onset of obesity. Such a study generally is not feasible on a large-scale population basis. More often, a population of non-obese individuals completes a BMI assessment at the beginning and end of a specified observation period. The proportion of the initial population defined as obese at the second assessment gives an approximation of the underlying incidence of obesity during the follow-up interval. This approximation likely does not reflect the true rate of obesity because individuals who become obese during follow-up and revert to being non-obese before the second BMI assessment are not counted as cases.

One of the first such studies was conducted by Williamson et al. [31] among 9,862 U.S. adults ages 25 to 74 years who underwent initial BMI assessments in 1971 through 1975 and then again from 1982 through 1984 as part of the HANES. Obesity cases were defined as a BMI ≥27.3 kg/m$^2$ which was consistent with national criteria at the time. During a follow-up interval of 10 years, an average weight gain of 2 pounds was seen in the combined population of women and men. The incidence of obesity among initially non-obese study participants ranged from 4 to 13 cases per 100 population at risk in women, and from 5 to 16 cases per 100 population at risk in men, depending on age and race–ethnicity. In both sexes, the highest obesity incidence was among those aged 35 to 44 years. Interestingly, the incidence of major weight gain (defined as an increase of >5 BMI units above baseline) was 2-fold higher in women (1.0 to 8.5 cases per 100 women) than men (0.5 to 3.75 cases per 100 men), which broadly aligns with the sex differences in overall and abdominal obesity shown in Table 3.2.

A recent report from the Framingham Heart Study [32] expands on the earlier work of Williamson and colleagues. BMI was assessed among study participants at two examinations separated on average by 8 years within each of 5 decades between 1950 and 1990. The incidence of obesity (BMI ≥30 kg/m$^2$) was 5.8 cases per 100 men at risk in the 1950s and 14.8 cases per 100 men at risk in the 1990s. Among women, obesity incidence was 3.9 and 14.0 cases per 100 at risk in the 1950s and 1990s, respectively. These findings suggest a progressive 3-fold increase in obesity incidence during the past 50 years, and like the findings reported by Williamson et al. [31], align with the secular changes in obesity prevalence among U.S. adults during the same interval (Table 3.2).

## 3.5 HEALTH CONSEQUENCES OF OBESITY

Obesity is associated with a variety of prevalent medical conditions and with increased risk of premature morbidity, disability, and mortality. An extensive review of the extant literature is not possible here; rather a selective overview of major aspects of the obesity-related disease burden is presented.

### 3.5.1 Coronary Heart Disease and Metabolic Risk Factors

The prevalence of major modifiable CHD risk factors (smoking, hypercholesterolemia, hypertension, diabetes mellitus, and physical inactivity) is substantially higher among obese compared with normal weight adults, despite a recent trend for declining risk factor prevalence, except for diabetes and physical inactivity, in each of the major BMI groups [33,34]. The prevalence of non-traditional CHD risk factors such as inflammatory biomarkers [35], metabolic syndrome [36], obstructive sleep apnea [37], non-alcoholic fatty liver disease [38], and measures of subclinical CHD such as coronary artery calcium scores, carotid artery wall thickness, and left ventricular hypertrophy [39] also is greater among obese than non-obese adults.

Epidemiological evidence for a link between inflammation and obesity is seen in recent data from the U.S. HANES [35]. The age-adjusted prevalence of elevated C-reactive protein (CRP; exceeding the sex-specific 85th percentile) was 8.7%, 8.8%, 15.3%, 25.9%, 38.0%, and 51.3% across BMI categories of <18.5 (underweight), 18.5 to 24.9 (normal weight), 25.0 to 29.9 (overweight), 30.0 to 34.9 (class 1 obesity), 35.0 to 39.9 (class 2 obesity), and ≥40 kg/m$^2$ (class 3 obesity). After adjusting for several biological and lifestyle factors and for anti-inflammatory medication use, the odds of having elevated CRP was 2- to 4-fold higher in obese men and 4- to 11-fold higher in obese women, compared with their same-sex non-obese counterparts.

Interestingly, a recent cross-sectional report on 5,440 U.S. adults in the HANES showed that one in three obese participants had clinically normal values for a variety of traditional and non-traditional CHD risk factors (including CRP); likewise, about one in five adults of normal weight exhibited clinically abnormal risk factor profiles [40]. These data suggest that environmental or behavioral factors may influence phenotypic risk factor expression in susceptible individuals beyond the effect of weight status alone.

### 3.5.2 Mortality

Several prospective studies have shown a significant increased risk of total and cause-specific mortality among obese compared with normal weight adults. Analysis of pooled data from 388,622 adults (60,374 deaths) from 26 large prospective cohort studies [41] showed that after adjusting for age and smoking status, compared with normal weight adults (BMI 18.5 to 25.0), overweight (BMI 25 to 30) was not associated with increased risk of total or cancer mortality, but was associated with a significant 10 to 16% greater risk of CHD mortality. Obesity (BMI ≥30) was associated with a significant 20 to 27% greater risk for total mortality and 51 to 62% greater risk for CHD mortality; obesity was not associated with cancer mortality. These findings generally are consistent with two recent reports on U.S. adults in the HANES that showed overweight was not associated with increased risk of total or cancer mortality, and that class 2 obesity (BMI >35) was associated with significantly higher risk of total and CHD mortality in adults aged 25 to 69 years, but not in those >70 years [42,43].

Other studies have shown that measures of total body fat and fat distribution are important mortality predictors. After 8 years of follow-up on 21,925 men aged 20 to 83, age-adjusted death rates (per 10,000 man years) across percent body fat levels <16.7%, 16.7 to 25.0%, and >25.0% were 23, 20, and 30 for total mortality (trend, $P < 0.05$), and 6, 8, and 13 for cardiovascular mortality (trend, $P < 0.05$) [44].

Similar patterns of mortality were seen across incremental tertiles of waist circumference. In a 10-year follow-up study on 39,756 U.S. men aged 40 to 75, after adjusting for age, smoking habits, and other potential confounders, the relative risk for total mortality was 1.37 (P <0.05), and for cardiovascular mortality was 2.94 (P <0.05) in men whose waist circumference was >102 cm compared with <87.6 cm [45]. Waist circumference was not associated with cancer mortality, and when analyses were stratified by age groups (<65 versus ≥65 yr), waist circumference generally was not associated with mortality risk in the older men. Another study in 57,053 Danish adults aged 50 to 64 years showed that greater waist circumference and higher levels of fat mass and fat-free mass (determined by bioelectrical impedance) were associated with greater risk of total mortality (P <0.05 each) in both sexes [46]. However, adjusting for waist circumference eliminated the risk associated with fat mass and fat-free mass, whereas adjusting for either of the former factors did not influence the significant mortality risk associated with a large waist circumference.

The amount of excess mortality associated with obesity has been a matter of recent debate [42,43,47,48]. Based on analysis of the U.S. HANES, an initial report indicated that of the total 2.3 million adult deaths in 2000, 400,000 (16.6%) were attributed to a combination of overweight and obesity (BMI ≥25) [48]. However, a subsequent report by a different investigator using the same data but better analytic methodology indicated no excess deaths attributed to overweight and only 111,909 deaths (4.8%) attributable to obesity [42]. The more modest latter estimate of excess deaths related to obesity should not mask the substantial health risks of obesity; rather it provides a more accurate scale of the public health burden of obesity in comparison to other behavioral health risks such as smoking, alcoholism, poor diet, and physical inactivity [47].

### 3.5.3 Morbidity

The association between obesity and new onset morbidity varies in magnitude from relative risks ≥3 for type 2 diabetes mellitus, hypertension, gallbladder disease, and sleep apnea, to relative risks of 2 to 3 for CHD and musculoskeletal disorders, to relative risks of <2 for breast, endometrial, and colon cancers, polycystic ovary syndrome, and impaired fertility [5,6,9]. Table 3.3 shows the relative risks associated with levels of BMI and waist circumference for incident cardiovascular disease, type 2 diabetes, hypertension, and breast cancer in three large cohorts of U.S. adults [49,50]. Significantly higher risks of each outcome are seen with greater levels of BMI or waist circumference.

In the study of cardiovascular disease [50], adjusting the BMI-CVD association for waist girth and vice versa substantially attenuated the relative risks. The value for BMI or waist at which risk becomes significantly elevated varies by sex and outcome and depends on the value assigned for the referent category. A doubling of cardiovascular risk occurs at a BMI of about 35 kg/m$^2$ in both women and men, whereas it occurs at a waist circumference of about 122 cm in men and about 111 cm in women. Likewise, diabetes risk doubles at BMI and waist measurements of about 23 kg/m$^2$ and 75 cm, respectively, but hypertension risk doubles for a BMI and waist of 30 kg/m$^2$ and 88 cm, respectively. In the study of Iowa women [49], the investigators specifically evaluated the WHO and NHLBI criteria for evaluating health risk based jointly on BMI and waist circumference. Although there were too few women

## TABLE 3.3
## Relative Risk of Chronic Disease Outcomes Based on BMI and Waist Circumference

| | Study Characteristics | Findings[a] | | | |
|---|---|---|---|---|---|
| Ref.[50] | 16,332 U.S. men, 32,700 U.S. women; mean age 61; self-reported HT, WT, WC<br>Outcome: fatal and nonfatal CVD)<br>Follow-up: men: 14 yr (1,505 cases), women: 6 yr (414 cases) | BMI, kg/m$^2$ | | Waist, cm | |
| | | Men: | RR (95% CI) | | RR (95% CI) |
| | | <20.0: | 0.83 (0.55–1.24) | <80.0 | 0.60 (0.33–1.06) |
| | | 20.0–22.4: | 0.95 (0.80–1.12) | 80.0–86.9 | 0.96 (0.80–1.15) |
| | | 22.5–24.9: | 1.00 | 87.0–94.6 | 1.00 |
| | | 25.0–27.4: | 1.18 (1.04–1.35) | 94.5–103.1 | 1.22 (1.07–1.40) |
| | | 27.5–29.9: | 1.39 (1.18–1.64) | 103.2–110.5 | 1.48 (1.27–1.74) |
| | | 30.0–34.9: | 1.97 (1.61–2.41) | 110.6–121.9 | 1.67 (1.37–2.05) |
| | | ≥35.0: | 2.12 (1.36–3.30) | ≥122.0 | 2.15 (1.44–3.21) |
| | | Women: | RR (95% CI) | | RR (95% CI) |
| | | <20.0: | 0.89 (0.54–1.46) | <69.2 | 0.85 (0.43–1.65) |
| | | 20.0–22.4: | 0.70 (0.49–1.02) | 69.2–76.2 | 1.00 (0.68–1.49) |
| | | 22.5–24.9: | 1.00 | 76.3–84.5 | 1.00 |
| | | 25.0–27.4: | 1.08 (0.80–1.47) | 84.6–92.7 | 1.49 (1.09–2.05) |
| | | 27.5–29.9: | 1.47 (1.06–2.03) | 92.3–98.4 | 1.55 (1.08–2.23) |
| | | 30.0–34.9: | 1.46 (1.06–2.01) | 98.5–111.1 | 1.53 (1.09–2.14) |
| | | ≥35.0: | 2.11 (1.46–3.05) | ≥111.2 | 2.78 (1.95–3.97) |
| Ref.[49] | 31,702 U.S. women age 55–69; self-reported HT, WT, WC<br>Follow-up: ≈12 yr<br>Outcomes: diabetes (1,578 cases), hypertension (4,077 cases), breast cancer (1,299 cases) | BMI, kg/m$^2$ | Diabetes | Hypertension | Br. Cancer |
| | | | RR (95% CI) | RR (95% CI) | RR (95% CI) |
| | | <22.8: | 1.0 | 1.0 | 1.0 |
| | | 22.8–24.8: | 1.9 (1.4–2.5) | 1.1 (1.0–1.3) | 1.2 (0.9–1.4) |
| | | 24.9–27.1: | 2.9 (2.2–3.8) | 1.4 (1.2–1.5) | 1.2 (0.9–1.4) |
| | | 27.2–30.2: | 6.6 (5.0–8.5) | 1.7 (1.6–1.9) | 1.3 (1.1–1.5) |
| | | >30.2: | 13.8 (10.6–17.8) | 2.2 (2.0–2.4) | 1.4 (1.2–1.7) |
| | | Waist, cm | | | |
| | | <74.3: | 1.0 | 1.0 | 1.0 |
| | | 74.3–79.9: | 1.7 (1.2–2.3) | 1.1 (1.0–1.3) | 1.4 (1.1–1.7) |
| | | 80.0–87.2: | 3.4 (2.5–4.6) | 1.5 (1.3–1.6) | 1.2 (1.0–1.5) |
| | | 87.3–96.0: | 7.2 (5.5–9.6) | 1.8 (1.6–2.0) | 1.4 (1.2–1.7) |
| | | >96.0: | 16.5 (12.6–21.7) | 2.3 (2.1–2.5) | 1.7 (1.4–2.0) |
| | | | RR (95% CI) of Diabetes, Hypertension, CVD | | |
| | | BMI, kg/m$^2$ | | WC <88 cm | WC ≥88 cm |
| | | <18.5 (underweight): | | 0.92 (0.7–1.3) | — |
| | | 18.5–24.9 (normal weight): | | 1.00 | 1.2 (1.1–1.3) |
| | | 25.0–29.9 (overweight): | | 1.3 (1.2–1.4) | 1.9 (1.7–2.1) |
| | | 30.0–34.9 (Class 1 obesity) | | 1.5 (1.1–2.0) | 2.4 (2.2–2.7) |
| | | 35.0–39.0 (Class 2 obesity) | | — | 3.1 (2.6–3.7) |
| | | ≥40 (Class 3 obesity) | | — | 4.2 (3.1–5.7) |

[a] RR = relative risk. CI = confidence interval. Multivariable adjusted RRs and CIs reported. Covariables include age, smoking, physical activity, alcohol and diet intake, family history of disease, and other risk predictors.

cross-classified as low BMI–high waist and vice versa, the overall findings seem to support improved risk assessment for diabetes, hypertension, and CVD when based jointly on BMI and waist rather than on BMI alone. A similar finding was reported for diabetes risk in men [51].

## 3.6 WEIGHT CHANGE AND HEALTH RISK

Observational data on the associations of body weight and fat distribution generally derive from a single baseline exposure of weight status or adiposity and the subsequent occurrence of morbid and mortal events. Stronger evidence for a causal relationship comes from studies that obtain serial adiposity assessments and then relate patterns of change in adiposity with subsequent disease risk. Intentional weight loss of even modest amounts (e.g., 5 to 10%) appears beneficial in improving risk factor levels [52], but recidivism is high and adverse risk factors often re-emerge even after significant intentional weight loss [53].

Using economic forecasting models, it has been estimated that a sustained 10% weight loss in an initially obese individual would reduce expected lifetime health-care costs associated with cardiovascular-related diseases by $22,000 to $53,000 [54]. Prospective data regarding weight loss and weight gain, subsequent risk of mortality, and incident nonfatal disease are sparse. Among 5,608 British men aged 40 to 59 years compared with men whose weight remained stable over 12 years, those who gained >4% of baseline weight did not have increased mortality risk, whereas those who lost >4% of baseline weight or whose weight cycled by >4% had a significant 34 to 44% greater risk in total and cardiovascular mortality [55]. Most of the excess mortality associated with weight loss or cycling was accounted for by differences in socioeconomic status or pre-existing disease.

In a follow-up of U.S. nurses who reported weight changes between ages 18 and ≈40, weight loss was not associated with mortality risk, but weight gain of 10 to 20 kg was associated with a significant 32 to 51% greater risk of total, cardiovascular, and cancer mortality; weight gain of 40 kg more than doubled mortality risk from each cause [56]. In the same study, each 4-kg weight gain was associated with a 20% greater risk ($P < 0.05$) of hypertension, whereas weight loss and cycling were not associated with hypertension occurrence [57]. In a large cohort of U.S. men who reported 10-year changes in body weight and waist circumference, each 1-kg weight gain was associated with a 7% greater risk ($P < 0.05$) of diabetes; each 14-cm increase in waist circumference was associated with a 70% greater risk ($P < 0.05$) of diabetes, independent of overall weight gain [58]. These data provide little evidence of excess risk associated with adult weight loss, but clearly confirm the adverse effect of weight gain.

## 3.7 METHODOLOGICAL ISSUES IN STUDY OF OBESITY AND HEALTH

A variety of methodological issues can lead to inconsistent study findings on obesity-related health risks [42,59]. The potential for errors in recall and the resulting misclassification of obesity status is a concern in epidemiological studies that rely on questionnaire assessments of height, weight, BMI, and waist circumference [60]. However, prospective studies in which obesity status was assessed objectively using BMI from measured height and weights (61), using fat and fat-free mass from hydrodensitometry [44] and bioelectrical impedance [46], and using abdominal fat mass from CT scans [62] yielded findings similar to studies that used questionnaire exposure assessments [41,49–51]. The dose–response between BMI and disease risk has been reported to be J-shaped, U-shaped, and linear [1,59]. Two major methodological considerations underlying discrepancies in the patterns of dose-response are failure to adequately account for smoking habits and the presence of subclinical disease.

Since both characteristics likely are prevalent among those in the lower part of the BMI distribution (BMI <20), insufficient control for these factors could lead to a spuriously high disease risk at lower BMI values. Studies that excluded past and current smokers and the cases of disease identified early in follow-up (e.g., during the first 3 to 5 years) generally have shown that the higher risks associated with lower BMI are eliminated [1,59]. Furthermore, adjusting associations between adiposity and disease outcomes for factors that are likely intermediate in the causal pathway (e.g., dyslipidemia, hyperglycemia, hypertension) may spuriously lower the risk estimated for BMI, waist circumference, or other measures of adiposity [59,64].

Growing evidence indicates that greater adiposity may not confer increased disease risk among older populations [42,43,45,63]. Simply controlling for age in multivariable analysis may obscure the true age-specific association between obesity and disease; when possible, analyses stratified by age groups are preferable. A recent 25-year follow-up on BMI and cardiovascular mortality risk revealed an important issue related to follow-up duration [64]. In 38,379 adults, age- and smoking-adjusted relative risks were computed in four separate age strata for women and men and deaths occurring in early follow-up were excluded. During follow-ups earlier than 15 years, associations were weak and often non-significant (7 to 24% higher risk per one SD increase in BMI). However, when follow-up was extended to 25 years, significant and stronger associations were seen in all age–sex strata (10 to 42% higher risk per one SD increase in BMI). These findings suggest that follow-up on studies of shorter duration may be insufficient to detect the adverse health effects of excess body weight and adiposity. This methodological limitation may contribute to inconsistency among reported study findings.

## 3.8 DETERMINANTS OF OBESITY

Consistent with the definition of epidemic, the prevalence of BMI-defined obesity is well in excess of expected population levels and a plausible propagating force underlies this excess phenotypic expression. An average positive caloric balance of 5.5 kcal/day is all that is required to produce the average weight gain of 24 pounds in 42 years recently reported in the U.S. [28]. If sustained over time, even this small positive caloric balance results in a sizable amount of fat accretion and subsequent increases in average body mass and BMI at the population level. Energy balance is a complex issue and the causes of positive caloric balance are unclear [65,66]. The genetically susceptible individual is becoming more clearly identified [67].

However, the human genetic constitution has not changed significantly in the past 10,000 years [68]; thus, maladaptive homeostatic changes in body mass regulation are likely the results of environmental and lifestyle forces interacting with a given level of genetic susceptibility. Environmental potentiators include a more mechanized built environment and lifestyle potentiators include readily available and low cost energy-dense foods and declines in daily physical activity-related energy expenditure. Cultural and economic factors are other likely converging forces toward net energy storage in specific population subgroups, particularly from an international vantage [2,66]. Data from the 1996 U.S. Behavioral Risk Factor Surveillance Survey indicate that less than half of obese adults receive advice to lose weight from their healthcare providers [69].

Secular trends in dietary habits suggest a paradox between energy intake and obesity patterns among U.S. adults [70]. During the interval from 1971 to 1974 and 1999 to 2000, daily total energy intake increased modestly from 2,450 kcal in men and 1,542 kcal in women to 2,618 kcal and 1,877 kcal, respectively. Paralleling these changes in energy intake, were small increases in the percentage of total kilocalories from carbohydrates (from 42 to 45% to 49 to 51%) and small decreases in the percent of kilocalories from total fat (36 to 32%) and from saturated fats (13 to 10%). It is unlikely that this pattern of shifts in dietary intake fully accounts for the substantial increase in obesity prevalence during the same time interval. More likely, the marked reductions in daily physical activity-related energy expenditure (proportion of population reporting sedentary lifestyles: 1960 to 1962, 13%; 2003 to 2004, 33%) made it near impossible for most adults to restrict energy intake to a level that matches their true energy requirements. Thus, sustained positive energy balance gives rise to population levels of obesity of epidemic proportions.

Disentangling the complex multifactorial etiology of obesity and its relationships with disease occurrence present major challenges. For example, physical inactivity may be an antecedent or a consequence of obesity, and higher levels of activity may attenuate whereas physical inactivity may accentuate obesity-related disease risk [71]. Several large prospective studies have shown that obese individuals who are physically active and have adequate functional capacity have lower mortality risk than their non-obese sedentary peers [71].

In a study of 462 obese men at high risk for cardiovascular disease because of existing type 2 diabetes, the multivariable adjusted risk of cardiovascular mortality was 2.8 ($P < 0.05$) in men with low ( <6 METs) compared with moderate and higher (>6 METs) levels of exercise tolerance measured during a treadmill exercise stress test [61]. In another study among men whose percent body fat was >25%, cancer mortality rates (per 10,000 man years) were 25 and 16 in men with low compared to moderate and higher exercise tolerance; among men whose waist circumference was >102 cm, cancer mortality rates were 21 and 14, respectively [72]. Among 1,358 obese women who survived initial diagnoses of breast cancer, a 48% lower risk of breast cancer mortality was seen across incremental quintiles (trend, $P = 0.01$) of reported physical activity habits prior to diagnosis [73].

These findings do not dismiss the adverse consequences of obesity, but rather emphasize the favorable influence of physical activity and functional capacity on these risks among individuals in chronic positive energy balance. Considerably more research is needed to better define the specific roles of energy intake and energy expenditure in relation to sustained positive energy balance. Because currently available data do not identify a single (or most important) causal agent of the recent positive energy balance that has afflicted many, it is crucial for both the scientific and lay communities to be vigilant in promoting sensible healthy eating habits and regular physical activity as attainable strategies for long-term weight management in the population.

## REFERENCES

1. National Heart, Lung, and Blood Institute. *Clinical Guidelines on the Identification, Evaluation, and Treatment of Overweight and Obesity in Adults: The Evidence Report.* Rockville, MD: National Institutes of Health, 1998.
2. World Health Organization. *Obesity: Preventing and Managing the Global Epidemic. Report of a World Health Organization Consultation.* Geneva, World Health Organization, Technical Report 894. 2000.
3. Eckel RH et al. Prevention Conference VII: Obesity, a worldwide epidemic related to heart disease and stroke: executive summary. *Circulation.* 2004; 110: 2968–2975.
4. Vasan RS et al. Estimated risks for developing obesity in the Framingham Heart Study. *Ann Intern Med.* 2005; 143: 473–480.
5. Caterson ID et al. Prevention Conference VII: Obesity, a worldwide epidemic related to heart disease and stroke. Group III: worldwide comorbidities of obesity. *Circulation.* 2004; 110: e476–e483.
6. Pi-Sunyer FX. Comorbidities of overweight and obesity: current evidence and research issues. *Med Sci Sports Exerc.* 1999; 31: S602–S608.
7. Peeters A et al. Obesity in adulthood and its consequences for life expectancy: a life-table analysis. *Ann Intern Med.* 2003; 138: 24–32.
8. Wolf AM and Colditz GA. Current estimates of the economic cost of obesity in the United States. *Obes Res.* 1998; 6: 97–106.
9. Bray GA and Gray DS. Obesity. Part I: Pathogenesis. *West J Med.* 1988; 149: 429–441.
10. Kuczmarski RJ and Flegal KM. Criteria for definition of overweight in transition: background and recommendations for the United States. *Am J Clin Nutr.* 2000; 72: 1074–1081.
11. Wilmore JH and Haskell WL. Body composition and endurance capacity of professional football players. *J Appl Physiol.* 1972; 33: 564–567.
12. Chumlea WC, LaMonte MJ, and Bray GA. Anthropometric assessment: stature, weight, and the body mass index in adults. In: Berdanier CD et al., Eds. *Handbook of Nutrition and Food.* 2nd ed. Boca Raton, CRC Press, 2007, pp. 581–587.
13. Adams TD et al. The relationship between body mass index and percent body fat in the severely obese. *Diabetes Obes Metab.* 2007; 9: 498–505.
14. Deurenberg P, Weststrate JA, and Seidell JC. Body mass index as a measure of body fatness: age- and sex-specific prediction formulas. *Br J Nutr.* 1991; 65: 105–114.
15. Jackson AS et al. The effect of sex, age and race on estimating percentage body fat from body mass index: Heritage Family Study. *Int J Obes Relat Metab Disord.* 2002; 26: 789–796.
16. Harp JB and Hecht L. Obesity in the National Football League. *JAMA.* 2005; 293: 1061–1062.

17. Bouchard C. BMI, fat mass, abdominal adiposity and visceral fat: where is the beef? *Int J Obes.* 2007; 31: 1552–1553.
18. Lemieux S et al. A single threshold value of waist girth identifies normal weight and overweight subjects with excess visceral adipose tissue. *Am J Clin Nutr.* 1996; 64: 685–693.
19. Pouliot MC et al. Waist circumference and abdominal sagittal diameter: best simple anthropometric indexes of abdominal visceral adipose tissue accumulation and related cardiovascular risk in men and women. *Am J Cardiol.* 1994; 73: 460–468.
20. Bray GA. Don't throw the baby out with the bath water. *Am J Clin Nutr.* 2004; 79: 347–349.
21. Bray GA. Fat distribution and body weight. *Obes Res.* 1993; 1: 203–205.
22. Zhu S et al. Percentage body fat ranges associated with metabolic syndrome risk: results based on the third National Health and Nutrition Examination Survey (1988–1994). *Am J Clin Nutr.* 2003; 78: 228–235.
23. Molarius A et al. Educational level, relative body weight, and changes in their association over 10 years: an international perspective from the WHO MONICA Project. *Am J Public Health.* 2000; 90: 1260–1268.
24. Martorell R et al. Obesity in women from developing countries. *Eur J Clin Nutr.* 2000; 54: 247–252.
25. Wardle J and Boniface D. Changes in the distributions of body mass index and waist circumference in English adults, 1993/1994 to 2002/2003. *Int J Obes.* 2008; 32: 527–532.
26. Ogden CL et al. *Mean Body Weight, Height, and Body Mass Index, United States 1960–2002. Advance Data from Vital and Health Statistics, No. 347.* Hyattsville, MD: National Center for Health Statistics; 2004.
27. Flegal KM et al. Overweight and obesity in the United States: prevalence and trends, 1960–1994. *Int J Obes Relat Metab Disord.* 1998; 22: 39–47.
28. Ogden CL et al. Prevalence of overweight and obesity in the United States, 1999–2004. *JAMA.* 2006; 295: 1549–1555.
29. Li C et al. Increasing trends in waist circumference and abdominal obesity among U.S. adults. *Obesity.* 2007; 15: 216–224.
30. Molarius A et al. Waist and hip circumferences and waist–hip ratios in 19 populations of the WHO MONICA Project. *Int J Obes Relat Metab Disord.* 1999; 23: 116–125.
31. Williamson DF et al. The 10-year incidence of overweight and major weight gain in US adults. *Arch Intern Med.* 1990; 150: 665–672.
32. Parikh NI et al. Increasing trends in incidence of overweight and obesity over 5 decades. *Am J Med.* 2007; 120: 242–250.
33. Gregg EW et al. Secular trends in cardiovascular disease risk factors according to body mass index in U.S. adults. *JAMA.* 2005; 293: 1868–1874.
34. Kruger J, Yore MM, and Kohl HW. Leisure time physical activity patterns by weight control status: 1999–2002 NHANES. *Med Sci Sports Exerc.* 2007; 39: 788–795.
35. Ford ES. Body mass index, diabetes, and C-reactive protein among U.S. adults. *Diabetes Care.* 1999; 22: 1971–1977.
36. Park YW et al. The metabolic syndrome: prevalence and associated risk factor findings in the U.S. population from the Third National Health and Nutrition Examination Survey, 1988–1994. *Arch Intern Med.* 24 2003; 163: 427–436.
37. Grunstein RR et al. Impact of obstructive sleep apnea and sleepiness on metabolic and cardiovascular risk factors in the Swedish Obese Subjects (SOS) Study. *Int J Obes Relat Metab Disord.* 1995; 19: 410–418.
38. Clark JM, Brancati FL, and Diehl AM. Nonalcoholic fatty liver disease. *Gastroenterology.* 2002; 122: 1649–1657.

39. Burke GL et al. The impact of obesity on cardiovascular disease risk factors and subclinical vascular disease: the Multi-Ethnic Study of Atherosclerosis. *Arch Intern Med.* 2008; 168: 928–935.
40. Wildman RP et al. The obese without cardiometabolic risk factor clustering and the normal weight with cardiometabolic risk factor clustering: prevalence and correlates of two phenotypes among the U.S. population (NHANES 1999–2004). *Arch Intern Med.* 2008; 168: 1617–1624.
41. McGee DL. Body mass index and mortality: a meta-analysis based on person-level data from twenty-six observational studies. *Ann Epidemiol.* 2005; 15: 87–97.
42. Flegal KM et al. Excess deaths associated with underweight, overweight, and obesity. *JAMA.* 2005; 293: 1861–1867.
43. Flegal KM et al. Cause-specific excess deaths associated with underweight, overweight, and obesity. *JAMA.* 2007; 298: 2028–2037.
44. Lee CD, Blair SN, and Jackson AS. Cardiorespiratory fitness, body composition, and all-cause and cardiovascular disease mortality in men. *Am J Clin Nutr.* 1999; 69: 373–380.
45. Baik I et al. Adiposity and mortality in men. *Am J Epidemiol.* 2000; 152: 264–271.
46. Bigaard J et al. Waist circumference and body composition in relation to all-cause mortality in middle-aged men and women. *Int J Obes.* 2005; 29: 778–784.
47. Blair SN, LaMonte MJ, and Nichaman MZ. Modifiable behavioral factors as causes of death. *JAMA.* 2004; 291: 2943.
48. Mokdad AH et al. Actual causes of death in the United States, 2000. *JAMA.* 2004; 291: 1238–1245.
49. Folsom AR et al. Associations of general and abdominal obesity with multiple health outcomes in older women: Iowa Women's Health Study. *Arch Intern Med.* 2000; 160: 2117–2128.
50. Gelber RP et al. Measures of obesity and cardiovascular risk among men and women. *J Am Coll Cardiol.* 2008; 52: 605–615.
51. Wang Y et al. Comparison of abdominal adiposity and overall obesity in predicting risk of type 2 diabetes among men. *Am J Clin Nutr.* 2005; 81: 555–563.
52. Goldstein DJ. Beneficial health effects of modest weight loss. *Int J Obes Relat Metab Disord.* 1992; 16: 397–415.
53. Sjostrom L et al. Lifestyle, diabetes, and cardiovascular risk factors 10 years after bariatric surgery. *New Engl J Med.* 2004; 351: 2683–2693.
54. Oster G et al. Lifetime health and economic benefits of weight loss among obese persons. *Am J Public Health.* 1999; 89: 1536–1542.
55. Wannamethee SG, Shaper AG, and Walker M. Weight change, weight fluctuation, and mortality. *Arch Intern Med.* 2002; 162: 2575–2580.
56. Hu FB et al. Adiposity as compared with physical activity in predicting mortality among women. *New Engl J Med.* 23 2004; 351: 2694–2703.
57. Field AE et al. Weight cycling, weight gain, and risk of hypertension in women. *Am J Epidemiol.* 1999; 150: 573–579.
58. Koh-Banerjee P et al. Changes in body weight and body fat distribution as risk factors for clinical diabetes in US men. *Am J Epidemiol.* 2004; 159: 1150–1159.
59. Manson JE et al. Body weight and longevity: a reassessment. *JAMA.* 1987; 257: 353–358.
60. Nieto-Garcia FJ, Bush TL, and Keyl PM. Body mass definitions of obesity: sensitivity and specificity using self-reported weight and height. *Epidemiology.* 1990; 1: 146–152.
61. Church TS et al. Cardiorespiratory fitness and body mass index as predictors of cardiovascular disease mortality among men with diabetes. *Arch Intern Med.* 2005; 165: 2114–2120.

62. Nicklas BJ et al. et al. Association of visceral adipose tissue with incident myocardial infarction in older men and women: Health, Aging and Body Composition Study. *Am J Epidemiol.* 2004; 160: 741–749.
63. Heiat A, Vaccarino V, and Krumholz HM. An evidence-based assessment of federal guidelines for overweight and obesity as they apply to elderly persons. *Arch Intern Med.* 2001; 161: 1194–1203.
64. Dyer AR et al. Long-term consequences of body mass index for cardiovascular mortality: the Chicago Heart Association Detection Project in Industry Study. *Ann Epidemiol.* 2004; 14: 101–108.
65. Bouchard C. Current understanding of the etiology of obesity: genetic and non-genetic factors. *Am J Clin Nutr.* 1991; 53: 1561S–1565S.
66. Hill JO and Melanson EL. Overview of the determinants of overweight and obesity: current evidence and research issues. *Med Sci Sports Exerc.* 1999; 31: S515–521.
67. Rankinen T et al. The human obesity gene map: 2005 update. *Obesity.* 2006; 14: 529–644.
68. Macaulay V et al. The emerging tree of West Eurasian mtDNAs: a synthesis of control region sequences and RFLPs. *Am J Hum Genet.* 1999; 64: 232–249.
69. Galuska DA et al. Are health care professionals advising obese patients to lose weight? *JAMA.* 1999; 282: 1576–1578.
70. Wright JD et al. Trends in intake of energy and macronutrients: United States, 1971–2000. *MMWR* 2004; 53: 80–82.
71. LaMonte MJ and Chumlea WC. Obesity and health: survival of the fittest? *Int J Body Comp Res.* 2006; 4: 19–25.
72. Farrell SW et al. Cardiorespiratory fitness, different measures of adiposity, and cancer mortality in men. *Obesity.* 2007; 15: 3140–3149.
73. Holmes MD et al. Physical activity and survival after breast cancer diagnosis. *JAMA.* 2005; 293: 2479–2486.
74. Okosun IS et al. Abdominal adiposity in U.S. adults: prevalence and trends, 1960–2000. *Prev Med.* 2004; 39: 197–206.

# 4 Inflammatory and Anti-Inflammatory Mediators Secreted by Adipose Tissue

*Herbert Tilg and Alexander R. Moschen*

## CONTENTS

## 4.1 INTRODUCTION

The number of obese and overweight individuals has risen dramatically over recent years. Obesity exposes individuals to increased risks of developing many diseases such as atherosclerosis, type 2 diabetes (T2D), nonalcoholic fatty liver disease (NAFLD), and certain cancers. Obesity, in particular visceral obesity—the accumulation of adipose tissue inside the abdominal cavity—is associated with resistance to the effects of insulin (insulin resistance; IR) often leading to the development of T2D. Insulin resistance is frequently associated with a state of low-grade inflammation and therefore it is assumed that inflammation contributes to its development [1,2]. Furthermore, although evidence is limited, obesity may be associated with certain immune-mediated disorders such as asthma, psoriasis, and certain cancers.

Adipose tissue exists in mammals in two forms: white adipose tissue (WAT) and brown adipose tissue (BAT). WAT is the fat in which triglycerides are stored and from which lipids are mobilized for systemic utilization when energy is needed; it is often divided into subcutaneous and abdominal depots, whose physiologies may be different and whose roles in diseases are distinct. In addition to adipocytes, adipose tissue also contains pre-adipocytes, endothelial cells, fibroblasts, and various leukocytes including macrophages, T lymphocytes, and neutrophils. These various infiltrating leukocytes are additional sources of soluble mediators (cytokines, adipocytokines) in the adipose tissue. Adipose tissue macrophages are commonly bone-marrow derived and the number of these cells present in WAT correlates with obesity [3].

Research in recent years has identified important pathways that link metabolism with the immune system and *vice versa.* Many of these interactions between the metabolic and immune systems seem to be mediated by a complex group of soluble mediators derived from immune cells and adipocytes called adipocytokines [4]. Adipocytokines are defined as soluble mediators that are mainly produced by adipocytes and exert their biological function either in an autocrine, paracrine, or systemic manner. This group of mediators is growing rapidly and is currently thought to represent a major link between the adipose tissue and the immune system.

The family of adipocytokines includes adiponectin, leptin, pre-B cell enhancing factor (PBEF; also known as Nampt and visfatin), resistin, tumor necrosis factor-alpha (TNF-$\alpha$), interleukin-6 (IL-6), C-C chemokine ligand 2 (CCL2), plasminogen activator-1, angiotensinogen, retinol-binding protein-4 (RPB4), serum amyloid A, and others [2,5–10]. Although macrophages in adipose tissue seem to be the major sources of TNF-$\alpha$, adipocytes contribute almost one third of circulating IL-6 in obese patients [11]. CCL2, also produced by adipocytes, has recently been identified as one potential factor contributing to macrophage infiltration into adipose tissue [12]. Obesity is associated with a chronic inflammatory response characterized by abnormal cytokine production, increased synthesis of acute phase reactants, such as C-reactive protein (CRP), and activation of inflammatory signalling pathways [2]. This chapter provides an overview of recent advances in the understanding of adipose tissue with its secreted mediators and will focus on classical cytokines and transcription factors involved in regulating adipose tissue inflammation (see Table 4.1).

**TABLE 4.1**
**Mediators of Immune Cells and Adipocytes Involved in Regulation of Adipose Tissue Inflammation**

| Cytokines and Chemokines | Transcription Factors | Others |
|---|---|---|
| TNFα[26,27] | NF-κB/IKKβ[53,56,57] | Osteopontin[74] |
| IL-1αβ[33,35] | JNK-1[58,59] | SAA[90] |
| Gp130 family (IL-6, CNTF)[42,43,91] | PPARγ_[19,64] | CRP[65,68,70] |
| IL-10[92,93] | SREBP-1c[94-97] | FABP-4[72,73] |
| IL-18[49-51] | LXR[98-101] | Oxidative stress[76-78] |
| MCP-1[12,15] | | ER stress[61,62] |
| MIP-1αβ[102] | | iNOS[75] |
| RANTES[22] | | E-selectin[103,104], P-selectin[105] |
| | | ICAM-1[103,106], VCAM-1[75,103] |
| | | TLR-4[107,108], PKC-θ[109], C3[71] |

*Note:* Refences are indicated by superscript numbers.

## 4.2 INFILTRATION OF ADIPOSE TISSUE WITH LEUKOCYTES: MACROPHAGES

Obesity is commonly associated with a chronic inflammatory syndrome characterized by abnormal cytokine and acute phase reactant synthesis and activation of inflammation in adipose tissue. Macrophage recruitment to the adipose tissue in obesity contributes to enhanced tissue inflammatory activity, and therefore may contribute to obesity-associated metabolic dysfunction [13].

Two reports presented evidence that adipose tissue is infiltrated by macrophages [3,14]. A large number of gene transcripts found in "inflamed" adipose tissue are well defined macrophage genes [3]. Macrophages may differentiate from pre-adipocytes or may enter the adipose tissue attracted by certain chemokines. Furthermore, macrophages may differentiate from pre-adipocytes and mesenchymal stem cells in adipose tissue. Adipocytes secrete several chemoattractants that drive monocytes into this tissue. It has been demonstrated that obese adipose tissue exhibits increased CCL2 expression, a key chemokine that recruits macrophages [12].

The absence of CCL2 in mice, however, does not limit obesity-associated infiltration of macrophages into adipose tissue [15]. Adipose tissue from CCL2-deficient mice was collected for analysis of macrophage infiltration after a high-fat diet. $CCL2^{-/-}$ mice on a high-fat diet showed no reductions in adipose tissue macrophages although they were glucose intolerant, suggesting that CCL2 is not critical for adipose tissue macrophage recruitment. Several other candidates may play a role in the recruitment of monocytes/macrophages such as migration inhibitory factor and macrophage inflammatory protein-1-alpha [16,17].

Macrophages have been recently characterized as exhibiting either a more proinflammatory M1 or a more anti-inflammatory M2 phenotype. In obesity, as adipose tissue is loaded with macrophages, resulting in local inflammation, IR aggravates. Different types of macrophages reside in the adipose tissue [18]. Resident macrophages present in adipose tissues of lean mice display the alternatively activated phenotype (M2 or "alternatively activated" macrophages characterized by activated genes for Ym1, arginase 1, and IL-10) [18]. Pro-inflammatory classically activated macrophages are recruited to sites of tissue damage in the adipose tissue as in obesity (M1 or "classically activated" macrophages producing enhanced levels of TNF-$\alpha$ and iNOS).

Diet-induced obesity influences the state of adipose tissue macrophages from an M2 polarized state that protects adipocytes from inflammation in lean animals to an M1 pro-inflammatory state leading to IR. This obesity-induced phenotypic switch in adipose tissue macrophage polarization has been recently demonstrated to be orchestrated by PPAR$\gamma$ [19]. Using mice with specific macrophage deletion of PPAR$\gamma$, these authors demonstrated that PPAR$\gamma$ is required for the maturation of alternatively activated (M2) macrophages. It is therefore expected that a predominantly M2 phenotype may improve insulin sensitivity by suppressing the synthesis of pro-inflammatory cytokines.

Kang et al. recently demonstrated that adipocytes synthesize the IL-13 Th2 cytokine that induces macrophage PPAR$\delta$ expression through a STAT6 binding site [20]. Interestingly, ablation of PPAR$\delta$ makes macrophages incapable of transition into an M2 phenotype, thereby leading to inflammation. More importantly, these authors demonstrated that hepatocyte-derived IL-13 and macrophage PPAR$\delta$ in the liver control hepatic lipid metabolism as myeloid-specific PPAR$\delta^{-/-}$ mice develop severe steatohepatitis. However, other adipocyte-derived mediators may regulate the development of alternatively activated M2 macrophages. Stienstra and colleagues recently presented evidence that PPAR$\gamma$ may also be involved in this process as treatment with rosiglitazone, a PPAR$\gamma$ ligand, results in an increase of M2 macrophages in the adipose tissue [21]. All these studies convincingly show how adipocytes and macrophages may interact in the adipose tissue.

## 4.3 INFILTRATION OF ADIPOSE TISSUE WITH LEUKOCYTES: T CELLS AND NEUTROPHILS

Besides macrophages, other immune cells such as T cells or neutrophils may infiltrate adipose tissue [22]. Wu and colleagues showed that adipose tissue from diet-induced obese insulin-resistant mice is infiltrated by T cells. This infiltration was accompanied by an increased expression of the RANTES T cell chemoattractant; furthermore, adiponectin$^{-/-}$ mice showed higher RANTES expression compared to wild-type mice. Obese humans with metabolic syndrome had higher mRNA levels of RANTES and CCR5 in subcutaneous adipose tissue than lean humans. RANTES expression was higher in human visceral fat and expression correlated with CD3 and CD11b staining in human visceral adipose tissue. Pro-inflammatory T cells are present in visceral adipose tissues of patients with T2D. In a mouse model of obesity-mediated IR, a marked T lymphocyte infiltration preceded recruitment of macrophages, suggesting that T lymphocytes may contribute to obesity-associated inflammation even at a rather early stage [23].

Interestingly, T cells have also been demonstrated to play a role in the generation of angiotensin II-induced hypertension and vascular dysfunction [24]. Therefore, T cells may represent a novel therapeutic target for the treatment of obesity-related inflammation and even hypertension.

Even neutrophils may become a relevant leukocyte population in relation to adipose tissue inflammation [25]. Early (3 to 7 days) after starting a high-fat diet, C57BL/6J mice showed an increase in neutrophilic infiltration in the intra-abdominal adipose tissue. As expected, this neutrophilic infiltration was followed by macrophage infiltration. Human data in this area, however, are lacking. It may not be a surprise if other immune cells such as natural killer (NK) cells or dendritic cells infiltrate adipose tissue and participate in immune dysfunctions observed in states of obesity.

## 4.4 PRO- AND ANTI-INFLAMMATORY MEDIATORS SECRETED BY ADIPOSE TISSUE

### 4.4.1 Tumor Necrosis Factor-α (TNF-α)

In 1993, Hotamisligil and colleagues first described an increase in the expression of a pro-inflammatory cytokine, namely TNF-α, in adipose tissue and its interference with insulin action [26]. Their findings led to the concept of inflammation in obesity and demonstrated that adipocytes are potential sources of TNF-α. Expression of this cytokine in obese animals (*fa/fa* rat and *ob/ob* mice) was increased and shown to regulate insulin action [26]. Further evidence supporting a key role for TNF-α in IR came from studies published by Uysal et al. who observed that mice lacking TNF-α or TNF receptors showed improved insulin sensitivity in both dietary and genetic (*ob/ob*) models of obesity [27]. Similar findings were noted in humans [28] with increased adipose tissue TNF-α expression in obesity and improvement in cytokine expression following weight loss [29]. Exposure of cells to TNF-α stimulated inhibitory phosphorylation of serine residues of IRS-1 [30,31]. TNF-α is "the" classical pro-inflammatory cytokine that links inflammation, obesity and IR.

### 4.4.2 Interleukin-1

IL-1α and IL-1β, among the first identified cytokines, exert strong pro-inflammatory functions [32]. IL-1$\beta^{-/-}$ mice exhibited lower fasting glucose and insulin levels and improved insulin sensitivity as determined by insulin tolerance testing, compared with wild-type controls [33]. IL-1β and IL-6 serum levels predict risk for T2D in humans [34]. IL-1β is able to reduce IRS-1 expression at a transcriptional level through a mechanism that is ERK-dependent and at a post-transcriptional level independent of ERK activation [35].

Neutralization of endogenous IL-1 by an IL-1β antibody improves glycemic control in a mouse model of diet-induced obesity [36]. These effects were paralleled by a decrease in the levels of serum amyloid A. Further strengthening a role for IL-1 in obesity-related diseases, it has been demonstrated that certain IL-1 gene polymorphisms are associated with central obesity and the metabolic syndrome [37].

IL-1ra$^{-/-}$ mice show an increased fat-to-carbohydrate oxidation ratio on a high-fat diet that results in increased sympathetic tone [38]. Therefore, a defect in a regulatory anti-inflammatory system like IL-1ra leads to increased energy expenditure, heart rate, and catecholamine production. By targeting IRS-1, IL-1β is capable of impairing insulin signalling and action, and may thus participate in concert with other cytokines in the development of IR.

### 4.4.3 Interleukin-6

The IL-6 cytokine family (or gp130 cytokines) consists of IL-6, ciliary neurotrophic factor, IL-11, leukemia inhibitory factor, oncostatin M, and cardiotrophin 1. IL-6 signals via induction of a gp130 homodimer after binding to the IL-6 receptor [39]. IL-6 was one of the first cytokines considered as a predictor of IR and cardiovascular disease. Serum levels of IL-6 decrease in parallel with weight loss and improvement of IR in patients undergoing bariatric surgery [40].

Visceral fat has been demonstrated as a major site for IL-6 production in humans [41]. IL-6 synthesis in abdominal adipose tissue is several times higher compared with subcutaneous adipose tissue, thereby potentially contributing to hepatic IR. This cytokine may be indeed involved in the pathogenesis of hepatic IR as insulin sensitivity increases in diet-induced obese mice treated with anti-IL-6 antibodies [42]. IL-6$^{-/-}$ mice are insulin-resistant and develop mature onset obesity [43]. These results were not reproducible in another IL-6$^{-/-}$ mouse model as these authors did not observe age-related obesity [44]. This discrepancy has been addressed by the authors initially describing obesity in IL-6$^{-/-}$ mice showing decreased energy expenditures and thermogenesis compared to wild-type mice [45].

Gp130 cytokines signal through a receptor that shows many similarities to leptin signalling, and leptin is known to activate signal transduction pathways that promote increased energy expenditure and insulin sensitivity [46]. IL-6 has been shown (like leptin) to activate AMP-activated protein kinase (AMPK) in both skeletal muscle and adipose tissue. Consistent with activation of AMPK, IL-6 has also been shown to increase fat oxidation *in vitro* and *ex vivo* and in humans *in vivo* [47]. The definite role of IL-6 in IR and obesity-related diseases must still be clarified and will be possible only when patients with T2D and/or IR receive treatment with an IL-6 neutralizing antibody.

### 4.4.4 Interleukin-18

IL-18 is another pro-inflammatory cytokine that plays a role in septic shock, joint inflammation, and inflammatory bowel diseases [48]. An intracellular pool of IL-18 is released after activation of caspase-1 which cleaves pro-IL-18. Its bioactivity on the other side is under tight control of its physiologic antagonist, the IL-18 binding protein. As IL-18 concentrations are increased in patients with T2D, this may reflect a role in the regulation of IR [49,50]. Indeed, as demonstrated, IL-18$^{-/-}$ mice and IL-18R$^{-/-}$ mice exhibited increased body weight accompanied by IR, hyperglycemia, lipid abnormalities, and atherosclerosis compared to wild-type mice [51].

Intracerebral administration of recombinant IL-18 inhibited food intake and reversed hyperglycemia in these mice by activation of STAT3 phosphorylation. Increased IL-18 levels in patients with T2D may reflect either an attempt of IL-18 to counteract hyperglycemia or resistance to this cytokine as observed for others such as insulin or leptin. Such an explanation could also be offered for the increased IL-6 levels observed in stages of IR. IL-18 resistance has also been found by Zilverschoon et al. Leukocytes isolated from patients with obesity and T2D showed lower IFN-γ responses after stimulation with IL-18 [52]. The authors concluded that IL-18 resistance may be a relevant mechanism explaining the increased susceptibility of these patients to various infections.

### 4.4.5 CCL2

Adipocytes secrete various chemoattractants that attract monocytes. Obese adipose tissue shows increased CCL2 expression, a key factor in the recruitment of macrophages [12]. These authors found that $CCL2^{-/-}$ mice exhibited reduced macrophage infiltration in adipose tissues and reduced IR. Conversely, they observed an increase in macrophage infiltration when CCL2 was overexpressed. Another study also revealed reduced macrophage infiltration in the adipose tissue and decreased IR in $CCL2^{-/-}$ mice [7].

In contrast, Inouye et al. recently demonstrated that the absence of CCL2 in mice does not limit obesity-associated infiltration of macrophages into adipose tissue [15]. The authors used $CCL2^{-/-}$ mice and adipose tissue was collected for analysis of macrophage infiltration. Surprisingly, $CCL2^{-/-}$ mice on a high-fat diet showed no reductions in adipose tissue macrophages although they were glucose-intolerant and had mildly increased plasma glucose and decreased serum adiponectin levels compared with wild-type mice. These data suggest that this chemokine is not the only critical mediator for adipose tissue macrophage recruitment. Several other candidates may play a role in the recruitment of monocytes/macrophages into adipose tissue.

## 4.5 TRANSCRIPTION FACTORS REGULATING INFLAMMATORY RESPONSE

### 4.5.1 Role of IKKβ–NF-κB pathway

Yuan et al. reported in 2001 that the inhibitor of κB kinase-β (IKKβ), a kinase located proximal of nuclear factor-κB (NF-κB), is a target for TNF-α-induced IR [53]. Yin et al. demonstrated a few years earlier that aspirin and salicylates inhibited the activity of IKKβ [54]. High doses of salicylates were demonstrated to lower high blood glucose concentrations 130 years ago by William Ebstein [55]. Yuan et al. demonstrated in their work that high doses of salicylates reverse hyperglycemia, hyperinsulinemia, and dyslipidemia in *fa/fa* rats and *ob/ob* mice, and overexpression of IKKβ attenuates insulin signalling in cultured cells. These findings clearly suggest the involvement of inflammatory pathways in IR, highlighting the important role of IKKβ.

Two other groups have elucidated the relationship between IKKβ expression in the liver and IR [56,57]. Cai et al. created a stage of chronic subacute inflammation in the liver in a transgenic mouse model by selective hepatocellular activation of NF-κB, causing continuous low level expression of IKKβ. These mice exhibited a T2D phenotype with evidence of moderate systemic IR that was improved by systemic neutralization of IL-6 or oral salicylate therapy. Arkan et al. presented similar findings in mice lacking IKKβ in hepatocytes or myeloid cells [57]. Liver-specific abrogation of IKKβ resulted in relative insulin sensitivity in the liver in mice placed on a high fat diet or intercrossed with *ob/ob* mice, but they developed IR in muscle and fat. In contrast, mice deficient in myeloid IKKβ exhibited increased insulin sensitivity and were partially protected from IR.

### 4.5.2 c-Jun N-terminal kinase (JNK)

Several serine and threonine kinases are activated by inflammatory stimuli contributing to IR including JNK, IKK and others. Activation of these kinases takes place in situations that activate inflammatory and metabolic pathways. JNK has emerged as an important regulator of IR in obesity [58]. The JNK group belongs to the group of MAPKs and controls many cellular functions through regulation of activator protein-1 (AP-1), including c-Jun and JunB. In obesity, JNK activity is increased in the liver, muscle, and fat tissues, probably due to the increase of free fatty acids and TNF-α. The loss of JNK1 prevents the development of IR in both genetic and dietary models of obesity. Liver-specific knockdown of JNK1 lowers circulating glucose and insulin levels, supporting a role for this pathway in the development of IR [59].

### 4.5.3 Endoplasmatic Reticulum (ER) Response

Experimental evidence suggests that ER stress is important in the initiation and regulation of inflammation and insulin action as observed in IR [1]. Folding, maturation, storage, and transport of most proteins take place in the ER. When folding is disturbed, an unfolded protein response (UPR) is initiated to restore this organelle that involves three key molecules: inositol-requiring enzyme-1 (IRE-1), PKR-like endoplasmatic reticulum kinase (PERK), and activating transcription factor 6 (ATF6) [60].

Two important pathways in the regulation of IR (discussed above), namely NF-κB/IKKβ and JNK-AP-1, are linked to activation of IRE-1 and PERK [61]. Indeed, ER stress may be involved in both dietary and genetic models of obesity and regulation of IR [62]. ATF6 and X-box binding protein-1 (XBP-1) are critical regulators of ER function and its adaptive responses as gain- and loss-of-function studies with XPB-1 demonstrated the close interaction with insulin action in vitro and in vivo [62]. Interestingly, hepatic-specific deletion of XBP-1 results in marked hypocholesteremia and hypotriglyceridemia, suggesting that this important transcription factor is critically involved in hepatic lipogenesis [63].

### 4.5.4 PPARγ

PPARγ, a transcription factor and genetic sensor of fatty acids, is a member of the nuclear receptor superfamily of ligand-dependent transcription factors. PPARγ is required for fat cell development and is the molecular target of thiazolidinediones (TZDs) that exert insulin sensitizing effects in adipose tissue, skeletal muscle, and the liver. TZDs suppress the production of various pro-inflammatory cytokines that promote IR.

In adipocytes, TZDs suppress the synthesis of IL-6, TNF-α, PAI-1, CCL2, and angiotensinogen; in macrophages that also express PPARγ, TZDs inhibit Toll-like receptor (TLR)- and IFN-γ-mediated inflammatory responses. Because macrophages invade adipose tissue in obesity, the role of macrophage-derived PPARγ has been studied. Hevener et al. demonstrated that macrophage PPARγ is essential for normal skeletal muscle and liver insulin sensitivity [64]. This suggests that that macrophage PPARγ expression is needed for a proper TZD response.

## 4.6 OTHER MEDIATORS AND PATHWAYS INVOLVED IN ADIPOSE TISSUE INFLAMMATION

### 4.6.1 Acute Phase Proteins

CRP, the best known and studied human acute phase protein, correlates with states of IR [65]. CRP in most instances is considered as an inflammatory marker related to atherosclerosis and cardiovascular diseases [66,67]. Human CRP, however, may also have some anti-inflammatory properties as it reduces atherosclerosis development in a mouse model of human-like hypercholesteremia [68]. CRP has been demonstrated to upregulate the synthesis of anti-inflammatory cytokines such as IL-1 receptor antagonist, which may help to explain the above phenomenon [69]. It has been suggested that CRP may play a role in leptin resistance by acting as a serum leptin-interacting protein [70]. Serum C3 levels are also significantly related to the presence of IR [71]. In this study primarily investigating IR in elderly people, C3 levels were better correlated with IR compared to CRP, blood sedimentation rate, and leukocyte numbers.

### 4.6.2 Adipocyte Fatty Acid Binding Protein (aP2, FABP4)

Fatty acid binding proteins are a family of small proteins that bind with high affinity to hydrophobic ligands such as saturated and unsaturated long-chain fatty acids and eicosanoids. Adipocyte fatty-acid binding protein, aP2 (FABP4), is expressed in adipocytes and macrophages, links inflammatory and metabolic processes, and is mainly regulated by PPARγ agonists, insulin, and fatty acids. Deficiency of aP2 protects mice against the development of IR associated with genetic or diet-induced obesity [72]. The macrophage is the critical site of FABP action, and total or macrophage-specific aP2 deficiency leads to a marked protection against early and advanced atherosclerosis in apolipoprotein-deficient mice. Furuhashi et al.

demonstrated that an orally active small-molecule inhibitor of aP2 improved severe atherosclerosis and T2DM in various mouse models [73].

### 4.6.3 Osteopontin (OPN)

OPN is a secreted matrix glycoprotein and pro-inflammatory cytokine playing a relevant role in cell-mediated immunity. Tissue infiltration of macrophages as observed in the adipose tissues in cases of obesity is dependent on the expression of OPN, which regulates monocyte chemotaxis and motility. Nomiyama et al. demonstrated that mice after a high-fat diet showed increased circulating OPN levels [74]. Obese mice lacking OPN exhibited improved insulin sensitivity and decreased macrophage infiltration into adipose tissue.

### 4.6.4 Inducible Nitric Oxide Synthase (iNOS)

Inducible nitric oxide synthase (iNOS) and its deletion are associated with improvement of high-fat diet induced IR. Blockade of iNOS by N(G)-nitro-l-arginine methyl ester (L-NAME) improved high fat diet-induced obesity and glucose intolerance. These effects were accompanied by a reduction of inflammation in the adipose tissue and improved signalling in skeletal muscle [75].

### 4.6.5 Oxidative Stress

One of the final common mediators of IR seems to be oxidative stress due to generation of reactive oxygen species (ROS) and/or decreased antioxidant defenses [76]. In both non-alcoholic steatohepatitis and experimental steatohepatitis, hepatic expression of CYP2E1 increased, leading to oxidative stress and this enhanced expression was shown to impair insulin signalling [77]. Xu and colleagues recently demonstrated a key role for the Nrf1 gene in NASH [78]. Mice with liver-specific deletions of Nrf1, a gene mediating activation of oxidative stress response genes, develop all features of non-alcoholic fatty liver disease including steatosis, apoptosis, necrosis, inflammation, fibrosis, and finally liver cancer, highlighting the importance of oxidative stress in this disease.

## 4.7 ANTI-INFLAMMATORY STRATEGIES TO COUNTERACT ADIPOSE TISSUE INFLAMMATION

### 4.7.1 Anti-TNF Approaches

Infliximab, a chimeric human–mouse anti-TNF antibody, has been demonstrated to reverse steatosis and to improve insulin signalling in a rat model of high-fat diet-induced IR, suggesting that neutralization of this key cytokine may improve liver inflammation, steatosis, and fibrosis and insulin signalling [79]. It is important to state studies to date using neutralizing anti-TNF antibodies in humans have not shown improved insulin sensitivity [80,81]. Single doses of infliximab failed to improve IR in diabetic and obese subjects [80,82].

Subsequent placebo-controlled studies conducted over a treatment period of 4 weeks provided no improvements in insulin sensitivity in obese diabetic subjects [83] nor in obese insulin-resistant subjects without diabetes [81]. Lo et al. demonstrated in a study of metabolic syndrome that etanercept, a TNFRp75-IgG fusion protein neutralizing TNF-α, increased total adiponectin concentration but concentrations of the HMW form thought to mediate insulin sensitivity were unchanged [84].

Another recent study investigated the effects of adalimumab, a fully human anti-TNF-α antibody, on IR in rheumatoid arthritis patients. These patients with active disease showed marked IR that was not influenced by anti-TNF therapy despite reductions in systemic inflammation (IL-6, CRP serum levels) during treatment [85]. Thus, the role of TNF-α blockade as a treatment strategy to improve IR in humans remains unclear.

### 4.7.2 Chemical Chaperones

ER stress is a key link of obesity, IR, and T2D [1]. Pharmaceutical chaperones such as 4-phenyl butyric acid (PBA) and endogenous bile acids and derivates such as ursodeoxycholic acid (UDCA) including its taurine-conjugated derivates (TUDCAs) are able to modulate ER function. Ozcan et al. recently demonstrated that chaperone treatment of obese and diabetic mice improved hyperglycemia, insulin sensitivity, and fatty liver disease [86]. Importantly, the concentrations, e.g., for UDCA, used in these studies were approximately ten times higher than those achieved in humans. UDCA treatment of patients with non-alcoholic steatohepatitis has not been successful to date [87].

### 4.7.3 IL-1 Receptor Antagonist (IL-1Ra)

IL-1Ra is markedly upregulated in the sera of obese patients, is correlated with BMI and IR, and is overexpressed in the white adipose tissues of obese humans [88]. Treatment of T2D patients with recombinant human IL-1Ra indeed improved glycemic control, supporting the concept of inflammation in T2D and IR [89]. This is the first evidence that an anti-inflammatory strategy may indeed improve glycemic control and IR.

## 4.8 CONCLUSIONS

Various pro-inflammatory cytokines, adipocytokines, and transcription factors are critically involved in adipose tissue inflammation and systemic inflammation associated with obesity. This concept is supported by many studies in patients with T2D and NAFLD in which obesity is correlated with a state of low-grade chronic inflammation. Adipose tissue as evolved as a major immune organ secreting large numbers of various cytokines and adipocytokines and also contributing significantly to the overall synthesis of these mediators that exert such major control over health and disease. Therefore, adipose tissue can be added to the list of major organs affecting many inflammatory and immune-mediated processes other than obesity-related disorders in humans.

## REFERENCES


1. Hotamisligil GS. Inflammation and metabolic disorders. *Nature* 2006; 444: 860–867.
2. Wellen KE and Hotamisligil GS. Inflammation, stress, and diabetes. *J Clin Invest* 2005; 115: 1111–1119.
3. Weisberg SP et al. Obesity is associated with macrophage accumulation in adipose tissue. *J Clin Invest* 2003; 112: 1796–1808.
4. Tilg H and Moschen AR. Adipocytokines: mediators linking adipose tissue, inflammation and immunity. *Nat Rev Immunol* 2006; 6: 772–783.
5. La Cava A and Matarese G. The weight of leptin in immunity. *Nat Rev Immunol* 2004; 4: 371–379.
6. Kusminski CM, McTernan PG, and Kumar S. Role of resistin in obesity, insulin resistance and Type II diabetes. *Clin Sci* 2005; 109: 243–256.
7. Weisberg SP et al. CCR2 modulates inflammatory and metabolic effects of high fat feeding. *J Clin Invest* 2006; 116: 115–124.
8. Mannino DM et al. Boys with high body masses have an increased risk of developing asthma: findings from the National Longitudinal Survey of Youth (NLSY). *Int J Obes* 2006; 30: 6–13.
9. Calle EE and Kaaks R. Overweight, obesity and cancer: epidemiological evidence and proposed mechanisms. *Nat Rev Cancer* 2004; 4: 579–591.
10. Graham TE et al. Retinol-binding protein 4 and insulin resistance in lean, obese, and diabetic subjects. *New Engl J Med* 2006; 354: 2552–2563.
11. Fantuzzi G. Adipose tissue, adipokines, and inflammation. *J Allergy Clin Immunol* 2005; 115: 911–919.
12. Kanda H et al. MCP-1 contributes to macrophage infiltration into adipose tissue, insulin resistance, and hepatic steatosis in obesity. *J Clin Invest* 2006; 116: 1494–1505.
13. Wellen KE and Hotamisligil, GS. Obesity-induced inflammatory changes in adipose tissue. *J Clin Invest* 2003; 112: 1785–1788.
14. Xu H et al. Chronic inflammation in fat plays a crucial role in the development of obesity-related insulin resistance. *J Clin Invest* 2003; 112: 1821–1830.
15. Inouye KE et al. Absence of CC chemokine ligand 2 does not limit obesity-associated infiltration of macrophages into adipose tissue. *Diabetes* 2007; 56: 2242–2250.
16. Sartipy P and Loskutoff DJ. Monocyte chemoattractant protein 1 in obesity and insulin resistance. *Proc Natl Acad Sci USA* 2003; 100: 7265–7270.
17. Skurk T et al. Production and release of macrophage migration inhibitory factor from human adipocytes. *Endocrinology* 2005; 146: 1006–1011.
18. Lumeng CN, Bodzin JL, and Saltiel AR. Obesity induces a phenotypic switch in adipose tissue macrophage polarization. *J Clin Invest* 2007; 117: 175–184.
19. Odegaard JI et al. Macrophage-specific PPARγ controls alternative activation and improves insulin resistance. *Nature* 2007; 447: 1116–1120.
20. Kang K et al. Adipocyt-derived Th2 cytokines and myeloid PPARδ regulate macrophage polarization and insulin sensitivity. *Cell Metab* 2008; 7: 485–495.
21. Stienstra R et al. PPARγ activation promotes infiltration of alternatively activated macrophages into adipose tissue. *J Biol Chem* 2008; 283: 22620–22627.
22. Wu H et al. T cell accumulation and regulated on activation, normal T cell expressed and secreted upregulation in adipose tissue in obesity. *Circulation* 2007; 115: 1029–1038.
23. Kintscher U et al. T lymphocyte infiltration in visceral adipose tissue: a primary event in adipose tissue inflammation and the development of obesity-mediated insulin resistance. *Arterioscler Thromb Vasc Biol* 2008; 28: 1304–1310.
24. Guzik TJ et al. Role of the T cell in the genesis of angiotensin I-induced hypertension and vascular dysfunction. *J Exp Med* 2007; 204: 2449–2460.

25. Elgazar-Carmon V et al. Neutrophils transiently infiltrate intra-abdominal fat early in the course of high fat feeding. *J Lipid Res* 2008; 49: 1894–1903.
26. Hotamisligil GS, Shargill NS, and Spiegelman BM. Adipose expression of tumor necrosis factor-α: direct role in obesity-linked insulin resistance. *Science* 1993; 259: 87–91.
27. Uysal KT et al. Protection from obesity-induced insulin resistance in mice lacking TNF-α function. *Nature* 1997; 389: 610–614.
28. Kern PA et al. The expression of tumor necrosis factor in human adipose tissue: regulation by obesity, weight loss, and relationship to lipoprotein lipase. *J Clin Invest* 1995; 95: 2111–2119.
29. Dandona P et al. Tumor necrosis factor-α in sera of obese patients: fall with weight loss. *J Clin Endocrinol Metab* 1998; 83: 2907–2910.
30. Paz K et al. A molecular basis for insulin resistance. Elevated serine/threonine phosphorylation of IRS-1 and IRS-2 inhibits their binding to the juxtamembrane region of the insulin receptor and impairs their ability to undergo insulin-induced tyrosine phosphorylation. *J Biol Chem* 1997; 272: 29911–29918.
31. Aguirre V et al. The c-Jun NH(2) terminal kinase promotes insulin resistance during association with insulin receptor substrate-1 and phosphorylation of Ser(307). *J Biol Chem* 2000; 275: 9047–9054.
32. Dinarello CA. Blocking IL-1 in systemic inflammation. *J Exp Med* 2005; 201: 1355–1359.
33. Matsuki T et al. IL-1 plays an important role in lipid metabolism by regulating insulin levels under physiological conditions. *J Exp Med* 2003; 198: 877–888.
34. Spranger J et al. Inflammatory cytokines and the risk to develop type 2 diabetes: results of the prospective population-based European Prospective Investigation into Cancer and Nutrition (EPIC) Potsdam Study. *Diabetes* 2003; 52: 812–817.
35. Jager J et al. Interleukin-1β-induced insulin resistance in adipocytes through down-regulation of insulin receptor substrate-1 expression. *Endocrinology* 2007; 148: 241–251.
36. Osborn O et al. Treatment with an Interleukin-1β antibody improves glycemic control in diet-induced obesity. *Cytokine* 2008; 44: 141–148.
37. Carter KW et al. Association of interleukin-1 gene polymorphisms with central obesity and metabolic syndrome in a coronary heart disease population. *Hum Genet* 2008; 124: 199–206.
38. Chida D et al. Increased fat:carbohydrate oxidation ratio in Il1ra (–/–) mice on a high fat diet is associated with increased sympathetic tone. *Diabetologia* 2008; 51: 1698–1706.
39. Kishimoto T et al. Interleukin-6 family of cytokines and gp130. *Blood* 1995; 86: 1243–1254.
40. Kopp HP et al. Impact of weight loss on inflammatory proteins and their association with the insulin resistance syndrome in morbidly obese patients. *Arterioscler Thromb Vasc Biol* 2003; 23: 1042–1047.
41. Fontana L et al. Visceral fat adipokine secretion is associated with systemic inflammation in obese humans. *Diabetes* 2007; 56: 1010–1013.
42. Klover PJ, Clementi AH, and Mooney RA. Interleukin-6 depletion selectively improves hepatic insulin action in obesity. *Endocrinology* 2005; 146: 3417–3427.
43. Wallenius V et al. Interleukin-6-deficient mice develop mature onset obesity. *Nat Med* 2002; 8: 75–79.
44. Di Gregorio GB et al. Lipid and carbohydrate metabolism in mice with a targeted mutation in the IL-gene: absence of development of age-related obesity. *Am J Physiol Endocrinol Metab* 2004; 287: E182–E187.

45. Wernstedt I et al. Reduced stress- and cold-induced increase in energy expenditure in interleukin-6-deficient mice. *Am J Physiol Regul Integr Comp Physiol* 2006; 291: R551–R557.
46. Flier JS. Obesity wars: molecular progress confronts an expanding epidemic. *Cell* 2004; 116: 337–350.
47. Febbraio MA. gp130 receptor ligands as potential therapeutic targets for obesity. *J Clin Invest* 2007; 117: 841–849.
48. Dinarello CA. Interleukin-18 and the pathogenesis of inflammatory diseases. *Semin Nephrol* 2007; 27: 98–114.
49. Zirlik A et al. Interleukin-18, the metabolic syndrome, and subclinical atherosclerosis: results from the Dallas Heart Study. *Arterioscler Thromb Vasc Biol* 2007; 27: 2043–2049.
50. Kim HJ et al. Effects of rosiglitazone and metformin on inflammatory markers and adipokines: decrease in interleukin-18 is an independent factor for the improvement of homeostasis model assessment-β in type 2 diabetes mellitus. *Clin Endocrinol* 2007; 66: 282–289.
51. Netea MG et al. Deficiency of interleukin-18 in mice leads to hyperphagia, obesity and insulin resistance. *Nat Med* 2006; 12: 650–656.
52. Zilverschoon GR et al. Interleukin-18 resistance in patients with obesity and type 2 diabetes mellitus. *Int J Obes* 2008; 32: 1407–1414.
53. Yuan M et al. Reversal of obesity- and diet-induced insulin resistance with salicylates or targeted disruption of IKKβ. *Science* 2001; 293: 1673–1677.
54. Yin MJ, Yamamoto Y, and Gaynor RB. The anti-inflammatory agents aspirin and salicylate inhibit the activity of I(κ)B kinase-β. *Nature* 1998; 396: 77–80.
55. Ebstein W. Zur Therapie des Diabetes mellitus, insbesondere über die Anwendung des salicylsauren Natron bei demselben. *Berl. Klin. Wochs.* 1876; 13: 337–340.
56. Cai D et al. Local and systemic insulin resistance resulting from hepatic activation of IKKβ and NF-κB. *Nat Med* 2005; 11: 183–190.
57. Arkan MC et al. IKKβ links inflammation to obesity-induced insulin resistance. *Nat Med* 2005; 11: 191–198.
58. Hirosumi J et al. A central role for JNK in obesity and insulin resistance. *Nature* 2002; 420: 333–336.
59. Yang R et al. Liver-specific knockdown of JNK1 upregulates proliferator-activated receptor γ coactivator 1β and increases plasma triglyceride despite reduced glucose and insulin levels in diet-induced obese mice. *J Biol Chem* 2007; 282: 22765–22774.
60. Harding HP, Zhang Y, and Ron D. Protein translation and folding are coupled by an endoplasmic reticulum-resident kinase. *Nature* 1999; 397: 271–274.
61. Urano F et al. Coupling of stress in the ER to activation of JNK protein kinases by transmembrane protein kinase IRE1. *Science* 2000; 287: 664–666.
62. Ozcan U et al. Endoplasmic reticulum stress links obesity, insulin action, and type 2 diabetes. *Science* 2004; 306: 457–461.
63. Lee AH et al. Regulation of hepatic lipogenesis by the transcription factor XBP1. *Science* 2008; 320: 1492–1496.
64. Hevener AL et al. Macrophage PPARγ is required for normal skeletal muscle and hepatic insulin sensitivity and full antidiabetic effects of thiazolidinediones. *J Clin Invest* 2007; 117: 1658–1669.
65. Yoneda M et al. High-sensitivity C-reactive protein is an independent clinical feature of nonalcoholic steatohepatitis (NASH) and also of the severity of fibrosis in NASH. *J Gastroenterol* 2007; 42: 573–582.
66. Ridker PM et al. C-reactive protein and other markers of inflammation in the prediction of cardiovascular disease in women. *New Engl J Med* 2000; 342: 836–843.
67. Lind L. Circulating markers of inflammation and atherosclerosis. *Atherosclerosis* 2003; 169: 203–214.

68. Kovacs A et al. Human C-reactive protein slows atherosclerosis development in a mouse model with human-like hypercholesterolemia. *Proc Natl Acad Sci USA* 2007; 104: 13768–13773.
69. Tilg H et al. Anti-inflammatory properties of hepatic acute phase proteins: preferential induction of interleukin 1 (IL-1) receptor antagonist over IL-1β synthesis by human peripheral blood mononuclear cells. *J Exp Med* 1993; 178: 1629–1636.
70. Chen K et al. Induction of leptin resistance through direct interaction of C-reactive protein with leptin. *Nat Med* 2006; 12: 425–432.
71. Muscari A et al. Serum C3 is a stronger inflammatory marker of insulin resistance than C-reactive protein, leukocyte count, and erythrocyte sedimentation rate: comparison study in an elderly population. *Diabetes Care* 2007; 30: 2362–2368.
72. Newberry EP et al. Protection against Western diet-induced obesity and hepatic steatosis in liver fatty acid-binding protein knockout mice. *Hepatology* 2006; 44: 1191–1205.
73. Furuhashi M et al. Treatment of diabetes and atherosclerosis by inhibiting fatty acid-binding protein aP2. *Nature* 2007; 447: 959–965.
74. Nomiyama T et al. Osteopontin mediates obesity-induced adipose tissue macrophage infiltration and insulin resistance in mice. *J Clin Invest* 2007; 117: 2877–2888.
75. Tsuchiya K et al. Chronic blockade of nitric oxide synthesis reduces adiposity and improves insulin resistance in high fat-induced obese mice. *Endocrinology* 2007; 148: 4548–4556.
76. Sanyal AJ et al. Nonalcoholic steatohepatitis: association of insulin resistance and mitochondrial abnormalities. *Gastroenterology* 2001; 120: 1183–1192.
77. Schattenberg JM et al. Hepatocyte CYP2E1 overexpression and steatohepatitis lead to impaired hepatic insulin signaling. *J Biol Chem* 2005; 280: 9887–9894.
78. Xu Z, Chen L, Leung L, et al. Liver-specific inactivation of the Nrf1 gene in adult mouse leads to nonalcoholic steatohepatitis and hepatic neoplasia. *Proc Natl Acad Sci USA* 2005; 102: 4120–4125.
79. Barbuio R et al. Infliximab reverses steatosis and improves insulin signal transduction in liver of rats fed a high-fat diet. *J Endocrinol* 2007; 194: 539–550.
80. Ofei F et al. Effects of an engineered human anti-TNF-α antibody (CDP571) on insulin sensitivity and glycemic control in patients with NIDDM. *Diabetes* 1996; 45: 881–885.
81. Bernstein LE et al. Effects of etanercept in patients with the metabolic syndrome. *Arch Intern Med* 2006; 166: 902–908.
82. Paquot N et al. No increased insulin sensitivity after a single intravenous administration of a recombinant human tumor necrosis factor receptor: Fc fusion protein in obese insulin-resistant patients. *J Clin Endocrinol Metab* 2000; 85: 1316–1319.
83. Dominguez H et al. Metabolic and vascular effects of tumor necrosis factor-α blockade with etanercept in obese patients with type 2 diabetes. *J Vasc Res* 2005; 42: 517–525.
84. Lo J et al. Effects of TNF-α neutralization on adipocytokines and skeletal muscle adiposity in metabolic syndrome. *Am J Physiol Endocrinol Metab* 2007; 293: E102–109.
85. Rosenvinge A et al. Insulin resistance in patients with rheumatoid arthritis: effect of anti-TNF-α therapy. *Scand J Rheumatol* 2007; 36: 91–96.
86. Ozcan U et al. Chemical chaperones reduce ER stress and restore glucose homeostasis in a mouse model of type 2 diabetes. *Science* 2006; 313: 1137–1140.
87. Lindor KD et al. Ursodeoxycholic acid for treatment of nonalcoholic steatohepatitis: results of a randomized trial. *Hepatology* 2004; 39: 770–778.
88. Somm E et al. Interleukin-1 receptor antagonist is upregulated during diet-induced obesity and regulates insulin sensitivity in rodents. *Diabetologia* 2006; 49: 387–393.

89. Larsen CM et al. Interleukin-1 receptor antagonist in type 2 diabetes mellitus. *New Engl J Med* 2007; 356: 1517–1526.
90. Yang RZ et al. Acute phase serum amyloid A: an inflammatory adipokine and potential link between obesity and its metabolic complications. *PLoS Med* 2006; 3: e287.
91. ALS CNTF Treatment Study Group. A double-blind placebo-controlled clinical trial of subcutaneous recombinant human ciliary neurotrophic factor (rHCNTF) in amyotrophic lateral sclerosis. *Neurology* 1996; 46: 1244–1249.
92. van Exel E et al. Low production capacity of interleukin-10 associates with the metabolic syndrome and type 2 diabetes: Leiden 85 Plus Study. *Diabetes* 2002; 51: 1088–1092.
93. den Boer MA et al. Endogenous interleukin-10 protects against hepatic steatosis but does not improve insulin sensitivity during high fat feeding in mice. *Endocrinology* 2006; 147: 4553–4558.
94. Nakayama H et al. Transgenic mice expressing nuclear sterol regulatory element-binding protein 1c in adipose tissue exhibit liver histology similar to nonalcoholic steatohepatitis. *Metabolism* 2007; 56: 470–475.
95. Yahagi N et al. Absence of sterol regulatory element-binding protein-1 (SREBP-1) ameliorates fatty livers but not obesity or insulin resistance in Lep(ob)/Lep(ob) mice. *J Biol Chem* 2002; 277: 19353–19357.
96. Takahashi A et al. Transgenic mice overexpressing SREBP-1a under control of the PEPCK promoter exhibit insulin resistance, but not diabetes. *Biochim Biophys Acta* 2005; 1740: 427–433.
97. Shimano H. SREBP-1c and TFE3: energy transcription factors that regulate hepatic insulin signaling. *J Mol Med* 2007; 85: 437–444.
98. Joseph SB et al. Reciprocal regulation of inflammation and lipid metabolism by liver X receptors. *Nat Med* 2003; 9: 213–219.
99. Laffitte BA et al. Activation of liver X receptor improves glucose tolerance through coordinate regulation of glucose metabolism in liver and adipose tissue. *Proc Natl Acad Sci USA* 2003; 100: 5419–5424.
100. Commerford SR et al. Dissection of the insulin-sensitizing effect of liver X receptor ligands. *Mol Endocrinol* 2007; 21: 3002–3012.
101. Kalaany NY et al. LXRs regulate the balance between fat storage and oxidation. *Cell Metab* 2005; 1: 231–244.
102. Sell H et al. Cytokine secretion by human adipocytes is differentially regulated by adiponectin, AICAR, and troglitazone. *Biochem Biophys Res Commun* 2006; 343: 700–706.
103. Song Y et al. Circulating levels of endothelial adhesion molecules and risk of diabetes in an ethnically diverse cohort of women. *Diabetes* 2007; 56: 1898–1904.
104. Matsumoto K et al. High serum concentrations of soluble E-selectin in patients with impaired glucose tolerance with hyperinsulinemia. *Atherosclerosis* 2000; 152: 415–420.
105. De Pergola G et al. sP-selectin plasma levels in obesity: association with insulin resistance and related metabolic and prothrombotic factors. *Nutr Metab Cardiovasc Dis* 2007; 18: 227–232.
106. Kent JW, Jr et al. Intercellular adhesion molecule-1 concentration is genetically correlated with insulin resistance, obesity, and HDL concentration in Mexican Americans. *Diabetes* 2004; 53: 2691–2695.
107. Kim F et al. Toll-like receptor-4 mediates vascular inflammation and insulin resistance in diet-induced obesity. *Circ Res* 2007; 100: 1589–1596.

108. Shi H et al. TLR4 links innate immunity and fatty acid-induced insulin resistance. *J Clin Invest* 2006; 116: 3015–3025.
109. Kim JK et al. PKC–θ knockout mice are protected from fat-induced insulin resistance. *J Clin Invest* 2004; 114: 823–827.

# 5 Adipokines and Inflammation

*Melissa E. Gove and Giamila Fantuzzi*

## CONTENTS



## 5.1 ADIPOSE TISSUE

The traditional function of white adipose tissue (WAT) as an organ exclusively devoted to storing energy during times of abundance and releasing it during fasting has recently been expanded. WAT is now considered a complex and highly active secretory organ, sending out and responding to signals that modulate appetite, energy expenditure, insulin sensitivity, endocrine and reproductive system functions, bone metabolism, inflammation, and immunity (Fantuzzi 2005; Kershaw and Flier 2004). These characteristics link WAT to the whole body and make it an essential tissue in the regulation of several physiological functions (Wang et al. 2008).

Adipokines are bioactive peptides predominantly secreted by adipocytes capable of acting at both the local and systemic levels (Kershaw and Flier 2004). Adipokines include leptin, adiponectin (APN), resistin, and visfatin—proteins that provide an important link of obesity, insulin resistance (IR), and inflammatory disorders (Tilg and Moschen 2006). Other factors, including cytokines such as interleukin (IL)-6, tumor necrosis factor (TNF)-α, chemokines, and others are also secreted by WAT; although some are produced by adipocytes, they are not strictly classified as adipokines because adipocytes are not the main sources of these mediators (Tilg and Moschen 2006). Production of adipokines by WAT is one of the most important ways for this tissue to influence physiological and pathological processes throughout the body and is the topic of this chapter.

## 5.2 LEPTIN

A genetic defect that caused a severely obese phenotype due to overeating and decreased energy expenditure in mice was identified at the Jackson Laboratories in the 1950s; mice carrying the mutation were called *ob/ob* (Ingalls, Dickie, and Snell 1950). A series of parabiotic experiments suggested that *ob/ob* mice did not produce satiety factors, but were able to respond to satiety factors from parabiotic mice (Coleman 1973). Similar parabiotic experiments performed with *db/db* mice, also identified at the Jackson Laboratories and displaying a similar phenotype, led to the hypothesis that the *db* gene encoded for the *ob* receptor (Coleman and Hummel 1973). Nearly three decades later, leptin was discovered in 1994 by Friedman et al. (Zhang et al. 1994). This 16-kDa polypeptide encoded by the *ob* gene was termed leptin from the Greek *leptos,* meaning thin.

Leptin is predominantly produced by adipocytes, but can also be synthesized by cells in the fundus of the stomach, skeletal muscle, liver, lymphocytes. and placenta, although at much lower levels than adipocytes (Juge-Aubry, Henrichot, and Meier 2005; Meier and Gressner 2004). Circulating levels and mRNA expression of leptin in adipocytes are strongly associated with body mass index (BMI) and with fat mass. Leptin expression is higher in subcutaneous than in visceral adipose tissue in humans (Bastard et al. 2006; Kershaw and Flier 2004), leading to serum levels of leptin two to three times higher in women—who have higher percentages of subcutaneous WAT than men—even when adjusted for age and BMI (Tilg and Moschen 2006).

Leptin exerts its biological activities by binding to its receptor (Ob-R). Five alternatively spliced isoforms of Ob-R have been identified and designated Ob-Ra, Ob-Rb, Ob-Rc, Ob-Rd, and Ob-Re. Each isoform is capable of forming homodimers in the presence and absence of leptin (Fantuzzi and Faggioni 2000; Meier and Gressner 2004). However, only the long form, Ob-Rb, contains an intracellular region capable of downstream signaling upon leptin binding.

Similar to other class 1 cytokine receptors, Ob-Rb requires the activation of Janus tyrosine kinase 2 (JAK2) for propagation of leptin signaling. Binding of leptin to Ob-Rb induces autophosphorylation of JAK2, leading to phosphorylation of three intracellular tyrosine residues of Ob-Rb, specifically in positions 985, 1077, and 1138 (Buettner et al. 2006; Robertson, Leinninger, and Myers 2008). When activated, each of these tyrosine residues recruits specific downstream signaling proteins, such as SH2-domain-containing phosphatase-2 (SHP-2), signal transducer and activator of transcription (STAT)-5 and STAT-3 that mediate intracellular signaling by leptin (Gong et al. 2007).

The best characterized and probably the most important function of leptin is its inhibitory effect on appetite; therefore leptin is classified as an anorexigenic molecule (Fantuzzi and Faggioni 2000). Through its effects on appetite, leptin helps maintain long-term control of adiposity and regulates adaptive metabolic changes in response to nutritional modifications. Leptin also regulates short-term energy intake by modulating meal size according to changes in energy balance (Valassi, Scacchi, and Cavagnini 2008).

In agreement with these functions of leptin, very high expression of Ob-Rb is present in the feeding centers of the hypothalamus, including the arcuate, dorsomedial, ventromedial, and premammillary nuclei (Meier and Gressner 2004; Munzberg and Myers 2005). Within the arcuate nucleus, Ob-Rb is found in two distinct populations of neurons (Valassi, Scacchi, and Cavagnini 2008). One population produces orexigenic neuropeptide Y (NPY) and agouti-related peptide (AgRP), while the other synthesizes anorexigenic pro-opiomelanocortin (POMC) (Munzberg and Myers 2005; Valassi, Scacchi, and Cavagnini 2008). Leptin decreases appetite and increases energy expenditure by activating POMC neurons and inhibiting NPY/AgRP neurons (Robertson, Leinninger, and Myers 2008). In addition to its critical role in the homeostatic control of food intake, mostly controlled by the hypothalamus, leptin also acts in the cortex and limbic areas of the brain where it regulates cognitive and hedonic responses to feeding (Rosenbaum et al. 2008).

Mice and humans with leptin or leptin receptor deficiency, in addition to being obese, suffer from a series of other endocrine abnormalities including reduced fertility, alterations in bone metabolism, and dysfunction of the immune system. The alterations secondary to lack of leptin or its receptor resemble the adaptive response to starvation, a situation in which leptin levels fall dramatically, out of proportion with fat mass (Chan and Mantzoros 2005; Fantuzzi and Faggioni 2000). Most neuroendocrine and immune alterations associated with fasting and/or starvation are normalized by the administration of leptin.

Studies on the role of leptin in regulating the immune system have been fueled largely by early observations of thymus atrophy in *db/db* mice, suggesting a role for leptin in regulation of T lymphocyte function and on T cell survival. Lord et al. (1998) have shown that leptin increases the activation and proliferation of $CD4^+$ T lymphocytes, having a more pronounced impact on the proliferation of naïve $CD4^+$ T lymphocytes compared to $CD4^+$ memory T lymphocytes. Subsequent studies demonstrated that leptin increases proliferation and activation of T lymphocytes when used as a co-stimulus with lectins such as phytohemagglutinin (PHA) and concanavalin A (ConA) (Martin-Romero et al. 2000).

De Rosa et al. demonstrated leptin's ability to enhance proliferation of effector $CD4^+$ T cells after anti-CD3 and anti-CD28 stimulation while inhibiting responsiveness of T regulatory (Treg) cells (De Rosa et al. 2007). Interestingly, in vitro neutralization of leptin with a monoclonal antibody during anti-CD3 and anti-CD28 stimulation resulted in Treg cell proliferation. In vivo data from *ob/ob* and *db/db* mice support this finding; these mice exhibited enhanced Treg proliferation compared to wild-type (WT) mice (De Rosa et al. 2007). In addition, *ob/ob* and *db/db* mice had decreased thymus size and cellularity due to high thymocyte apoptosis, which was normalized by peripheral administration of leptin in *ob/ob* mice (Howard et al. 1999). WT mice treated with leptin throughout a 48-hour fast were protected from thymic atrophy (Howard et al. 1999). Mansour et al. (2006) indicate that leptin protects lymphocytes from apoptosis in the thymus through an IRS-1/PI3-kinase signaling cascade, independent of JAK activation. Leptin also promotes $CD4^+$ T lymphocyte survival in vitro by suppressing Fas-mediated apoptosis (Papathanassoglou et al. 2006).

In vitro leptin enhances T lymphocyte activation and promotes T helper (TH)-1 cytokine production by stimulating the synthesis of IL-2 and IFN-$\gamma$ and suppressing production of TH-2 cytokines (Tilg and Moschen 2006; Juge-Aubry, Henrichot, and Meier 2005). In addition, leptin promotes the activation of monocytes and macrophages and their secretion of leukotriene $B_4$ ($LTB_4$), cycloxygenase 2 (COX2), nitric oxide (NO), and pro-inflammatory cytokines such as TNF-α, IL-6 and IL-12 (Fantuzzi and Faggioni 2000; La Cava and Matarese 2004). Leptin also initiates neutrophil chemotaxis and stimulates the production of reactive oxygen species (ROS) by these cells (Tilg and Moschen 2006). Natural killer (NK) cell differentiation, proliferation, and activation are also regulated by leptin (La Cava and Matarese 2004; Tilg and Moschen 2006).

Studies have demonstrated that *ob/ob*, and in some cases *db/db*, mice show resistance or less susceptibility in models of innate and adaptive immune-mediated inflammatory diseases (Otero, Lago, Gomez, Dieguez et al. 2006). It has been speculated that this is due to the reduced secretion of IFN-γ, IL-2, TNF-α and IL-18 and to increases in Th2 cytokines, such as IL-4 and IL-10 after stimulation in *ob/ob* mice (La Cava and Matarese 2004).

For example, in experimental autoimmune encephalomyelitis (EAE), an animal model for multiple sclerosis, *ob/ob* mice did not develop neurological impairment. However, leptin replacement in *ob/ob* mice reversed their resistance, indicating that leptin deficiency is the reason for this effect (Matarese et al. 2001). In experimental intestinal inflammation, *ob/ob* and *db/db* mice displayed decreased disease severity in models of chemically induced colitis (Siegmund, Lehr, and Fantuzzi 2002; Gove et al. *submitted*). Interestingly, the transfer of $CD4^+CD45RB^{high}$ from *db/db* mice into *scid* mice (immune-mediated models of colitis) induced delayed disease compared with *scid* mice that received WT cells (Siegmund, Sennello, Jones-Carson et al. 2004). However, as disease progressed, differences between mice receiving WT and *db/db* cells were no longer apparent, suggesting that leptin affects the immune system partly by acting on Ob-Rb expressed on T lymphocytes (Siegmund, Sennello, Jones-Carson et al. 2004). However, the generation of double IL-10 and leptin-deficient mice (models of spontaneous colitis mediated by $CD4^+$ T cells) showed no difference in the development of colitis between single and double IL-10 and leptin-deficient mice (Siegmund, Sennello, Lehr et al. 2004).

In antigen-induced arthritis (AIA), an immune-mediated model of joint inflammation, *ob/ob* mice developed less severe arthritis compared to WT mice, with much lower expression of pro-inflammatory TNF-α, IL-1β and IFN-γ paired with higher levels of anti-inflammatory IL-10 production (Busso et al. 2002). In contrast, zymosan-induced arthritis (ZIA), an arthritis model independent of the adaptive immune system, is not impaired in *ob/ob* and *db/db* mice (Bernotiene et al. 2004). In fact, the acute phase response remained elevated for a longer period in the *ob/ob* and *db/db* mice compared to WT mice (Bernotiene et al. 2004). This may be partially explained by the observation that *ob/ob* and *db/db* mice are more sensitive to agents stimulating innate immune responses than agents stimulating adaptive immune responses (Fantuzzi and Faggioni 2000; Bernotiene et al. 2004).

Leptin is an important factor in modulating immune and inflammatory reactions, both in humans and experimental animals. However, a better understanding of the mechanisms through which leptin participates in the complex network of immune and inflammatory mediators is needed.

## 5.3 ADIPONECTIN

Adiponectin (APN), a protein secreted predominantly by adipocytes, is present at a concentration of 5 to 30 μg/ml in blood of healthy humans, making it the highest circulating adipokine (Fantuzzi 2005). Adiponectin is also expressed and secreted from other cell types, such as cytokine-stimulated hepatocytes, osteoclasts, synovial fibroblasts, and skeletal muscle cells, although these cells produce much lower amounts of APN than adipocytes (Ehling et al. 2006; Berner et al. 2004; Delaigle et al. 2004).

Adiponectin was identified in 1995 and 1996 by four independent groups (Kershaw and Flier 2004). Adiponectin is a 247-amino acid protein consisting of an amino terminal signal sequence, a variable region, a collagenous domain, and a carboxy terminal globular domain (Oh, Ciaraldi, and Henry 2007). Adiponectin is a close homologue of C1q, a complement protein, and TNF-α, sharing structural but not sequence similarities with these molecules. Adiponectin forms low molecular weight (LMW) trimers that further associate to form middle molecular weight (MMW) hexamers and high molecular weight (HMW) oligomers.

These HMW oligomers form bouquet-like structures through disulfide bonds located within the collagenous domains of each monomer (Oh, Ciaraldi, and Henry 2007). All three forms of APN are present in serum, but whether the different forms exhibit differential biological actions remains unclear. A globular form of APN may also exist (Fruebis et al. 2001). Leukocyte elastase, secreted by activated monocytes and neutrophils, cleaves APN and generates a globular domain capable of forming a trimer (Waki et al. 2005). Globular APN, however, remains to be shown to exist physiologically in vivo. To add to the complexity, hydroxylation and glycosylation of lysines in the collagenous domain are necessary for complete biologic activity of APN (Chandran et al. 2003).

Part of the biological activity of APN is mediated by binding through its receptors, whereas other functions appear to be receptor-independent. Two APN receptors have been identified to date: AdipoR1 and AdipoR2. AdipoR1 is ubiquitously expressed but predominantly found in skeletal muscle, whereas AdipoR2 is abundantly expressed in the liver (Guzik, Mangalat, and Korbut 2006). AdipoR1 and AdipoR2 contain seven transmembrane domains that are structurally and functionally different from other G protein receptors, having the N terminus located in the cytoplasm and the C terminus located externally (Kadowaki and Yamauchi 2005). In vitro studies have shown AdipoR1 to be a high affinity receptor for globular APN, with a low affinity for full-length APN. AdipoR2 is an intermediate affinity receptor for both globular and full-length APNs.

Adiponectin binding to its receptors activates signaling molecules such as peroxisome proliferator-activated receptor (PPAR)-α, AMP-activated protein kinase (AMPK), and mitogen-activated protein kinase (p38 MAPK) (Yamauchi et al. 2003). The activation of PPAR-α is important in APN-stimulated fatty acid (FA) oxidation, but not for glucose uptake, whereas AMPK and MAPK activation is likely involved in both FA oxidation and glucose uptake (Yamauchi et al. 2003). However, as stated earlier, whether the different forms of APN initiate the same signals remains controversial (Oh, Ciaraldi, and Henry 2007). In addition, T-cadherin is a receptor for hexameric and HMW forms of APN (Hug et al. 2004). More recently, APN has been shown to bind calreticulin and to opsonize apoptotic cells to facilitate their clearance by macrophages (Takemura et al. 2007).

In addition to receptor-dependent effects, the high circulating concentration of APN suggests that low-affinity receptor-independent activities may be critical for the biological functions of this protein. Adiponectin binds to and inhibits the activity of heparin-binding epidermal growth factor (HB-EGF), basic fibroblast growth factor (bFGF), and platelet-derived growth factor (PDGF), all of which are involved in tissue repair (Fayad et al. 2007; Wang et al. 2005). APN also binds to certain chemokines (such as CCL2, CCL5 and CXCL12), displacing them from their interactions with proteoglycans in the extracellular matrix (Masaie et al. 2007).

Finally, APN interacts with collagens I, III, and IV in injured vascular walls and myocardium (Okamoto et al. 2002). Therefore, APN acts in a complex way, through both receptor-dependent and -independent effects, to regulate a variety of biological functions involved in metabolism, tissue repair, cell proliferation, and inflammation.

Unlike leptin, APN levels decrease with the increase in fat mass observed in obesity. In fact, the current hypothesis states that chronic inflammation associated with obesity and cardiovascular disease (CVD) inhibits production of APN, leading to the subsequent perpetuation of inflammation (Fantuzzi 2008). An inverse correlation between APN and IR has been established both in animal and human studies (Chandran et al. 2003; Mlinar et al. 2007). Human studies in obese and IR individuals, in which chronic low-grade inflammation of adipose tissue is common, also demonstrated a negative relationship between APN and classic markers of inflammation, such as CRP and IL-6 (Fantuzzi 2008). Adiponectin improves insulin sensitivity by various mechanisms (Mlinar et al. 2007). In the liver, APN activates AMPK, which enhances insulin sensitivity by increasing FA oxidation, down-regulating gluconeogenic enzymes and increasing glucose-6-phosphate biosynthesis (Mlinar et al. 2007; Combs et al. 2001).

Activation of PPAR-α by APN further increases insulin sensitivity by increasing FA oxidation, thereby decreasing FA synthesis in the liver (Kadowaki and Yamauchi 2005). In muscle, APN stimulates the phosphorylation of acetyl-CoA carboxylase, FA oxidation, glucose utilization, and lactate production (Mlinar et al. 2007). These effects are observed in parallel with activation of AMPK and PPAR-α by APN and are blocked by inhibition of AMPK and PPAR-α signaling (Kadowaki and Yamauchi 2005; Combs et al. 2001; Yamauchi et al. 2002; Diez and Iglesias 2003). Treatment with thaizolindinediones (TZDs), drugs that activate PPAR-γ and are commonly used to treat patients with type 2 diabetes (T2D) and IR, increases APN levels and improves glucose tolerance and insulin sensitivity. PPAR-γ is a ligand-activated transcription factor involved in the regulation of adipocyte differentiation. Studies of mice and adipocyte cultures have shown that activation of PPAR-γ increases APN mRNA levels in adipocytes (Boden et al. 2003; Chandran et al. 2003), further strengthening the link between adipocyte biology and IR.

Reduced levels of APN in obesity also contribute to endothelial dysfunction observed in subjects with CVD. In the early stages of CVD, endothelial cell activation occurs via inflammatory stimuli such as TNF-α that induce expression of vascular adhesion molecule-1 (VCAM-1), endothelial leukocyte adhesion molecule-1

(E-selectin), and intracellular adhesion molecule-1 (ICAM-1)—a crucial event in the development of vascular disease (Ouchi et al. 1999; Ouchi et al. 2000). Adiponectin reverses these deleterious effects of TNF-α on endothelial dysfunction by hindering its ability to activate the nuclear factor κB (NFκB) pro-inflammatory transcription factor and the subsequent induction of adhesion molecules (Ouchi et al. 1999; Goldstein and Scalia 2004).

In addition, APN inhibits foam cell formation and smooth muscle cell migration, both of which play an important role in the development of atherosclerosis (Arita et al. 2002). These effects of APN are also present in vivo in apolipoprotein E (apoE)-deficient mice, a well established model for atherosclerosis. When treated with adenovirus-derived APN, apoE-deficient mice did not develop severe atherosclerosis due to APN's ability to attenuate the endothelial inflammatory response and macrophage foam cell formation by inhibiting lipoprotein lipase and class A scavenger receptor (Okamoto et al. 2002). In addition, APN protects the heart from ischemia–reperfusion injury in vivo in mice by inhibiting NO and ROS production, thus protecting tissues from nitrative and oxidative stress (Tao et al. 2007). In vitro studies demonstrated that APN decreases IFNγ, IL-6 and TNF-α production in lipopolysaccharide (LPS)-activated primary macrophages and upregulates IL-10 and IL-1RA production (Wolf et al. 2004; Wulster-Radcliffe et al. 2004).

Although most studies have shown that APN acts as an anti-inflammatory molecule, evidence also indicates that it can be pro-inflammatory. Outside the context of diseases associated with obesity, inflammatory status is not necessarily associated with low APN levels. In fact, the opposite is observed, with APN levels increasing during inflammatory conditions that are unrelated to an increase in fat mass (Fantuzzi 2008). This trend is particularly pronounced in chronic autoimmune diseases.

In inflammatory bowel disease (IBD), serum APN levels increase significantly in patients with ulcerative colitis (UC) and Crohn's disease (CD) when compared to healthy controls (HCs) (Karmiris et al. 2006; Yamamoto et al. 2005). When comparing UC and CD patients to each other, patients with UC have higher APN levels (Karmiris et al. 2006). Interestingly, APN mRNA and protein levels are increased in hypertrophic mesenteric adipose tissue in CD compared to tissues obtained from HC and UC subjects (Yamamoto et al. 2005). In addition, this increase in APN secretion and expression among CD patients is higher in the inflamed hypertrophied mesenteric adipose tissue compared to non-inflamed adipose tissue (Paul et al. 2006; Yamamoto et al. 2005).

This increase in APN levels in the presence of inflammation has also been observed in patients with systemic lupus erythematosus, a chronic autoimmune disease. In these patients, plasma APN levels increase compared to HC, particularly in subjects with inflammatory renal flare (Sada et al. 2006; Rovin et al. 2005). Interestingly, high serum levels of APN are observed despite the simultaneous presence of high TNF-α, which is considered a major inhibitor of APN production (Fantuzzi 2008).

Patients with rheumatoid arthritis (RA), another autoimmune disease, also exhibit increased APN levels in serum and in the synovial fluid of inflamed joints (Senolt et al. 2006; Schaffler et al. 2003; Otero, Lago, Gomez, Lago et al. 2006). Ehling et al. demonstrated that APN is present in the synovium and stimulates synovial fibroblasts to produce IL-6 and pro-matrix metalloproteinase-1, two key mediators in the pathogenesis of RA (Ehling et al. 2006).

Adiponectin also stimulates production of receptor activator of NFκB ligand (RANKL) while inhibiting osteoprotegrin expression in human osteoblasts, indicating that APN can indirectly increase osteoclast formation and therefore augment bone resorption (Luo et al. 2006)—a property APN shares with its TNF-α structural homologue (Berner et al. 2004).

In vitro studies have shown APN to have pro-inflammatory effects by inducing secretion of IL-8 in colonic epithelial cancer cells (Ogunwobi and Beales 2006). Finally, although it has commonly been accepted that APN inhibits TNF-α, APN can activate the TNF-α promoter and induce TNF-α production in RAW264.7 macrophages (Park et al. 2007). Thus, APN is emerging as an important mediator in the pathogenesis of obesity-related, autoimmune, and chronic inflammatory disorders. How APN regulates inflammation may be context- and tissue-specific and warrants further investigation.

## 5.4 RESISTIN

Three groups independently discovered resistin (Steppan et al. 2001; Kim et al. 2001; Holcomb et al. 2000). This 12-kDA polypeptide is the product of the *retn* gene and belongs to a unique family of cysteine-rich C terminal domain proteins called resistin-like molecules (RELMs), also known as "found in the inflammatory zone" (FIZZ) (Fantuzzi 2005; Kershaw and Flier 2004). In rodents, resistin is produced almost exclusively by adipocytes (Steppan and Lazar 2004). However, a degree of controversy surrounds this adipokine because some studies have shown human adipose tissue to express resistin and others did not find it or detected it only at very low levels in human adipocytes. In fact, it is believed that monocytes and macrophages, rather than adipocytes, are the predominant producers of resistin in humans (Bastard et al. 2006), thus questioning its classification as an adipokine.

Holcomb et al. (2000) described the first member of the gene family, FIZZ-1, and its specificity to adipose tissue. Soon thereafter, Steppan et al. (2001) discovered elevated serum levels of resistin (FIZZ-3) in rodent models of genetic and diet-induced obesity, suggesting a role in obesity-related dysfunction. Further investigation indicated that administration of recombinant resistin to lean mice impaired glucose homeostasis and insulin sensitivity. In addition, neutralization of resistin by injection of antibodies into diet-induced obese mice decreased blood glucose levels and improved insulin sensitivity (Steppan and Lazar 2004).

Further animal studies using resistin-deficient mice showed decreases in fasting glucose, improved glucose tolerance, and enhanced insulin sensitivity compared to wild-type mice. These resistin-deficient mice exhibited reduced fasting glucose after a high fat diet compared to their weight matched controls (Kershaw and Flier 2004; Antuna-Puente et al. 2008). In vitro studies demonstrated the ability of resistin to inhibit insulin-stimulated glucose uptake (Meier and Gressner 2004; Kershaw and Flier 2004). Despite conflicting data, resistin in rodents is generally considered to play an important role in development of hyperglycemia and IR observed in obesity (Bastard et al. 2006).

Most of the work linking resistin to obesity-related disorders was conducted in mouse models. However, the human resistin protein is only 55% homologous to its mouse counterpart, suggesting that resistin may not be evolutionarily well conserved across species (Tilg and Moschen 2006). As a result, the role of resistin in humans remains largely unclear. No correlation of resistin mRNA expression, body weight, obesity, and IR was observed in humans (Meier and Gressner 2004). However, the pro-inflammatory properties of resistin suggest a role for it in regulating inflammatory processes (Tilg and Moschen 2008).

Data demonstrating plasma resistin levels associated with markers of inflammation, and not IR, in both T2D and non-diabetic populations support this notion (Reilly et al. 2005). As a result, many studies have been conducted to determine how resistin is involved in the inflammatory response. In primary human macrophages, resistin expression is induced by stimulation with the IL-1, IL-6, and TNF-α cytokines in combination with LPS (Lago et al. 2007). Treatment with LPS also increases resistin mRNA expression in murine 3T3-L1 adipocytes and human monocytes (Lu et al. 2002). In human peripheral blood mononuclear cells, resistin induces production of TNF-α, IL-6, and IL-1β through activation of the NFκB pathway (Bokarewa et al. 2005; Lago et al. 2007). Thus resistin is part of the network of inflammatory mediators that regulate innate immunity in humans.

Levels of resistin have been associated with chronic and autoimmune diseases. Plasma levels of resistin have been correlated with reduced high-density lipoprotein (HDL) and high levels of coronary artery calcification, measures of coronary atherosclerosis, even after controlling for established risk factors such as metabolic syndrome and CRP levels (Reilly et al. 2005). Resistin also increases expression of the VCAM-1 and ICAM-1 adhesion molecules in human aortic endothelial cells. This induction of adhesion molecules by resistin is inhibited by APN, which also inhibits the ability of TNF-α to induce adhesion molecules on the vascular endothelium (Ouchi et al. 1999; Ouchi et al. 2000). Resistin stimulates smooth muscle cell proliferation in the human aorta as well, another physiological process necessary in the development of atherosclerosis (Antuna-Puente et al. 2008). Thus, resistin in humans may be more critical to regulation of vascular function and inflammatory responses than as a factor controlling metabolic processes.

Resistin has been associated with the pathogenesis of RA. In fact, patients with RA have significantly higher serum resistin levels compared to healthy controls that correlated with circulating CRP and TNF-α levels as well (Migita et al. 2006). Interestingly, serum resistin levels are rapidly reduced by anti-TNF-α therapy in patients with RA (Gonzalez-Gay et al. 2008). Furthermore, resistin accumulates in the synovial fluid of RA patients at levels significantly higher than in patients with osteoarthritis, a non-autoimmune joint disease (Bokarewa et al. 2005), further suggesting a potential role for resistin in the inflammatory cascade of RA. In IBD, circulating levels of resistin increase significantly in CD and UC patients compared to HCs (Karmiris et al. 2006). This increase in resistin is significantly associated with leukocyte blood count, CRP, and disease activity (Konrad et al. 2007). In conclusion, although adipocyte-derived resistin in rodents is likely implicated in metabolic control, in humans this protein should be more properly classified as a "classical" pro-inflammatory cytokine primarily produced by immune cells.

## 5.5 VISFATIN

Visfatin was originally identified by Samal et al. (1994) as a protein highly expressed in bone marrow, liver, and muscle. This 52-kDa protein was isolated and characterized from a human peripheral blood lymphocyte cDNA library and termed pre-B cell colony-enhancing factor (PBEF) (Samal et al. 1994). The molecule was rediscovered in 2005 as visfatin (Fukuhara et al. 2005) and was renamed because it was expressed predominantly in visceral adipose tissue. In both genetic and diet-induced animal models of obesity, visfatin expression is increased in visceral adipose tissue (Antuna-Puente et al. 2008).

With the rediscovery of visfatin, Fukuhara et al. demonstrated insulin-like effects of this molecule in vitro and in vivo when administered to lean mice by binding to and activating the insulin receptor (Fukuhara et al. 2005). Since then, the authors retracted some of their findings (Tilg and Moschen 2008). More recent data suggest that serum visfatin increases with progressive β cell deterioration observed in T2D (Lopez-Bermejo et al. 2006). Interestingly, circulating levels of visfatin increase in overweight and obese individuals, with even larger increases seen in those with metabolic syndrome compared to subjects that do not fulfill the criteria for complete metabolic syndrome diagnosis (Filippatos et al. 2007).

In addition to visceral WAT, visfatin expression is also upregulated in activated neutrophils and macrophages (Tilg and Moschen 2006). Visfatin inhibits apoptosis of neutrophils and upregulates expression of the IL-1β, IL-6, and TNF-α pro-inflammatory cytokines in human monocytes, as well as in mice treated with recombinant visfatin (Tilg and Moschen 2008). Visfatin is also expressed in human amniotic epithelial cell lines in vitro, and is upregulated by IL-6 and IL-8 pro-inflammatory cytokines (Ognjanovic and Bryant-Greenwood 2002).

Since the initial discovery, visfatin has been implicated in several inflammatory disorders such as acute lung injury (ALI), sepsis, IBD, and RA. Gene expression profiling of lung tissue from experimental models of ALI identified visfatin as a highly upregulated gene. Protein levels of visfatin increase in bronchoalveolar lavage, serum and lung tissue in both animal models, and in human ALI, suggesting visfatin as a potential biomarker (Ye et al. 2005). In neutrophils of critically ill septic patients expression of visfatin is markedly increased compared to resting neutrophils or LPS-stimulated neutrophils from HCs (Jia et al. 2004).

Studies using anti-sense oligonucleotides designed to target the translation initiation site of visfatin indicate that the increase in visfatin is partly responsible for the inhibition of neutrophil apoptosis observed in septic patients (Jia et al. 2004). Serum levels of visfatin also increase in IBD patients compared to HCs (Moschen et al. 2007). Significantly higher mRNA levels of visfatin are found in the inflamed colonic mucosa of IBD patients compared to HCs as well (Moschen et al. 2007). However, the cellular source responsible for this increase in visfatin remains unclear in IBD patients. A marked increase in visfatin was observed in patients with RA (Otero, Lago, Gomez, Lago et al. 2006). Nowell et al. showed that this upregulation of visfatin is, at least in part, regulated by IL-6 trans-signaling through STAT-3 (Nowell et al. 2006). Synovial fibroblasts are the major visfatin producing cells in the joints of RA patients (Nowell et al. 2006). These data suggest that visfatin is a pro-inflammatory mediator; whether it also plays a role in regulation of metabolism remains to be fully clarified.

## 5.6 CONCLUSIONS

It is no longer true that WAT exists solely as an energy source. Research has indicated that WAT is an important mediator of physiological and pathological processes, largely because of the release of adipokines. The deregulated release of adipokines by WAT during obesity and inflammatory conditions remains largely unexplored. Further understanding of the mechanisms of how adipokines are involved in the inflammatory response and the pathogenesis of metabolic, chronic and autoimmune disorders may lead to novel therapy options in the near future.

## REFERENCES

Antuna-Puente, B. et al. 2008. Adipokines: the missing link between insulin resistance and obesity. *Diabetes Metab* 34: 2–11.

Arita, Y. et al. 2002. Adipocyte-derived plasma protein adiponectin acts as a platelet-derived growth factor BB-binding protein and regulates growth factor-induced common post-receptor signal in vascular smooth muscle cell. *Circulation* 105: 2893–2898.

Bastard, J.P. et al. 2006. Recent advances in the relationship between obesity, inflammation, and insulin resistance. *Eur Cytokine Netw* 17: 4–12.

Berner, H.S. et al. 2004. Adiponectin and its receptors are expressed in bone-forming cells. *Bone* 35: 842–849.

Bernotiene, E. et al. 2004. Delayed resolution of acute inflammation during zymosan-induced arthritis in leptin-deficient mice. *Arthritis Res Ther* 6: R256–R263.

Boden, G. et al. 2003. Effect of thiazolidinediones on glucose and fatty acid metabolism in patients with type 2 diabetes. *Metabolism* 52: 753–759.

Bokarewa, M. et al. 2005. Resistin, an adipokine with potent pro-inflammatory properties. *J Immunol* 174: 5789–5795.

Buettner, C. et al. 2006. Critical role of STAT3 in leptin's metabolic actions. *Cell Metab* 4: 49–60.

Busso, N. et al. 2002. Leptin signaling deficiency impairs humoral and cellular immune responses and attenuates experimental arthritis. *J Immunol* 168: 875–882.

Chan, J.L. and C.S. Mantzoros. 2005. Role of leptin in energy deprivation states: normal human physiology and clinical implications for hypothalamic amenorrhoea and anorexia nervosa. *Lancet* 366: 74–85.

Chandran, M. et al. 2003. Adiponectin: more than just another fat cell hormone? *Diabetes Care* 26: 2442–2450.

Coleman, D.L. 1973. Effects of parabiosis of obese with diabetes and normal mice. *Diabetologia* 9: 294–298.

Coleman, D.L. and K.P. Hummel. 1973. The influence of genetic background on the expression of the obese (Ob) gene in the mouse. *Diabetologia* 9: 287–293.

Combs, T.P. et al. 2001. Endogenous glucose production is inhibited by the adipose-derived protein Acrp30. *J Clin Invest* 108: 1875–1881.

De Rosa, V. et al. 2007. A key role of leptin in the control of regulatory T cell proliferation. *Immunity* 26: 241–255.

Delaigle, A. M. et al. 2004. Induction of adiponectin in skeletal muscle by inflammatory cytokines: in vivo and in vitro studies. *Endocrinology* 145: 5589–5597.

Diez, J.J. and P. Iglesias. 2003. The role of the novel adipocyte-derived hormone adiponectin in human disease. *Eur J Endocrinol* 148: 293–300.

Ehling, A. et al. 2006. The potential of adiponectin in driving arthritis. *J Immunol* 176: 4468–4478.

Fantuzzi, G. 2005. Adipose tissue, adipokines, and inflammation. *J Allergy Clin Immunol* 115: 911–920.

Fantuzzi, G. 2008. Adiponectin and inflammation: consensus and controversy. *J Allergy Clin Immunol* 121: 326–330.

Fantuzzi, G. and R. Faggioni. 2000. Leptin in the regulation of immunity, inflammation, and hematopoiesis. *J Leukoc Biol* 68: 437–446.

Fayad, R. et al. 2007. Adiponectin deficiency protects mice from chemically induced colonic inflammation. *Gastroenterology* 132: 601–614.

Filippatos, T.D. et al. 2007. Increased plasma levels of visfatin/pre-B cell colony-enhancing factor in obese and overweight patients with metabolic syndrome. *J Endocrinol Invest* 30: 323–326.

Fruebis, J. et al. 2001. Proteolytic cleavage product of 300-kDa adipocyte complement-related protein increases fatty acid oxidation in muscle and causes weight loss in mice. *Proc Natl Acad Sci USA* 98: 2005–2010.

Fukuhara, A. et al. 2005. Visfatin: a protein secreted by visceral fat that mimics the effects of insulin. *Science* 307: 426–430.

Goldstein, B.J. and R. Scalia. 2004. Adiponectin: a novel adipokine linking adipocytes and vascular function. *J Clin Endocrinol Metab* 89: 2563–2568.

Gong, Y. et al. 2007. The long form of the leptin receptor regulates STAT5 and ribosomal protein S6 via alternate mechanisms. *J Biol Chem* 282: 31019–31027.

Gonzalez-Gay, M.A. et al. 2008. Anti-TNF-α therapy modulates resistin in patients with rheumatoid arthritis. *Clin Exp Rheumatol* 26: 311–316.

Gove, M.E. et al. 2009. Role of leptin receptor-induced STAT-3 signaling in modulation of intestinal and hepatic inflammation in mice. *J Leukoc Biol.*, submitted.

Guzik, T.J., D. Mangalat, and R. Korbut. 2006. Adipocytokines: novel link between inflammation and vascular function? *J Physiol Pharmacol* 57: 505–528.

Holcomb, I.N. et al. 2000. FIZZ1, a novel cysteine-rich secreted protein associated with pulmonary inflammation, defines a new gene family. *EMBO J* 19: 4046–4055.

Howard, J.K. et al. 1999. Leptin protects mice from starvation-induced lymphoid atrophy and increases thymic cellularity in *ob/ob* mice. *J Clin Invest* 104: 1051–1059.

Hug, C. et al. 2004. T-cadherin is a receptor for hexameric and high molecular weight forms of Acrp30/adiponectin. *Proc Natl Acad Sci USA* 101: 10308–10313.

Ingalls, A.M., M.M. Dickie, and G.D. Snell. 1950. Obese: a new mutation in the house mouse. *J Hered* 41: 317–318.

Jia, S.H. et al. 2004. Pre-B cell colony-enhancing factor inhibits neutrophil apoptosis in experimental inflammation and clinical sepsis. *J Clin Invest* 113: 1318–1327.

Juge-Aubry, C.E., E. Henrichot, and C.A. Meier. 2005. Adipose tissue: a regulator of inflammation. *Best Pract Res Clin Endocrinol Metab* 19: 547–566.

Kadowaki, T. and T. Yamauchi. 2005. Adiponectin and adiponectin receptors. *Endocr Rev* 26: 439–451.

Karmiris, K. et al. 2006. Circulating levels of leptin, adiponectin, resistin, and ghrelin in inflammatory bowel disease. *Inflamm Bowel Dis* 12: 100–105.

Kershaw, E.E. and J.S. Flier. 2004. Adipose tissue as an endocrine organ. *J Clin Endocrinol Metab* 89: 2548–2556.

Kim, K.H. et al. 2001. A cysteine-rich adipose tissue-specific secretory factor inhibits adipocyte differentiation. *J Biol Chem* 276: 11252–11256.

La Cava, A. and G. Matarese. 2004. The weight of leptin in immunity. *Nat Rev Immunol* 4: 371–379.

Lago, F. et al. 2007. Adipokines as emerging mediators of immune response and inflammation. *Nat Clin Pract Rheumatol* 3: 716–724.

Lopez-Bermejo, A. et al. 2006. Serum visfatin increases with progressive β cell deterioration. *Diabetes* 55: 2871–2875.

Lord, G.M. et al. 1998. Leptin modulates the T cell immune response and reverses starvation-induced immunosuppression. *Nature* 394: 897–901.

Lu, S.C. et al. 2002. Lipopolysaccharide increases resistin gene expression in vivo and in vitro. *FEBS Lett* 530: 158–162.

Luo, X.H. et al. 2006. Adiponectin stimulates RANKL and inhibits OPG expression in human osteoblasts through the MAPK signaling pathway. *J Bone Miner Res* 21: 1648–1656.

Mansour, E. et al. 2006. Leptin inhibits apoptosis in thymus through a janus kinase-2-independent, insulin receptor substrate-1/phosphatidylinositol-3 kinase-dependent pathway. *Endocrinology* 147: 5470–5479.

Martin-Romero, C. et al. 2000. Human leptin enhances activation and proliferation of human circulating T lymphocytes. *Cell Immunol* 199: 15–24.

Masaie, H. et al. 2007. Adiponectin binds to chemokines via the globular head and modulates interactions between chemokines and heparan sulfates. *Exp Hematol* 35: 947–956.

Matarese, G. et al. 2001. Requirement for leptin in the induction and progression of autoimmune encephalomyelitis. *J Immunol* 166: 5909–5916.

Meier, U. and A.M. Gressner. 2004. Endocrine regulation of energy metabolism: review of pathobiochemical and clinical chemical aspects of leptin, ghrelin, adiponectin, and resistin. *Clin Chem* 50: 1511–1525.

Migita, K. et al. 2006. The serum levels of resistin in rheumatoid arthritis patients. *Clin Exp Rheumatol* 24: 698–701.

Mlinar, B. et al. 2007. Molecular mechanisms of insulin resistance and associated diseases. *Clin Chim Acta* 375: 20–35.

Moschen, A.R. et al. 2007. Visfatin, an adipocytokine with pro-inflammatory and immunomodulating properties. *J Immunol* 178: 1748–1758.

Munzberg, H. and M.G. Myers, Jr. 2005. Molecular and anatomical determinants of central leptin resistance. *Nat Neurosci* 8: 566–570.

Nowell, M.A. et al. 2006. Regulation of pre-B cell colony-enhancing factor by STAT-3-dependent interleukin-6 trans-signaling: implications in the pathogenesis of rheumatoid arthritis. *Arthritis Rheum* 54: 2084–2095.

Ognjanovic, S. and G.D. Bryant-Greenwood. 2002. Pre-B-cell colony-enhancing factor: a novel cytokine of human fetal membranes. *Am J Obstet Gynecol* 187: 1051–1058.

Ogunwobi, O.O. and I.L. Beales. 2006. Adiponectin stimulates proliferation and cytokine secretion in colonic epithelial cells. *Regul Pept* 134: 105–113.

Oh, D.K., T. Ciaraldi, and R.R. Henry. 2007. Adiponectin in health and disease. *Diabetes Obes Metab* 9: 282–289.

Okamoto, Y. et al. 2002. Adiponectin reduces atherosclerosis in apolipoprotein E-deficient mice. *Circulation* 106: 2767–2770.

Otero, M. et al. 2006. Toward a pro-inflammatory and immunomodulatory emerging role of leptin. *Rheumatology* 45: 944–950.

Otero, M. et al. 2006. Changes in plasma levels of fat-derived hormones adiponectin, leptin, resistin and visfatin in patients with rheumatoid arthritis. *Ann Rheum Dis* 65: 1198–1201.

Ouchi, N. et al. 1999. Novel modulator for endothelial adhesion molecules: adipocyte-derived plasma protein adiponectin. *Circulation* 100: 2473–2476.

Ouchi, N. et al. 2000. Adiponectin, an adipocyte-derived plasma protein, inhibits endothelial NFκB signaling through a cAMP-dependent pathway. *Circulation* 102: 1296–1301.

Papathanassoglou, E. et al. 2006. Leptin receptor expression and signaling in lymphocytes: kinetics during lymphocyte activation, role in lymphocyte survival, and response to high fat diet in mice. *J Immunol* 176: 7745–7752.

Park, P.H. et al. 2007. Short-term treatment of RAW264.7 macrophages with adiponectin increases tumor necrosis factor-α (TNF-α) expression via ERK1/2 activation and Egr-1 expression: role of TNF-α in adiponectin-stimulated interleukin-10 production. *J Biol Chem* 282: 21695–21703.

Paul, G. et al. 2006. Profiling adipocytokine secretion from creeping fat in Crohn's disease. *Inflamm Bowel Dis* 12: 471–477.

Reilly, M.P. et al. 2005. Resistin is an inflammatory marker of atherosclerosis in humans. *Circulation* 111: 932–939.

Robertson, S.A., G.M. Leinninger, and M.G. Myers, Jr. 2008. Molecular and neural mediators of leptin action. *Physiol Behav* 94: 637–642.

Rosenbaum, M. et al. 2008. Leptin reverses weight loss-induced changes in regional neural activity responses to visual food stimuli. *J Clin Invest* 118: 2583–2591.

Rovin, B.H. et al. 2005. Plasma, urine, and renal expression of adiponectin in human systemic lupus erythematosus. *Kidney Int* 68: 1825–1833.

Sada, K.E. et al. 2006. Altered levels of adipocytokines in association with insulin resistance in patients with systemic lupus erythematosus. *J Rheumatol* 33: 1545–1552.

Samal, B. et al. 1994. Cloning and characterization of the cDNA encoding a novel human pre-B cell colony-enhancing factor. *Mol Cell Biol* 14: 1431–1437.

Schaffler, A. et al. 2003. Adipocytokines in synovial fluid. *JAMA* 290: 1709–1710.

Senolt, L. et al. 2006. Increased adiponectin is negatively linked to the local inflammatory process in patients with rheumatoid arthritis. *Cytokine* 35: 247–252.

Siegmund, B., H.A. Lehr, and G. Fantuzzi. 2002. Leptin: a pivotal mediator of intestinal inflammation in mice. *Gastroenterology* 122: 2011–2025.

Siegmund, B. et al. 2004. Leptin receptor expression on T lymphocytes modulates chronic intestinal inflammation in mice. *Gut* 53: 965–972.

Siegmund, B. et al. 2004. Development of intestinal inflammation in double IL-10- and leptin-deficient mice. *J Leukoc Biol* 76: 782–786.

Steppan, C.M. et al. 2001. The hormone resistin links obesity to diabetes. *Nature* 409: 307–312.

Steppan, C.M. and M.A. Lazar. 2004. The current biology of resistin. *J Intern Med* 255: 439–447.

Takemura, Y. et al. 2007. Adiponectin modulates inflammatory reactions via calreticulin receptor-dependent clearance of early apoptotic bodies. *J Clin Invest* 117: 375–386.

Tao, L. et al. 2007. Adiponectin cardioprotection after myocardial ischemia/reperfusion involves the reduction of oxidative/nitrative stress. *Circulation* 115: 1408–1416.

Tilg, H. and A.R. Moschen. 2006. Adipocytokines: mediators linking adipose tissue, inflammation and immunity. *Nat Rev Immunol* 6: 772–783.

Tilg, H. and A.R. Moschen. 2008. Inflammatory mechanisms in the regulation of insulin resistance. *Mol Med* 14: 222–231.

Valassi, E., M. Scacchi, and F. Cavagnini. 2008. Neuroendocrine control of food intake. *Nutr Metab Cardiovasc Dis* 18: 158–168.

Waki, H. et al. 2005. Generation of globular fragment of adiponectin by leukocyte elastase secreted by monocytic cell line THP-1. *Endocrinology* 146: 790–796.

Wang, P. et al. 2008. The secretory function of adipocytes in the physiology of white adipose tissue. *J Cell Physiol* 216: 3–13.

Wang, Y. et al. 2005. Adiponectin inhibits cell proliferation by interacting with several growth factors in an oligomerization-dependent manner. *J Biol Chem* 280: 18341–18347.

Wolf, A.M. et al. 2004. Adiponectin induces the anti-inflammatory cytokines IL-10 and IL-1RA in human leukocytes. *Biochem Biophys Res Commun* 323: 630–635.

Wulster-Radcliffe, M.C. et al. 2004. Adiponectin differentially regulates cytokines in porcine macrophages. *Biochem Biophys Res Commun* 316: 924–929.

Yamamoto, K. et al. 2005. Production of adiponectin, an anti-inflammatory protein, in mesenteric adipose tissue in Crohn's disease. *Gut* 54: 789–796.

Yamauchi, T. et al. 2003. Cloning of adiponectin receptors that mediate antidiabetic metabolic effects. *Nature* 423: 762–769.

Yamauchi, T. et al. 2002. Adiponectin stimulates glucose utilization and fatty acid oxidation by activating AMP-activated protein kinase. *Nat Med* 8: 1288–1295.

Ye, S.Q. et al. 2005. Pre-B cell colony-enhancing factor as a potential novel biomarker in acute lung injury. *Am J Respir Crit Care Med* 171: 361–370.

Zhang, Y. et al. 1994. Positional cloning of the mouse obese gene and its human homologue. *Nature* 372: 425–432.

# 6 Insulin as Modulator of Adipose Inflammation

*Joseph Doria and Ahmad Aljada*

## CONTENTS



## 6.1 INTRODUCTION

Diet-induced obesity and the subsequent development of features of the metabolic syndrome have become major worldwide health problems. According to the National Center for Health Statistics more than 34% of the U.S. population is obese and almost 70% of adults in the U.S. are overweight. Obesity has and continues to threaten and overload health care infrastructures due to association with various chronic diseases including increased risk of certain cancers, type II diabetes mellitus (T2DM) and cardiovascular complications and diseases [1]. Reduced life expectancy and degradation of elderly life quality due to obesity and its related diseases appear imminent. Obesity is associated with low-grade inflammation in adipose tissues resulting from increased production of pro-inflammatory cytokines that can subsequently contribute to the development of insulin resistance [2–4].

The development of the concept that obesity is an inflammatory condition is an exciting and novel approach to the understanding of this condition. It presents implications in terms of both the pathogenesis and the complications of this condition. Adipocytes are responsible for secreting hormones and adipokines that positively or negatively modulate inflammation. Conversely, insulin exerts anti-inflammatory effects in various experimental models of inflammation. The anti-inflammatory effects of insulin have been demonstrated in endotoxin-induced lung injury, carrageenin-induced inflammation in rats, suppression of tumor necrosis factor-α (TNF-α) by peritoneal exudate cells, inhibition of TNF-α-induced interstitial pneumonitis, prevention of periportal inflammation in the liver, acute ST segment-elevation myocardial infarction, inhibition of nuclear factor κB (NF-κB) and early growth response 1 (Egr-1) transcription factors in the obese and many other models [5–12]. In addition, insulin suppresses macrophage migration inhibitory factor (MIF) expression by adipocytes in vitro [10,13,14]. It is of interest that insulin sensitizers have also been shown to exert anti-inflammatory effects [15–20]. This represents a significant novel non-metabolic action that requires further elucidation.

## 6.2 CLASSICAL METABOLIC AND NOVEL NON-METABOLIC ACTIONS OF INSULIN

The original description of the metabolic syndrome consisted of obesity, insulin resistance, hypertension, impaired glucose tolerance or diabetes, hyperinsulinemia, and dyslipidemia characterized by elevated triglyceride and high density lipoprotein (HDL) concentrations [21]. As our understanding of the action of insulin evolved, we now perceive that insulin resistance is the basis of most if not all the features of the metabolic syndrome.

The classical biological effects of insulin are vital and dispersed throughout the body. These include homeostasis of glucose and lipid levels. In adipose tissue, insulin lowers glucose levels by increasing lipogenesis and by decreasing lipolysis. Insulin also increases glucose uptake in adipose tissue and muscle, processes facilitated by the GLUT-4 transporter. Insulin in liver tissue suppresses glucagon release from α cells and decreases glycogenolysis, which is primarily activated by epinephrine and glucagon. Studies by Cherrington et al. showed that with increased insulin concentrations there is a net decrease in hepatic glucose output, suggesting that hepatic glucose production is regulated by insulin [22].

The liver also assists in maintaining normal insulin concentration by decreasing gluconeogenesis and increasing glycogenesis via activity of glycogen synthase. The effect of insulin is to increase tissue and plasma glucose utilization. Insulin also decreases formation of uric acid, free fatty acids (FFAs), and triglycerides. Furthermore, insulin controls appetite which essentially lowers macronutrient intake. Collectively these mechanisms are essential for normal blood glucose homeostasis. As glucose is absorbed from digested carbohydrates, blood glucose levels rise and these aforementioned mechanisms quickly and efficiently bring blood glucose levels back to their normal range (Figure 6.1).

Novel insulin effects continue to be proposed, including chronic anti-inflammatory properties (Figure 6.2). Hotamisligil et al. first described the concept of the relationship of obesity and insulin resistance to inflammation in 1993, when they proposed that adipocytes express pro-inflammatory TNF-α and that expression in obese animals was significantly increased [4]. It was later shown that human adipocytes express TNF-α and this was associated with weight loss [23,24]. When overexpressed, TNF-α promotes insulin resistance by various means in different cells and tissues. Interestingly, insulin has been shown to inhibit TNF-α production [7]. Insulin stimulates nitric oxide (NO) production, leading to vasodilation and platelet inhibition [25]. Additionally, insulin exerts cardioprotective effects [10].

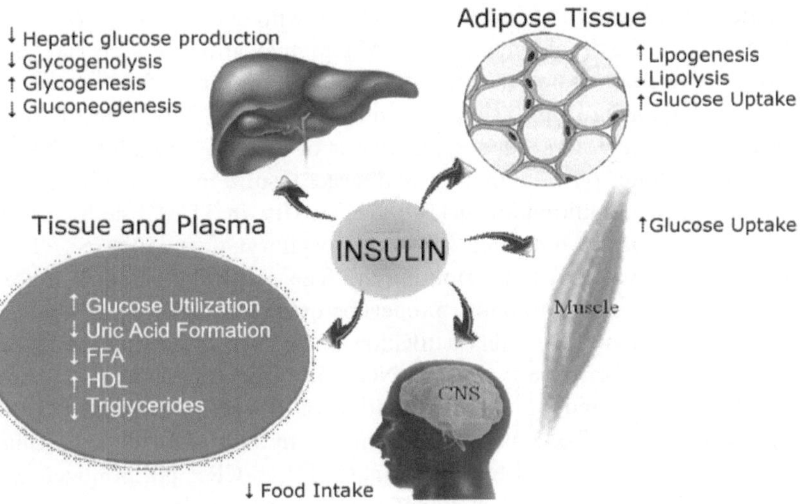


**FIGURE 6.1** Classical biological effects of insulin throughout the body, including increased uptake of blood glucose, lipogenesis, and decreased lipolysis in adipose tissue. In the liver, insulin decreases hepatic glucose production, glycogenolysis, and gluconeogensis, and increases glycogenesis. Insulin decreases uric acid formation, FFAs, and triglycerides, and increases HDL and glucose utilization. It facilitates glucose uptake in muscle and decreases food intake through the central nervous system.

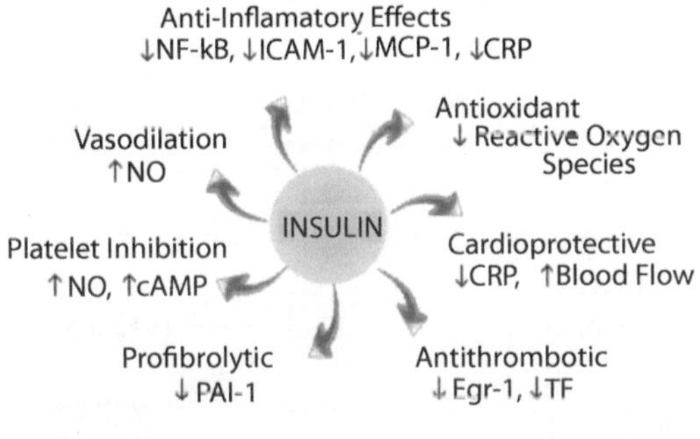


**FIGURE 6.2** Novel biological effects of insulin. In addition to lowering blood glucose and other classical biological effects, insulin also decreases atherosclerosis, acute myocardial infarction, inflammation, and ROS production and exerts cardioprotective effects.

An underlying mechanism may be through inhibition of TNF-α which is linked to acute myocardial infarction (AMI) and released in the early phase of AMI, directly decreasing contractility [26,27].

T2DM is a condition associated with insulin resistance in muscle and adipose tissue. Studies established T2DM as an inflammatory condition by showing elevated pro-inflammatory interleukin (IL)-6 and sialic acid in the plasma of patients [4,28,29]. Studies by Schmidt et al. showed that levels of inflammatory mediators such as IL-6, orosomucoid, sialic acid, and C-reactive protein (CRP) in the plasma may predict T2DM [30]. IL-6 levels are increased with obesity and IL-6 is an important marker for future development of T2DM and myocardial infarction [31,32].

Klover et al. demonstrated in mice that chronic exposure to IL-6 caused insulin resistance by impairing insulin signaling in liver cells [33]. Moreover, IL-6 reduces adiponectin expression and induces suppressor of cytokine signaling (SOCS) proteins induced in states of inflammation and target insulin receptor substrates (IRSs) for degradation by ubiquitination [34,35]. Conversely, in 3T3-L1 cells, insulin disrupts IL-6 pathways by reducing tyrosine phosphorylation of the STAT3 signal transducer and activator of transcription [34]. The anti-inflammatory, antioxidant, vasodilatory, and platelet inhibitory properties of insulin have been demonstrated in several studies. Insulin inhibits activation of NF-κB and formation of reactive oxygen species (ROS), while increasing NO and cyclic adenosine monophosphate (cAMP) (Figures 6.2 and 6.3). The inhibition of NF-κB activation by insulin is very significant since NF-κB plays a cardinal role in the transcription of inflammatory pathways such as c-Jun N-terminal kinase (JNK), CRP, plasminogen activator inhibitor-1 (PAI-1), TNF-α, and enzymes generating ROS. Insulin inhibits NF-κB activation by elevating the IκB that binds to cytosolic NF-κB, thus preventing translocation into the nucleus [36,37]. Moreover, the reduced level of NF-κB decreases ROS generation, preventing unwanted oxidative stress leading to further inflammation and cellular damage.

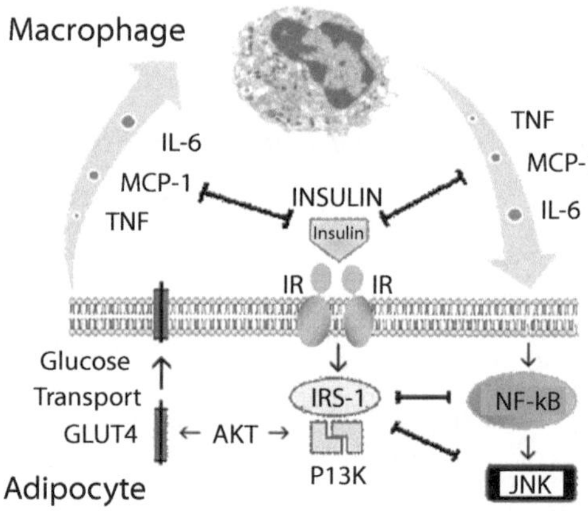


**FIGURE 6.3** Paracrine and autocrine actions and macrophage–adipocyte interactions in adipose tissue. Insulin stimulates phosphoinositide 3-kinase (PI3K) by binding insulin receptor (IR) and activating IRS-1, which can then bind to and activate PI3K. Once activated, PI3K induces expression of phosphatidylinositol (3,4,5)-trisphosphate (PIP3) leading to activation of Akt, and 3' phosphoinositide-dependent kinase-1 (PDK1), which ultimately promotes translocation of GLUT4 to the plasma membrane for use in glucose transport and uptake. Activated adipocytes secrete chemokines that attract macrophages to adipose tissue and adipokines that activate macrophages. Activated macrophages produce several cytokines that in turn may further activate adipocytes and eventually cause insulin resistance. Insulin inhibits TNF-α, MCP-1, IL-6, macrophage infiltration, NF-κB, and JNK pathways. Insulin, by inhibiting the synthesis and secretion of inflammatory mediators by adipocytes and macrophages, may reduce insulin resistance.

TNF-α also promotes insulin resistance in many cells and tissues by inhibiting phosphorylation on the tyrosine residue and/or promoting phosphorylation on serine residues of insulin receptor substrate-1 (IRS-1) which inhibits insulin signaling [38]. Strikingly, an overabundance of intracellular FFAs displays similar results, suggesting the FFAs reduce insulin activity by similar pathways. Ser-307 is another phosphorylation site on IRS-1 that is targeted by JNK which is increased by both FFAs and TNF-α. Ser-307 phosphorylation inhibits tyrosine phosphorylation, thus preventing transmission of insulin signals [38]. Neutralization of TNF-α by soluble TNF-α receptors leads to increased insulin sensitivity in obese animals [39]. Similar effects have not been demonstrated in humans. However, insulin does inhibit TNF-α production [7].

## 6.3 INSULIN MODULATION OF INFLAMMATORY SIGNALS IN ADIPOCYTES AND MACROPHAGES

In progressive stages of obesity in both mice and humans, macrophages increase in number and infiltrate adipose tissue [40]. It has been hypothesized that macrophage infiltration may promote the death of adipocytes in obesity based on the fact that macrophages have been observed surrounding dead adipocytes [41].

Expression analysis from adipose tissue shows that macrophages are accountable for the majority of pro-inflammatory secretions in obese adipose tissue and the secretion of almost all the TNF-α, inducible NOS (iNOS), and IL-6 observed in adipose tissue [40]. It has been proposed that accumulation of macrophages in adipose tissue may have a role in the inflammatory response of adipose tissue observed in obesity [42]. Insulin may prevent paracrine and autocrine inflammatory signals in adipocytes and macrophages by inhibiting production of TNF-α, monocyte chemotattractant protein-1 (MCP-1), and IL-6, factors that induce inflammation, macrophage infiltration, NF-κB and JNK pathways. As a result, insulin may repress this amplified inflammatory cascade.

## 6.4 INSULIN AND ADIPOKINES

Adipocytes influence the immune response through several cytokine-like mediators known as adipokines including leptin, resistin, and adiponectin. These adipokines exert profound effects on a number of physiologic processes including immune function and inflammation. By modulating these adipokines, insulin may exert an anti-inflammatory effect on adipocytes.

Leptin is produced by white adipose tissue and regulates adipose tissue mass by influencing food intake and energy expenditure [43]. The role of leptin in inflammation is less clear. Leptin may directly link nutritional status and pro-inflammatory T helper cell 1 immune responses [44,45]. Altered leptin production during infection and inflammation strongly suggests that leptin is a part of the cytokine cascade that orchestrates the innate immune response and host defense mechanisms [46,47]. Leptin stimulates T lymphocyte responses, thus filling a pro-inflammatory role in experimental models of autoimmune diseases. Leptin is a member of the helical cytokine family; its structure resembles those of the IL-2, IL-6, and IL-15 hematopoietic cytokines [44]. It exerts proliferative and anti-apoptotic activity in a variety of cell types including T lymphocytes, leukemia cells, and hematopoietic progenitors. It has also been implicated in the pathogenesis of chronic low grade inflammation present in obese individuals, increasing their risks of developing cardiovascular diseases, T2DM, and degenerative diseases including autoimmunity and cancer. On the other hand, leptin exerts protective anti-inflammatory effects in models of acute inflammation and during activation of innate immune responses.

Decreased plasma leptin concentration during food deprivation can lead to impaired immune function, reduced cell-mediated immune responses, and an overall increased risk of infection. Leptin deficiency after starvation in rodents is linked to increased glucocorticoid levels and decreased levels of thyroid and growth hormone, each of which may mediate immune suppression [48–50]. Additionally, leptin may be released in response to inflammatory cytokines and act to attenuate the responses to TNF-α, IL-6, and IL-1, all of which has been shown to play a protective role against the toxicity exerted by TNF-α [51]. Exogenous leptin can also reduce NO levels and CD40 expression in cerulean-induced acute pancreatitis, a non-infectious inflammatory reaction of the pancreas [52,53]. It is of interest that both the *ob/ob* mouse that lacks leptin and the *db/db* mouse that lacks the leptin receptor are susceptible to infections. Notably, insulin can stimulate leptin release via a phosphoinositide 3-kinase (PI3K)-independent mechanism and thus may modulate the immune system through leptin induction [54]. The definitive role of leptin in this context in humans has yet to be proved.

Resistin is an adipokine that is almost exclusively expressed in WAT. Resistin causes insulin resistance and diabetes in mice, while in humans it is linked to inflammation. In some pathophysiological conditions, plasma resistin levels were found to be associated with many inflammatory markers [55–60]. Insulin, glucose, many cytokines, and the anti-diabetic thiazolidinediones are regulators of resistin gene expression. Resistin levels increase in diet-induced and genetic forms of obesity (*ob/ob* and *db/db*) and decrease with the administration of the rosiglitazone anti-diabetic drug [61–66]. Insulin inhibits resistin expression and secretion in 3T3-L1 [64,66,67]. This inhibitory effect of insulin is independent of PI3K or p38-mitogen-activatied protein kinase pathways.

Remarkably, lipopolysaccharide (LPS) upregulates resistin levels in WAT and white blood cells in rats as well as in 3T3-L1 adipocytes. However, LPS was also reported to have no effect on resistin expression in 3T3-L1 adipocytes, while in a separate study LPS was shown to suppress resistin expression in adipose tissue [68,69]. Similarly, TNF-α increased resistin expression in human peripheral blood mononuclear cells (PBMCs) while it downregulated resistin in 3T3-L1 adipocytes [64,69,70]. IL-6 also increased resistin expression in PBMCs, but had no effect in 3T3-L1 adipocytes [69,70]. Resistin upregulation by IL-6 and TNF-α in human PBMCs is mediated via the NF-κB pathway [55]. Perhaps these conflicting results may be explained by the use of different techniques and conditions.

Adiponectin, one of the most abundantly synthesized hormones by adipocytes in adipose tissue, produces an anti-inflammatory effect by inhibiting phagocytic activity, TNF-α production, and modulating insulin sensitivity [71]. Adiponectin has been shown to increase sensitivity to insulin, glucose tolerance, and FFA oxidation in several tissues [72–75]. Administration of physiological doses of adiponectin in insulin-resistant mice increased insulin sensitivity. Adiponectin is found in lower concentrations in men than women and its levels decrease with obesity in humans, a trend similar to many anti-inflammatory hormones and cytokines.

Reduced adiponectin levels have been documented not only with obesity but also with T2DM, coronary artery disease (CAD), and insulin resistance [76–78]. Adiponectin may be linked to atherosclerosis because it inhibits TNF-α and the conversion of macrophages to foam cells and decreases cell adhesion molecules, thus lowering the number of macrophages attached to endothelial cells [76,79–81]. Weight loss is a general way to lower risk of T2DM and CAD. Yang et al. demonstrated that weight loss results in increased plasma levels of adiponectin [80]. By increasing insulin sensitivity, adiponectin positively enhances the anti-inflammatory effects of insulin.

The sirtuin 1 (SIRT1) enzyme, a human homologue of *Sir2*, is upregulated in calorie-restricted diets and has an anti-inflammatory response on adipocytes. SIRT1 can reduce TNF-α, and deacetylate NF-κB, thus inhibiting its transcription. In addition, a SIRT1 inhibitory compound showed induction of differentiation of adipocytes—a process often coupled with insulin sensitization [82]. Olefsky et al. indicated that a knockdown of SIRT1 increases IRS-1 serine phosphorylation and JNK phosphorylation and decreases insulin signaling and IRS-1 tyrosine phosphorylation, all inhibiting insulin signaling to glucose transport [83].

This study also displayed an increase in mRNA expression for TNF-α, IL-6, CRP, and JNK with the knockdown of SIRT1. Investigations by Sun et al. concluded that SIRT1 improves insulin signaling in insulin-resistant states by inhibiting the protein tyrosine phosphatase 1B (PTP1B) that directly inhibits insulin action [84,85]. However, it is important to note that SIRT1 improved insulin sensitivity and glucose transport only in resistant states and little to no change was noticed in insulin-sensitive conditions, suggesting that under normal conditions SIRT1 may have little effect in this regard.

SIRT1 has recently been linked to reversing inflammatory pathways and oxidative stress in adipose tissue associated with fatty acid overload through mechanisms that increase forkhead transcription factor FKHR (FOXO1), a key regulator of glucose homeostasis, cell cycle progression, and apoptosis [86]. It has also been shown that SIRT1 increases adiponectin transcription in adipocytes by activating FOXO1 and enhancing FOXO1 and CCAAT/enhancer binding protein (C/EBPβ) interaction; both FOXO1 and SIRT1 protein levels are significantly lower in epididymal fat tissues from *db/db* mice [87]. The addition of FFAs to adipocytes decreased FOXO1 protein levels which correlated with an increase in the production of ROS, IL-6, PAI-1, and MCP-1 mRNA expression levels and reduction in adiponectin [86]. Therefore, the observed anti-inflammatory properties of SIRT1 are possibly modulated by increased adiponectin levels through FOXO1, resulting in insulin sensitization. Increased insulin sensitization in return may account for these anti-inflammatory properties of SIRT1. However, the presence of other mechanisms is possible and has yet to be characterized.

FOXO transcription factors are subject to complex regulation. Growth factors inhibit FOXO via serine–threonine phosphorylation and nuclear exclusion while cellular stress causes FOXO acetylation, thus promoting the interaction of FOXO and SIRT1 [88–92]. However, the effect of SIRT1-dependent deacetylation on FOXO function remains somewhat controversial, with most studies suggesting that deacetylation increases FOXO-dependent transcription. Insulin negatively regulates FOXO1 activity via phosphorylation and ubiquitination-mediated degradation, resulting in repression of target gene expression [93,94]. Conversely, FOXO factors control upstream elements of insulin signaling although they are downstream targets of insulin. Treatment of 3T3-L1 adipocytes with TNF-α attenuated Akt-dependent phosphorylation of FOXO1 and enhanced transcriptional activity of FOXO1 which in turn increased the expression of C/EBPβ, a positive regulator of MCP-1 and IL-6 genes, through directly binding to its promoter [95].

Visfatin (also known as pre-B cell colony-enhancing factor or PBEF) is another adipokine that increases during the development of obesity and T2DM and exerts a pro-inflammatory effect [96]. It has an insulin-mimetic effect via binding to the insulin receptor [97]. However, insulin does not influence synthesis of this adipocytokine [98]. The binding of visfatin to insulin receptor may block insulin binding, resulting in increased inflammatory status.

## 6.5 INSULIN AND INFLAMMATION: PROPOSED MODEL

Several studies confirmed that the presence of inflammatory mediators predicts the development of T2DM in populations at risk, suggesting that IR and inflammatory processes are related and may involve a causal link [30,99–107]. What leads to inflammation in obesity and T2DM? What is the link between inflammation and insulin resistance? Our work and that of others confirmed the pro-inflammatory effects of food intake on oxidative stress and inflammation at the cellular and molecular levels [108–113].

These inflammatory mediators activate pro-inflammatory transcription factors, including NF-κB, that in turn promote the transcription of pro-inflammatory genes. On the other hand, levels of inflammatory mediators fall with dietary restriction and weight loss [114]. It is likely that the pro-inflammatory state of obesity is related to chronic excessive food intake. TNF-α has been shown to induce serine phosphorylation of IRS-1, thus inhibiting autophosphorylation of tyrosine residues of insulin receptor in adipocytes [115]. This effect has been shown to be related to SOCS family proteins that are increased by inflammatory cytokines [116]. SOCS proteins are elevated in insulin-resistant tissues and impair IR signaling by blocking IR-mediated tyrosine phosphorylation [116] or by promoting ubiquitination and degradation of IR [35]. SOCS-3 binds IR and inhibits its autophosphorylation and the tyrosine phosphorylation of IRS-1, the association of p85 subunit of PI3 kinase to IRS-1, and subsequent activation of Akt [117]. SOCS-3 may also be the mediator of leptin resistance in obese humans.

Plasma adiponectin concentration has an inverse relationship with adiposity, insulin resistance, diastolic pressure, triglyceride concentration, and TNF-α receptor concentration [118]. Leptin, on the other hand, is elevated in obese humans and can regulate immune function by stimulating responses to inflammatory challenge. Notably, CRP can directly interact with leptin, causing leptin resistance, and thus inhibiting leptin from regulating energy intake and energy expenditure, resulting in obesity [119]. CRP can also inhibit adiponectin. Resistin, on the other hand, may be activated by several inflammatory mediators and can itself activate these mediators, including the NF-κB pathway, potentiating the inflammatory status in obesity.

A chronic overload of food intake leads to chronic inflammation which leads to lower levels of secreted adiponectin and higher rates of lipolysis, CRP, and FFAs, and ultimately leads to leptin and insulin resistance (Figure 6.4). FFAs are key players of insulin resistance and inflammation in skeletal muscle, liver, and endothelial cells [120]. FFAs are elevated in obesity due to enlarged adipocytes releasing higher concentrations, but additionally FFA clearance will be reduced, along with an increase in lipolysis and a deficiency of perilipins, inhibiting lipases from hydrolyzing triglycerides and releasing FFAs [121].

FFAs exert pro-inflammatory effects. In diabetes, elevated FFAs increased mitochondrial ROS production and overall FFAs increased oxidative stress, activating many pro-inflammatory pathways and inflammatory cytokine production, and reducing circulating adiponectin levels [122,123]. It is also noteworthy that elevated circu-

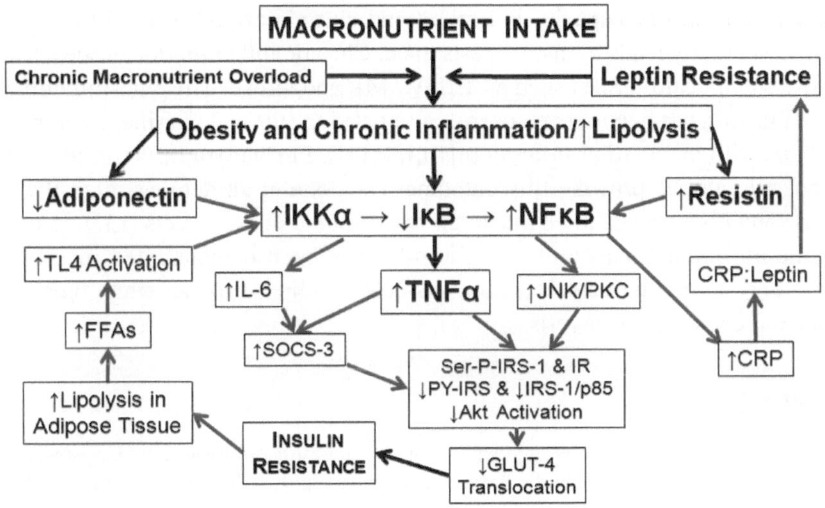


**FIGURE 6.4** Relationship of macronutrient intake, oxidative stress, inflammation, obesity, and impaired insulin signaling. Excessive caloric intake leads to chronic inflammation, reduced glucose tolerance, and insulin resistance through multiple pro-inflammatory pathways including NF-κB and TNF-α. These further increase pro-inflammatory pathways leading to decreased insulin sensitivity through impairment of IR. Inflammation in obesity leads to increased expression of TNF-α and SOCS-3 which impair Yp-IR and IRS-1-p85 PI3-kinase interaction which leads to IR. In addition to IR, adipokine levels are altered, adiponectin is inhibited, and leptin and resistin concentrations are increased, leading to resistance of leptin and ultimately increasing food intake.

lating FFAs in normal humans caused endothelial dysfunction, including decreases in blood flow and NO production through stimulation of NADPH oxidase [124,125].

The activation of NADPH and the consequent decrease in NO are likely due to an increase of ROS that is protein kinase C (PKC)-dependent [126]. Toll-like receptor 4 (TLR4) is an LPS receptor, also activated by FFAs, that transduces cytokine expression [127]. TLR4 activates NF-κB signaling in adipocytes, resulting in decreases in Akt and GSK3β phosphorylation, key mediators of insulin signaling and glucose uptake. FFAs and TNF-α can also induce Ser-307 phosphorylation on IRS-1, which inhibits tyrosine phosphorylation and prevents insulin signal transmission [38]. PKC, induced by the NF-κB pathway and FFAs, can inhibit insulin signaling by phosphorylating IRS-1. Lowering FFAs in non-diabetic and obese T2DM subjects improves insulin sensitivity and glucose tolerance, and reduces hyperinsulinemia, illustrating another possible correlation to insulin resistance, inflammation, and obesity [128].

However, contrary to typical thought, too little fat can result in decreased insulin sensitivity, fatty liver, hyperlipidemia, and hyperglycemia. Shulman et al. demonstrated that transplanting adipose tissue reversed hyperglycemia and diminished levels of insulin in fat-deficient mice, concluding that moderate levels of adipocytes are necessary for conventional insulin signaling and function [129].

Following high calorie food intake, NF-κB and ROS inflammatory pathways are activated. This pro-inflammatory state is counteracted by insulin. Nevertheless, insulin compensatory capacity is limited and chronic excessive food intake, as in obesity, eventually leads to insulin resistance. Chronic inflammation ensues, leading to disease and health issues such as T2DM, MI, and stroke. This anti-inflammatory effect of insulin has been proven useful in improving survival in critically ill patients and those with myocardial infarction [10,130,131]. Further studies have also shown that not all calories provoke this outcome. Dandona et al. demonstrated that high fructose and high orange juice intake do not stimulate these effects [132]. In conclusion, the anti-inflammatory effect of insulin has been demonstrated in numerous in vitro and in vivo models, including models of adipose tissue, white blood cells, endothelial cells, and liver cells.

## REFERENCES

1. Hubert HB et al. Obesity as an independent risk factor for cardiovascular disease: a 26-year follow-up of participants in the Framingham Heart Study. *Circulation* 67: 968–977, 1983.
2. Serrano-Rios M. Relationship between obesity and the increased risk of major complications in non-insulin-dependent diabetes mellitus. *Eur J Clin Invest* 28: 14–18, 1998.
3. Marette A. Mediators of cytokine-induced insulin resistance in obesity and other inflammatory settings. *Curr Opin Clin Nutr Metab Care* 5: 377–383, 2002.
4. Dandona P, Aljada A, and Bandyopadhyay A. Inflammation: the link between insulin resistance, obesity and diabetes. *Trends Immunol* 25: 4–7, 2004.
5. Chen HI et al. Insulin attenuates endotoxin-induced acute lung injury in conscious rats. *Crit Care Med* 34: 758–764, 2006.
6. Ottlecz A et al. Involvement of the kinin system in the insulin-induced inhibition of carrageenin oedema in rats. *Monog Allergy* 12: 131–137, 1977.

7. Satomi N, Sakurai A, and Haranaka K. Relationship of hypoglycemia to tumor necrosis factor production and antitumor activity: role of glucose, insulin, and macrophages. *J Natl Cancer Inst* 74: 1255–1260, 1985.
8. Fraker DL, Merino MJ, and Norton JA. Reversal of the toxic effects of cachectin by concurrent insulin administration. *Am J Physiol* 256: E725–E731, 1989.
9. Jeschke MG et al. Insulin attenuates the systemic inflammatory response to thermal trauma. *Mol Med* 8: 443–450, 2002.
10. Chaudhuri A et al. Anti-inflammatory and profibrinolytic effect of insulin in acute ST segment-elevation myocardial infarction. *Circulation* 109: 849–854, 2004.
11. Aljada A et al. Insulin inhibits the pro-inflammatory transcription factor early growth response gene-1 (Egr)-1 expression in mononuclear cells (MNC) and reduces plasma tissue factor (TF) and plasminogen activator inhibitor-1 (PAI-1) concentrations. *J Clin Endocrinol Metab* 87: 1419–1422, 2002.
12. Dandona P et al. Insulin inhibits intranuclear nuclear factor κB and stimulates IκB in mononuclear cells in obese subjects: evidence for an anti-inflammatory effect? *J Clin Endocrinol Metab* 86: 3257–3265, 2001.
13. Sakaue S et al. Regulation of macrophage migration inhibitory factor (MIF) expression by glucose and insulin in adipocytes in vitro. *Mol Med* 5: 361–371, 1999.
14. Ghanim H et al. Acute modulation of Toll-like receptors by insulin. *Diabetes Care* 31: 1827–1831, 2008.
15. Ricote M et al. The peroxisome proliferator-activated receptor-γ is a negative regulator of macrophage activation. *Nature* 391: 79–82, 1998.
16. Jiang C, Ting AT, and Seed B. PPAR-γ agonists inhibit production of monocyte inflammatory cytokines. *Nature* 391: 82–86, 1998.
17. Su GD et al. A novel therapy for colitis utilizing PPAR-γ ligands to inhibit the epithelial inflammatory response. *J Clin Invest* 104: 383–389, 1999.
18. Aljada A et al. Nuclear factor-κB suppressive and inhibitor-κB stimulatory effects of troglitazone in obese patients with type 2 diabetes: evidence of anti-inflammatory action? *J Clin Endocrinol Metab* 86: 3250–3256, 2001.
19. Ghanim H et al. Suppression of nuclear factor-κB and stimulation of inhibitor κB by troglitazone: evidence for an anti-inflammatory effect and a potential anti-atherosclerotic effect in the obese. *J Clin Endocrinol Metab* 86: 1306–1312, 2001.
20. Ghanim H et al. Low-dose rosiglitazone exerts an anti-inflammatory effect with an increase in adiponectin independently of free fatty acid fall and insulin sensitization in obese type 2 diabetics. *J Clin Endocrinol Metab* 91: 3553–3558, 2006.
21. Reaven GM. Banting lecture 1988. Role of insulin resistance in human disease. *Diabetes* 37: 1595–1607, 1988.
22. Edgerton DS et al. Small increases in insulin inhibit hepatic glucose production solely caused by an effect on glycogen metabolism. *Diabetes* 50: 1872–1882, 2001.
23. Kern PA et al. The expression of tumor necrosis factor in human adipose tissue: regulation by obesity, weight loss, and relationship to lipoprotein lipase. *J Clin Invest* 95: 2111–2119, 1995.
24. Dandona P et al. Tumor necrosis factor-α in sera of obese patients: fall with weight loss. *J Clin Endocrinol Metab* 83: 2907–2910, 1998.
25. Kuboki K et al. Regulation of endothelial constitutive nitric oxide synthase gene expression in endothelial cells and in vivo: a specific vascular action of insulin. *Circulation* 101: 676–681, 2000.
26. Cain BS, Harken AH, and Meldrum DR. Therapeutic strategies to reduce TNF-α mediated cardiac contractile depression following ischemia and reperfusion. *J Mol Cell Cardiol* 31: 931–947, 1999.

27. Li D et al. Kinetics of tumor necrosis factor-α in plasma and the cardioprotective effect of a monoclonal antibody to tumor necrosis factor-α in acute myocardial infarction. *Am Heart J* 137: 1145–1152, 1999.
28. Crook MA, Tutt P, and Pickup JC. Elevated serum sialic acid concentration in NIDDM and its relationship to blood pressure and retinopathy. *Diabetes Care* 16: 57–60, 1993.
29. Pickup JC et al. Higher serum sialic acid in women than in men with NIDDM: possible relevance to increased cardiovascular risk in NIDDM women. *Diabetes Care* 20: 1496, 1997.
30. Schmidt MI et al. Markers of inflammation and prediction of diabetes mellitus in adults (Atherosclerosis Risk in Communities): a cohort study. *Lancet* 353: 1649–1652, 1999.
31. Pradhan AD et al. C-reactive protein, interleukin-6, and risk of developing type 2 diabetes mellitus. *JAMA* 286: 327–334, 2001.
32. Ridker PM et al. Plasma concentration of interleukin-6 and the risk of future myocardial infarction among apparently healthy men. *Circulation* 101: 1767–1772, 2000.
33. Klover PJ et al. Chronic exposure to interleukin-6 causes hepatic insulin resistance in mice. *Diabetes* 52: 2784–2789, 2003.
34. Andersson CX et al. Insulin antagonizes interleukin-6 signaling and is anti-inflammatory in 3T3-L1 adipocytes. *J Biol Chem* 282: 9430–9435, 2007.
35. Rui L et al. SOCS-1 and SOCS-3 block insulin signaling by ubiquitin-mediated degradation of IRS1 and IRS2. *J Biol Chem* 277: 42394–42398, 2002.
36. Grilli M, Chiu JJ, and Lenardo MJ. NF-κB and Rel: participants in a multiform transcriptional regulatory system. *Int Rev Cytol* 143: 1–62, 1993.
37. Liou HC and Baltimore D. Regulation of the NF-κB/rel transcription factor and I κB inhibitor system. *Curr Opin Cell Biol* 5: 477–487, 1993.
38. Hotamisligil GS. Inflammatory pathways and insulin action. *Int J Obes Relat Metab Disord* 27: S53–S55, 2003.
39. Hotamisligil GS, Shargill NS, and Spiegelman BM. Adipose expression of tumor necrosis factor-α: direct role in obesity-linked insulin resistance. *Science* 259: 87–91, 1993.
40. Weisberg SP et al. Obesity is associated with macrophage accumulation in adipose tissue. *J Clin Invest* 112: 1796–1808, 2003.
41. Furuhashi M et al. Adipocyte/macrophage fatty acid-binding proteins contribute to metabolic deterioration through actions in both macrophages and adipocytes in mice. *J Clin Invest* 118: 2640–2650, 2008.
42. Furuhashi M and Hotamisligil GS. Fatty acid-binding proteins: role in metabolic diseases and potential as drug targets. *Nat Rev Drug Discov* 7: 489–503, 2008.
43. Zhang Y et al. Positional cloning of the mouse obese gene and its human homologue. *Nature* 372: 425–432, 1994.
44. Fantuzzi G and Faggioni R. Leptin in the regulation of immunity, inflammation, and hematopoiesis. *J Leukoc Biol* 68: 437–446, 2000.
45. Friedman JM. The function of leptin in nutrition, weight, and physiology. *Nutr Rev* 60: S1–S14; discussion S68–84, 2002.
46. Faggioni R, Feingold KR, and Grunfeld C. Leptin regulation of the immune response and the immunodeficiency of malnutrition. *FASEB J* 15: 2565–2571, 2001.
47. Pecoits-Filho R et al. Soluble leptin receptors and serum leptin in end-stage renal disease: relationship with inflammation and body composition. *Eur J Clin Invest* 32: 811–817, 2002.
48. Flier JS. Lowered leptin slims immune response. *Nat Med* 4: 1124–1125, 1998.
49. Guillaume-Gentil C et al. Abnormal regulation of the hypothalamo–pituitary–adrenal axis in the genetically obese fa/fa rat. *Endocrinology* 126: 1873–1879, 1990.
50. Takeshita N, Yoshino T, and Mutoh S. Possible involvement of corticosterone in bone loss of genetically diabetic db/db mice. *Horm Metab Res* 32: 147–151, 2000.

51. Takahashi N, Waelput W, and Guisez Y. Leptin is an endogenous protective protein against the toxicity exerted by tumor necrosis factor. *J Exp Med* 189: 207–212, 1999.
52. Gultekin FA et al. Leptin treatment ameliorates acute lung injury in rats with cerulein-induced acute pancreatitis. *World J Gastroenterol* 13: 2932–2938, 2007.
53. Cakir B et al. The anti-inflammatory effect of leptin on experimental colitis: involvement of endogenous glucocorticoids. *Peptides* 25: 95–104, 2004.
54. Zeigerer A et al. Insulin regulates leptin secretion from 3T3-L1 adipocytes by a PI 3 kinase independent mechanism. *Exp Cell Res* 314: 2249–2256, 2008.
55. Bokarewa M et al. Resistin, an adipokine with potent proinflammatory properties. *J Immunol* 174: 5789–5795, 2005.
56. Silswal N et al. Human resistin stimulates the pro-inflammatory cytokines TNF-α and IL-12 in macrophages by NF-κB-dependent pathway. *Biochem Biophys Res Commun* 334: 1092–1101, 2005.
57. Stejskal D et al. Resistin concentrations in persons with type 2 diabetes mellitus and in individuals with acute inflammatory disease. *Biomed Pap Med Fac Univ Palacky Olomouc Czech Repub* 147: 63–69, 2003.
58. Harsch IA et al. Resistin levels in patients with obstructive sleep apnoea syndrome: link to subclinical inflammation? *Med Sci Monit* 10: CR510–CR515, 2004.
59. Reilly MP et al. Resistin is an inflammatory marker of atherosclerosis in humans. *Circulation* 111: 932–939, 2005.
60. Shetty GK et al. Circulating adiponectin and resistin levels in relation to metabolic factors, inflammatory markers, and vascular reactivity in diabetic patients and subjects at risk for diabetes. *Diabetes Care* 27: 2450–2457, 2004.
61. Steppan CM et al. The hormone resistin links obesity to diabetes. *Nature* 409: 307–312, 2001.
62. Vidal-Puig A and O'Rahilly S. Resistin: a new link between obesity and insulin resistance? *Clin Endocrinol* 55: 437–438, 2001.
63. Johnson V, Clark T, and Fleming JW. Parental views of the development of children in a rural South African community. *ABNF J* 13: 18–22, 2002.
64. Shojima N et al. Humoral regulation of resistin expression in 3T3-L1 and mouse adipose cells. *Diabetes* 51: 1737–1744, 2002.
65. Hartman HB et al. Mechanisms regulating adipocyte expression of resistin. *J Biol Chem* 277: 19754–19761, 2002.
66. Haugen F et al. Inhibition by insulin of resistin gene expression in 3T3-L1 adipocytes. *FEBS Lett* 507: 105–108, 2001.
67. Kawashima J et al. Insulin down-regulates resistin mRNA through the synthesis of protein(s) that could accelerate the degradation of resistin mRNA in 3T3-L1 adipocytes. *Diabetologia* 46: 231–240, 2003.
68. Lu SC et al. Lipopolysaccharide increases resistin gene expression in vivo and in vitro. *FEBS Lett* 530: 158–162, 2002.
69. Rajala MW et al. Cell type-specific expression and coregulation of murine resistin and resistin-like molecule-α in adipose tissue. *Mol Endocrinol* 16: 1920–1930, 2002.
70. Kaser S et al. Resistin messenger RNA expression is increased by proinflammatory cytokines in vitro. *Biochem Biophys Res Commun* 309: 286–290, 2003.
71. Fu Y, Luo N, Klein RL, and Garvey WT. Adiponectin promotes adipocyte differentiation, insulin sensitivity, and lipid accumulation. *J Lipid Res* 46: 1369–1379, 2005.
72. Fruebis J et al. Proteolytic cleavage product of 30-kDa adipocyte complement-related protein increases fatty acid oxidation in muscle and causes weight loss in mice. *Proc Natl Acad Sci USA* 98: 2005–2010, 2001.
73. Yamauchi T et al. The fat-derived hormone adiponectin reverses insulin resistance associated with both lipoatrophy and obesity. *Nat Med* 7: 941–946, 2001.

74. Scherer PE et al. A novel serum protein similar to C1q, produced exclusively in adipocytes. *J Biol Chem* 270: 26746–26749, 1995.
75. Greenberg AS and Obin MS. Obesity and the role of adipose tissue in inflammation and metabolism. *Am J Clin Nutr* 83: 461S–465S, 2006.
76. Ouchi N et al. Novel modulator for endothelial adhesion molecules: adipocyte-derived plasma protein adiponectin. *Circulation* 100: 2473–2476, 1999.
77. Arita Y et al. Paradoxical decrease of an adipose-specific protein, adiponectin, in obesity. *Biochem Biophys Res Commun* 257: 79–83, 1999.
78. Hotta K et al. Plasma concentrations of a novel, adipose-specific protein, adiponectin, in type 2 diabetic patients. *Arterioscler Thromb Vasc Biol* 20: 1595–1599, 2000.
79. Ouchi N et al. Adipocyte-derived plasma protein, adiponectin, suppresses lipid accumulation and class A scavenger receptor expression in human monocyte-derived macrophages. *Circulation* 103: 1057–1063, 2001.
80. Yang WS et al. Weight reduction increases plasma levels of an adipose-derived anti-inflammatory protein, adiponectin. *J Clin Endocrinol Metab* 86: 3815–3819, 2001.
81. Okamoto Y et al. Adiponectin reduces atherosclerosis in apolipoprotein E-deficient mice. *Circulation* 106: 2767–2770, 2002.
82. Nayagam VM et al. SIRT1 modulating compounds from high-throughput screening as anti-inflammatory and insulin-sensitizing agents. *J Biomol Screen* 11: 959–967, 2006.
83. Yoshizaki T et al. SIRT1 exerts anti-inflammatory effects and improves insulin sensitivity in adipocytes. *Mol Cell Biol*, 2008.
84. Sun C et al. SIRT1 improves insulin sensitivity under insulin-resistant conditions by repressing PTP1B. *Cell Metab* 6: 307–319, 2007.
85. Zabolotny JM and Kim YB. Silencing insulin resistance through SIRT1. *Cell Metab* 6: 247–249, 2007.
86. Subauste AR and Burant CF. Role of FOXO1 in FFA-induced oxidative stress in adipocytes. *Am J Physiol Endocrinol Metab* 293: E159–E164, 2007.
87. Qiao L and Shao J. SIRT1 regulates adiponectin gene expression through FOXO1-C/enhancer-binding protein α transcriptional complex. *J Biol Chem* 281: 39915–39924, 2006.
88. Kops GJ and Burgering BM. Forkhead transcription factors: new insights into protein kinase B (c-akt) signaling. *J Mol Med* 77: 656–665, 1999.
89. van der Horst A et al. FOXO4 is acetylated upon peroxide stress and deacetylated by the longevity protein hSir2 (SIRT1). *J Biol Chem* 279: 28873–28879, 2004.
90. Brunet A et al. Stress-dependent regulation of FOXO transcription factors by the SIRT1 deacetylase. *Science* 303: 2011–2015, 2004.
91. Motta MC et al. Mammalian SIRT1 represses forkhead transcription factors. *Cell* 116: 551–563, 2004.
92. Daitoku H et al. Silent information regulator 2 potentiates FOXO1-mediated transcription through its deacetylase activity. *Proc Natl Acad Sci USA* 101: 10042–10047, 2004.
93. Zhang X et al. Phosphorylation of serine 256 suppresses transactivation by FKHR (FOXO1) by multiple mechanisms. Direct and indirect effects on nuclear/cytoplasmic shuttling and DNA binding. *J Biol Chem* 277: 45276–45284, 2002.
94. Matsuzaki H et al. Insulin-induced phosphorylation of FKHR (Foxo1) targets to proteasomal degradation. *Proc Natl Acad Sci USA* 100: 11285–11290, 2003.
95. Ito Y, Daitoku H, and Fukamizu A. Foxo1 increases pro-inflammatory gene expression by inducing C/EBPβ in TNF-α-treated adipocytes. *Biochem Biophys Res Commun* 378: 290–295, 2009.
96. Moschen AR et al. Visfatin, an adipocytokine with proinflammatory and immunomodulating properties. *J Immunol* 178: 1748–1758, 2007.
97. Fukuhara A et al. Visfatin: a protein secreted by visceral fat that mimics the effects of insulin. *Science* 307: 426–430, 2005.

98. Berndt J et al. Plasma visfatin concentrations and fat depot-specific mRNA expression in humans. *Diabetes* 54: 2911–2916, 2005.
99. Duncan BB et al. Low-grade systemic inflammation and the development of type 2 diabetes: the Atherosclerosis Risk in Communities Study. *Diabetes* 52: 1799–1805, 2003.
100. Duncan BB et al. Fibrinogen, other putative markers of inflammation, and weight gain in middle-aged adults: the Atherosclerosis Risk in Communities Study. *Obes Res* 8: 279–286, 2000.
101. Pradhan AD et al. C-reactive protein, interleukin 6, and risk of developing type 2 diabetes mellitus. *JAMA* 286: 327–334, 2001.
102. Barzilay JI et al. The relation of markers of inflammation to the development of glucose disorders in the elderly: the Cardiovascular Health Study. *Diabetes* 50: 2384–2389, 2001.
103. Han TS et al. Prospective study of C-reactive protein in relation to the development of diabetes and metabolic syndrome in the Mexico City Diabetes Study. *Diabetes Care* 25: 2016–2021, 2002.
104. Pradhan AD et al. C-reactive protein is independently associated with fasting insulin in nondiabetic women. *Arterioscler Thromb Vasc Biol* 23: 650–655, 2003.
105. Hak AE et al. Associations of C-reactive protein with measures of obesity, insulin resistance, and subclinical atherosclerosis in healthy, middle-aged women. *Arterioscler Thromb Vasc Biol* 19: 1986–1991, 1999.
106. Yudkin JS et al. C-reactive protein in healthy subjects: associations with obesity, insulin resistance, and endothelial dysfunction: a potential role for cytokines originating from adipose tissue? *Arterioscler Thromb Vasc Biol* 19: 972–978, 1999.
107. Lemieux I et al. Elevated C-reactive protein: another component of the atherothrombotic profile of abdominal obesity. *Arterioscler Thromb Vasc Biol* 21: 961–967, 2001.
108. Esposito K et al. Inflammatory cytokine concentrations are acutely increased by hyperglycemia in humans: role of oxidative stress. *Circulation* 106: 2067–2072, 2002.
109. Mohanty P et al. Glucose challenge stimulates reactive oxygen species (ROS) generation by leucocytes. *J Clin Endocrinol Metab* 85: 2970–2973, 2000.
110. Aljada A et al. Glucose ingestion induces an increase in intranuclear nuclear factor κB, a fall in cellular inhibitor κB, and an increase in tumor necrosis factor α messenger RNA by mononuclear cells in healthy human subjects. *Metabolism* 55: 1177–1185, 2006.
111. Aljada A et al. Glucose intake induces an increase in activator protein 1 and early growth response 1 binding activities, in the expression of tissue factor and matrix metalloproteinase in mononuclear cells, and in plasma tissue factor and matrix metalloproteinase concentrations. *Am J Clin Nutr* 80: 51–57, 2004.
112. Mohanty P et al. Both lipid and protein intakes stimulate increased generation of reactive oxygen species by polymorphonuclear leukocytes and mononuclear cells. *Am J Clin Nutr* 75: 767–772, 2002.
113. Aljada A et al. Increase in intranuclear nuclear factor κB and decrease in inhibitor κB in mononuclear cells after a mixed meal: evidence for a pro-inflammatory effect. *Am J Clin Nutr* 79: 682–690, 2004.
114. Dandona P et al. The suppressive effect of dietary restriction and weight loss in the obese on the generation of reactive oxygen species. *J Clin Endocrinol Metab* 86: 355–362, 2001.
115. Hotamisligil GS et al. Reduced tyrosine kinase activity of the insulin receptor in obesity-diabetes: central role of tumor necrosis factor-α. *J Clin Invest* 94: 1543–1549, 1994.
116. Ueki K, Kondo T, and Kahn CR. Suppressor of cytokine signaling 1 (SOCS-1) and SOCS-3 cause insulin resistance through inhibition of tyrosine phosphorylation of insulin receptor substrate proteins by discrete mechanisms. *Mol Cell Biol* 24: 5434–5446, 2004.

117. Senn JJ et al. Suppressor of cytokine signaling-3 (SOCS-3), a potential mediator of interleukin-6-dependent insulin resistance in hepatocytes. *J Biol Chem* 278: 13740–13746, 2003.
118. Fernandez-Real JM et al. Novel interactions of adiponectin with the endocrine system and inflammatory parameters. *J Clin Endocrinol Metab* 88: 2714–2718, 2003.
119. Chen K et al. Induction of leptin resistance through direct interaction of C-reactive protein with leptin. *Nat Med* 12: 425–432, 2006.
120. Shulman GI. Cellular mechanisms of insulin resistance. *J Clin Invest* 106: 171–176, 2000.
121. Zhang HH et al. Lipase-selective functional domains of perilipin A differentially regulate constitutive and protein kinase A-stimulated lipolysis. *J Biol Chem* 278: 51535–51542, 2003.
122. Hansen LL et al. Insulin signaling is inhibited by micromolar concentrations of H(2) O(2): evidence for a role of H(2)O(2) in tumor necrosis factor-α-mediated insulin resistance. *J Biol Chem* 274: 25078–25084, 1999.
123. Lionetti L et al. From chronic overnutrition to insulin resistance: role of fat-storing capacity and inflammation. *Nutr Metab Cardiovasc Dis*, 2009.
124. Steinberg HO et al. Elevated circulating free fatty acid levels impair endothelium-dependent vasodilation. *J Clin Invest* 100: 1230–1239, 1997.
125. Boden G. Obesity and free fatty acids. *Endocrinol Metab Clin North Am* 37: 635–646, viii–ix, 2008.
126. Inoguchi T et al. High glucose level and free fatty acid stimulate reactive oxygen species production through protein kinase C-dependent activation of NAD(P)H oxidase in cultured vascular cells. *Diabetes* 49: 1939–1945, 2000.
127. Tilg H and Moschen AR. Adipocytokines: mediators linking adipose tissue, inflammation and immunity. *Nat Rev Immunol* 6: 772–783, 2006.
128. Santomauro AT et al. Overnight lowering of free fatty acids with acipimox improves insulin resistance and glucose tolerance in obese diabetic and nondiabetic subjects. *Diabetes* 48: 1836–1841, 1999.
129. Gavrilova O et al. Surgical implantation of adipose tissue reverses diabetes in lipoatrophic mice. *J Clin Invest* 105: 271–278, 2000.
130. Malmberg K et al. Randomized trial of insulin–glucose infusion followed by subcutaneous insulin treatment in diabetic patients with acute myocardial infarction (DIGAMI study): effects on mortality at 1 year. *J Am Coll Cardiol* 26: 57–65, 1995.
131. van den Berghe G et al. Intensive insulin therapy in the critically ill patients. *New Engl J Med* 345: 1359–1367, 2001.
132. Ghanim H et al. Orange juice or fructose intake does not induce oxidative and inflammatory response. *Diabetes Care* 30: 1406–1411, 2007.

# 7 Growth Hormone as Modulator of Adipose Inflammation

*Hong-Biao Liu*

## CONTENTS



## 7.1 INFLAMMATORY PROCESSES OF ADIPOSE TISSUE

Adipose tissue is both a classic energy storage organ and also an endocrine-immune organ that regulates homeostasis by secreting a series of hormones and cytokines [1,2]. Inflammatory mediators were once regarded as mainly produced only by the immune system and liver. In 1993, Hotmisligil et al. described direct evidence of an inflammatory process involving adipose tissue by demonstrating the production of tumor necrosis factor-α (TNF-α) by adipose tissue [3].

More recent studies have shown clearly that adipose tissue synthesizes and secretes both pro-inflammatory and anti-inflammatory cytokines [4], including transforming growth factor β, interferon γ, interleukins (IL-1, IL-6, IL-8, and IL-10), plasminogen activator inhibitor-1, fibrinogen, angiopoietin-related proteins, metallothionein, complement factor 3, C-reactive protein, haptoglobin, and serum amyloid A protein [2]. Inflammation within white adipose tissue may be a crucial step contributing to the pathologic features of metabolic syndrome.

Human adipose tissue is a potent source of inflammatory cytokines. Two types of cells in adipose are involved in the cytokines secretion: fat cells (adipocytes) and nonfat cells. Leptin and adiponectin are primarily secreted by adipocytes. Other inflammatory cytokines are produced by both fat cells and nonfat cells. Although nonfat cells may contribute significantly to the profiles of inflammatory cytokines

[5], this review will focus on the interaction between growth hormone (GH) and the adipose tissue-specific cell, the adipocyte.

## 7.2 GENERAL FUNCTIONAL MECHANISMS OF GROWTH HORMONE (GH)

GH is synthesized and secreted by somatotropes in the anterior pituitary. GH is also referred to as somatotrophin (the natively synthesized animal hormone) and somatropin (a product of recombinant DNA technology). The stimulation of body growth was regarded as its essential role, which was first described at the beginning of the 20th century. GH was isolated from bovine pituitary in 1944 and thereafter its native 191-amino acid sequence determined [6,7]. Dual effector theory is regarded as the functional model at the cellular level: there are direct effects and indirect effects.

Direct effects refer to the result of GH binding to its GH receptor on target tissues, including adipose tissue. While indirect effects of GH are exerted through induction of secondary hormones such as insulin-like growth factors (IGFs) produced by hepatocytes [8], GH exerts pleiotropic effects divided into two main types: effects on the growth and differentiation of cells and effects on metabolism. Other major regulatory effects that GH exerts on adipocytes include the enhancement of lipolytic activity and reduction of triglyceride accumulation, both of which result in decrease of total fat mass [9].

## 7.3 DIRECT EFFECTS OF GROWTH HORMONE ON ADIPOCYTES

GH exerts modulating effects on both pre-adipocytes and adipocytes via the GH receptor (GHR). Heterogeneity is present in both the GHR mRNA transcript and the mature protein. The human GHR genetic locus is on chromosome 5pl.1-5p12 and spans more than 87 kb. The protein-coding portion contains nine exons (2–10). Transcripts from the GHR gene are characterized by a disparate 5' untranslated region (exon 1). After screening adult liver and cardiac muscle cDNA libraries, nine variants of human GHR mRNA (V1–V9) were reported. Each differed in its 5' untranslated region [10,11]. V6 was subsequently revealed as a false positive result [12]; while the other eight were confirmed by genomic localization and expression profiling.

Recently five new human GHR mRNA variants (VA–VE) were detected in adipose cDNA libraries by RT-PCR assay [13]. Even though no GHR mRNA variant exclusive to adipose tissue has yet been detected, there is still an interesting distribution pattern of GHR mRNA in adipose: 89% V2, 3% V3, and 5% novel variants (VA–VE). The V2 variant has been validated as the predominant form in adipocytes at all adipose development stages. However, whether specific splice variants may be associated with different pathologic conditions in adipose is still unknown [13].

Expression of GHR in adipocytes has been well investigated since its description in 1973, the purification of the protein in 1980, and the cloning of human and rabbit cDNAs in 1987 [14–24]. GHR is a single transmembrane protein of the class I cytokine receptor family [25], and it is expressed in most cell types. This receptor family comprises the receptors for I-L2, IL-7, IL-9, IL-12, prolactin, thrombopoietin, leukemia inhibitory factor, erythropoietin, granulocyte colony-stimulating factor, granulocyte–macrophage colony-stimulating factor, and interferons α, β, and γ. GHR contains 620 amino acids spread over three parts: an extracellular hormone-binding domain (ECD) of 246 amino acid residues, a transmembrane region of 24 amino acid residues, and a long cytoplasmic domain of 350 residues [15].

The extracellular domain contains seven cysteine residues and five potential N-linked glycosylation sites, and within this domain is the binding interface for GH. The cytoplasmic domain is deficient of known intrinsic kinase or ATP binding motifs. Box 1 (residues 280–287 in the GHR) is a juxtamembrane, proline-rich sequence; while box 2 (23 amino acid residues downstream) contains a serine-rich sequence.

The two forms of GHR are (1) the full-length form of 620 amino acids which is a membrane-bound receptor and (2) the shorter soluble GH binding protein (GHBP) corresponding to the extracellular domain of the full-length receptor. GHBP was detected in different biological fluids, such as serum, milk, and follicular fluid [26,27]. Two independent processes (with species differences) are involved in the production of GHBP. Simple proteolytic cleavage leads to formation of GHBP with only an extracelluar domain and this occurs in humans and rabbits. Alternative splicing of the GHR gene allows for the insertion of a unique hydrophilic tail in the GHBP and this has been observed in humans [28,29]. Besides the full-length GHR, two additional truncated splice-variants GHR-(1–277) and GHR-(1–279), have been detected in fat, liver and muscle [30]. These forms lack intracellular domains. An interesting proposal is that the truncated GHR may inhibit normal GHR action.

Both the full-length GHR and GHBP are expressed at the surfaces of rat adipocytes with identical binding affinities for GH [31] and fixed ratios [32]. The half-life is approximately 45 minutes [33]. GHBP appears to compete with full-length GHR to bind GH at the pre-adipocyte surfaces and this inhibits adipocyte differentiation [34]. Intracellular GHBP has been shown to serve as a transcriptional enhancer to affect function in both pre-adipocytes and adipocytes [35].

Although the in-depth investigation of the effects of human GH on adipose tissue through GHR has been achieved, the understanding of how GHR expression is regulated in human adipocytes is limited. It is well known that obese people have low GH levels due to the inhibitory effects of excess free fatty acids (FFAs) on GH secretion and increased GH clearance [36–38]. Gleenson et al. [37] observed that a single bolus administration of GH in obese individuals leads to increased levels of IGF-1 and GHBP; GHBP reflects an indirect estimate of GH receptor numbers. The positive correlation between IGF-1 and GHR may provide a clue that GHR is regulated by GH indirectly via the IGF-1 pathway.

## 7.4 BINDING MODEL OF GROWTH HORMONE TO GROWTH HORMONE RECEPTORS

Dimerization of two GHRs binding to one GH molecule (GHR2:GH) was described in 1991 [39]. X-ray crystallographic studies have shown that two different sites in the GH are linked to individual GHR to form a complex pattern as GHR: site1-GH-site2:GHR [40]. The binding model of GH to GHR is evidenced by activity of the GH analogue GH-G 120R that lacks the ability to bind the GHR via site 2 and does not stimulate GHR signal transduction and therefore serves as an inhibitor of GHR [41–43]. Formation of GH-induced dimerization of GHR is a critical step for the hormonal effect on proliferation and stimulation of lipogenesis in adipocytes [41].

## 7.5 GROWTH HORMONE SIGNAL TRANSDUCTION

GHR dimerization occurs after GH binds to GHR at the cell surface. The GHR dimer induces association of Janus kinase 2 (JAK2) to the membrane proximal proline-rich box 1 region and this consequently activates JAK2 [44,45], a 130-kDa tyrosine kinase, first identified in 1993 as responsible for the tyrosine phosphorylation of GHR in mouse 3T3-F442A cells [46,47]. Increased entry of calcium coincides with GHR stimulation. Activated JAK2 triggers the phosphorylation of GHR and activation of a series of proteins to transduce the cellular signal. Interestingly, autophosphorylation also occurs simultaneously while JAK2 phosphorylates GHR [48]. Activated JAK2 in turn activates various signaling Pathways [28] including:

1. Other receptors, including epidermal growth factor receptor and non-receptor kinases such as focal adhesion kinase (FAK), c-Src, and c-Fyn [44,45]
2. Other GH-activated signal-transducing proteins including members of the mitogen-activated protein (MAP) kinase family (p42/44 MAP kinase, stress-activated protein kinase [SAPK], and p38 MAP kinase)
3. Members of insulin receptor substrate (IRS) family (IRS-1, IRS-2 and IRS-3)
4. Signal transducers and activators of transcription (STAT) family members (STATs 1, 3, 5a, and 5b)

The GH signaling pathway is evidenced by the following observations [49,50]. More than fifty ECD mutations of GHR have been identified as pathogenesis factors in dwarfism. In addition, growth retardation was observed in mice with cytoplasmic domain mutations. Three different mice with targeted mutations in cytoplasmic domains have been created involving (1) truncation at residue 569 that lacks most STAT5 signaling function, (2) truncation at 391, deleting all STAT5 generation and other signals in the central region of the cytoplasmic domain, and (3) mutation of the box 1 sequence, abrogating JAK2 function. The body sizes of these homozygous mutation mice, compared with sizes of wild-type (WT) mice, demonstrated an interesting pattern: WT > 569 > 391 > box 1 [51].

In addition to potential regulators involved in GH signaling cascades, the suppressors of cytokine signaling (SOCS) protein family has also been implicated in controlling cellular responses to GH [52,53]. SOCS proteins are produced in response to signals from a diverse range of cytokines and growth factors that attenuate signal transduction to exert negative feedback effect by inhibiting the JAK–STAT signaling cascade. This is evidenced by the in vivo observation that mice lacking SOCS displayed gigantism accompanied by deregulated GH signaling [54,55].

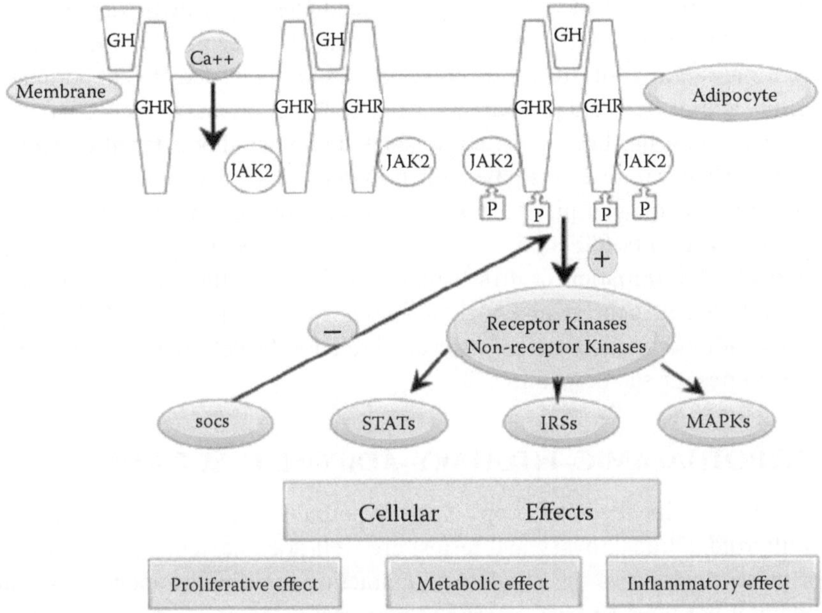


**FIGURE 7.1** Growth hormone signal transduction. GH binding to surface GHR results in dimerization of the receptor, phosphorylation of JAK2 and GHR, and subsequent phosphorylation of a series of proteins. The JAK–SAT pathway is a main route for GH effects on gene transcription. GH is also able to activate MAP kinases and utilize IRS-1 to induce insulin-like effect. In addition, the classic negative feedback loop via the SOCS protein family also plays a role in GH signal cascades.

Figure 7.1 illustrates signal transduction through the GH–GHR system. In summary, binding of GH to GHR results in dimerization of the receptor, phosphorylation of JAK2 and GHR, and subsequent phosphorylation of a series of proteins. Nuclear factors are involved in the effects of GH on gene transcription. The JAK–STAT pathway is a main pathway for GH effect on gene transcription. In addition, GH is able to activate MAP kinases and utilize IRS-1 to induce insulin-like effects. The classic negative feedback via SOCS protein family is also involved in GH signal cascades. Ultimate effects serve to regulate cell proliferation, metabolism, and tissue inflammation.

## 7.6 INDIRECT EFFECTS OF GROWTH HORMONE ON ADIPOCYTES

GH exerts its cellular effects directly through stimulating GHR and also indirectly through IGF-1. The liver is regarded as the major source of circulating IGF-1, but IGF-1 is produced by other tissues including adipose tissue. IGF-1 mRNA levels in adipose are as high as those in liver.

IGF-1 is a 7.6-kDa single chain polypeptide with a highly conserved sequence across species. For example, the human, bovine, and porcine IGF-1 sequences are identical [60]. IGF-1 interacts with two types of receptors: type I and type II.

Evidence indicating that the IGF-II receptor mediates any growth signaling function is very limited. The IGF-1 receptor, like the insulin receptor, is a member of the tyrosine kinase family [61]. IGF-1 receptor mRNA has been detected in porcine adipose tissue, adipocytes, and pre-adipocytes [62,63]. Additional evidence has shown that IGF-1 receptor has been detected in isolated human adipocytes after overnight culture [64]. However, several studies failed to detect IGF-1 receptor in mature adipocytes from humans and pigs [17,65,66]. The possible reason for the discrepancy may be the low level of IGF-1 receptor in the mature adipocytes compared to levels in cells before the initiation of differentiation [67,68]. Furthermore, recent studies using conditional gene targeting knockout strategies have shown that IGF-1 receptor signaling in adipocytes does not play a crucial role for the development and differentiation of adipose tissue in vivo [69,70].

## 7.7 HYPOTHALAMIC–PITUITARY–ADIPOSE TISSUE AXIS

The classic endocrine feedback loops of the hypothalamus, pituitary, and peripheral glands (thyroid, adrenal, ovary and testes) are well documented. The common feature of these networks is the regulatory interaction between hormone and source. The mediator hormone and its corresponding target receptor are the two fundamental factors involved in the interaction. It is reasonable to postulate that a hypothalamic–pituitary–adipose-axis also exists [71] (Figure 7.2).

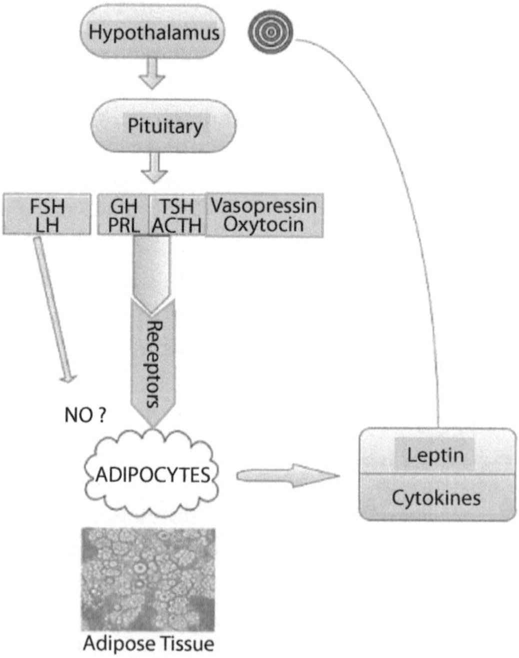


**FIGURE 7.2** Postulated hypothalamic–pituitary–adipose tissue axis.

Accumulating evidence indicates that adipose tissue functions as an endocrine/immune organ since it secretes several adipokines including leptin, adiponectin, resistin, TNF-α, IL-6, IL-8, haptoglobin, and adipsin [72,73]. These adipocyte-derived hormones and cytokines are involved in cellular proliferation, metabolism, and inflammatory processes. Besides GHR, other pituitary hormone receptors, such as those for ACTH, TSH, prolactin, and hypothalamic peptides (vasopressin and oxytocin), have been detected in mature adipocytes at the mRNA level [73]. Of note, the expression levels of some of these receptors in pre-adipocytes are low but appear to be induced in differentiated adipocytes. No evidence shows that receptors for LH and FSH exist in adipocytes.

GH is secreted by the pituitary under hypothalamic control and mediates cellular effects in adipose tissue. In turn, adipocytes secrete a series of hormones and cytokines. Adipocyte-derived leptin regulates GH secretion in the hypothalamus as evidenced directly by the widespread distribution and localization of leptin receptor in hypothalamus [74]. Indirect evidence of regulation also exists, e.g., the observation that genetically obese mice and rats with mutations of the leptin protein or its receptor exhibit decreased GH plasma levels [75]. Clinical evidence demonstrates that leptin receptor mutations in humans are associated with obesity and reductions of GH secretion [76,77].

GH is a modulator of inflammation in adipocytes. Adipocytes secrete hormones such as leptin to regulate GH secretion via the postulated hypothalamus–pituitary–adipose network [72]. The investigation of inflammatory processes in adipose tissue opens a new window to discover novel mechanisms and potential therapeutic targets involved in metabolic syndrome [78].

## REFERENCES


1. Rajala, M.W. and P.E. Scherer. The adipocyte: at the crossroads of energy homeostasis, inflammation, and atherosclerosis. *Endocrinology*, 2003. 144: 3765–3773.
2. Bluher, M. The inflammatory process of adipose tissue. *Pediatr Endocrinol Rev*, 2008. 6: 24–31.
3. Hotamisligil, G.S., N.S. Shargill, and B.M. Spiegelman. Adipose expression of tumor necrosis factor-α: direct role in obesity-linked insulin resistance. *Science*, 1993. 259: 87–91.
4. Wisse, B.E. The inflammatory syndrome: the role of adipose tissue cytokines in metabolic disorders linked to obesity. *J Am Soc Nephrol*, 2004. 15: 2792–2800.
5. Fain, J.N. Release of interleukins and other inflammatory cytokines by human adipose tissue is enhanced in obesity and primarily due to the nonfat cells. *Vitam Horm*, 2006. 74: 443–477.
6. Li, C.H. and H.M. Evans. The isolation of pituitary growth hormone. *Science*, 1944. 99: 183–184.
7. Li, C.H. and H. Papkoff. Preparation and properties of growth hormone from human and monkey pituitary glands. *Science*, 1956. 124: 1293–1294.
8. Green, H., M. Morikawa, and T. Nixon. A dual effector theory of growth hormone action. *Differentiation*, 1985. 29: 195–198.
9. Richelsen, B. Action of growth hormone in adipose tissue. *Horm Res*, 1997. 48: 105–110.

10. Pekhletsky, R.I., B.K. Chernov, and P.M. Rubtsov. Variants of the 5′ untranslated sequence of human growth hormone receptor mRNA. *Mol Cell Endocrinol*, 1992. 90: 103–109.
11. Goodyer, C.G. et al. Organization and evolution of the human growth hormone receptor gene 5' flanking region. *Endocrinology*, 2001. 142: 1923–1934.
12. Orlovskii, I.V., P.S. Sverdlova, and P.M. Rubtsov. Fine structure, expression and polymorphism of the human growth hormone receptor gene. *Mol Biol*, 2004. 38: 29–39.
13. Wei, Y., Z. Rhani, and C.G. Goodyer. Characterization of growth hormone receptor messenger ribonucleic acid variants in human adipocytes. *J Clin Endocrinol Metab*, 2006. 91: 1901–1908.
14. Tsushima, T. and H.G. Friesen. Radioreceptor assay for growth hormone. *J Clin Endocrinol Metab*, 1973. 37: 334–337.
15. Leung, D.W. et al. Growth hormone receptor and serum binding protein: purification, cloning and expression. *Nature*, 1987. 330: 537–543.
16. Carter-Su, C., J. Schwartz, and G. Kikuchi. Identification of a high affinity growth hormone receptor in rat adipocyte membranes. *J Biol Chem,* 1984. 259: 1099–1104.
17. DiGirolamo, M. et al. Specific binding of human growth hormone but not insulin-like growth factors by human adipocytes. *FEBS Lett*, 1986. 205: 15–19.
18. Fagin, K.D. et al. Specific binding of growth hormone by rat adipocytes. *Endocrinology*, 1980. 107: 608–615.
19. Gavin, J.R. 3rd, R.J. Saltman, and S.E. Tollefsen. Growth hormone receptors in isolated rat adipocytes. *Endocrinology*, 1982. 110: 637–643.
20. Grichting, G., L.K. Levy, and H.M. Goodman. Relationship between binding and biological effects of human growth hormone in rat adipocytes. *Endocrinology*, 1983. 113: 1111–1120.
21. Herington, A.C. Identification and characterization of growth hormone receptors on isolated rat adipocytes. *J Recept Res*, 1981. 2: 299–316.
22. Herington, A.C. et al. Growth hormone binding proteins in high-speed cytosols of multiple tissues of the rabbit. *Biochim Biophys Acta*, 1986. 881: 236–240.
23. Lobie, P.E. et al. Cellular localization of the growth hormone receptor/binding protein in the male and female reproductive systems. *Endocrinology,* 1990. 126: 2214–2221.
24. Lobie, P.E. et al. Localization of the growth hormone receptor/binding protein in skin. *J Endocrinol*, 1990. 126: 467–471.
25. Piwien-Pilipuk, G., J.S. Huo, and J. Schwartz. Growth hormone signal transduction. *J Pediatr Endocrinol Metab*, 2002. 15: 771–886.
26. Postel-Vinay, M.C. et al. Identification of prolactin and growth hormone binding proteins in rabbit milk. *Proc Natl Acad Sci USA*, 1991. 88: 6687–6690.
27. Amit, T. et al. Growth hormone-binding protein (GH-BP) levels in follicular fluid from human preovulatory follicles: correlation with serum GH-BP levels. *J Clin Endocrinol Metab*, 1993. 77: 33–39.
28. Nam, S.Y. and P.E. Lobie. The mechanism of effect of growth hormone on preadipocyte and adipocyte function. *Obes Rev*, 2000. 1: 73–86.
29. Ross, R.J. et al. A short isoform of the human growth hormone receptor functions as a dominant negative inhibitor of the full-length receptor and generates large amounts of binding protein. *Mol Endocrinol*, 1997. 11: 265–273.
30. Ballesteros, M. et al. Distribution and abundance of messenger ribonucleic acid for growth hormone receptor isoforms in human tissues. *J Clin Endocrinol Metab*, 2000. 85: 2865–2871.
31. Goodman, H.M., G.Frick, and L.R. Tai. The significance of the short form of the growth hormone receptor in rat adipocytes. *Proc Soc Exp Biol Med*, 1994. 206: 304–308.

32. Frick, G.P., L.R. Tai, and H.M. Goodman. Subcellular distribution of the long and short isoforms of the growth hormone (GH) receptor in rat adipocytes: both isoforms participate in specific binding of GH. *Endocrinology*, 1994. 134: 307–314.
33. Gorin, E. and H.M. Goodman. Turnover of growth hormone receptors in rat adipocytes. *Endocrinology*, 1985. 116: 1796–1805.
34. Lim, L. et al. Regulation of growth hormone (GH) bioactivity by a recombinant human GH-binding protein. *Endocrinology*, 1990. 127: 1287–1291.
35. Bick, T. et al. Overexpression of the short form of the growth hormone receptor in 3T3-L1 mouse preadipocytes. *Proc Soc Exp Biol Med*, 1994. 206: 185–199.
36. Franco, C. et al. Growth hormone treatment reduces abdominal visceral fat in postmenopausal women with abdominal obesity: a 12-month placebo-controlled trial. *J Clin Endocrinol Metab*, 2005. 90: 1466–1474.
37. Gleeson, H.K., C.A. Lissett, and S.M. Shalet. Insulin-like growth factor-I response to a single bolus of growth hormone is increased in obesity. *J Clin Endocrinol Metab*, 2005. 90: 1061–1067.
38. Miller, K.K. et al. Truncal adiposity, relative growth hormone deficiency, and cardiovascular risk. *J Clin Endocrinol Metab*, 2005. 90: 768–774.
39. Cunningham, B.C. et al. Dimerization of the extracellular domain of the human growth hormone receptor by a single hormone molecule. *Science,* 1991. 254: 821–825.
40. de Vos, A.M., M. Ultsch, and A.A. Kossiakoff. Human growth hormone and extracellular domain of its receptor: crystal structure of the complex. *Science*, 1992. 255: 306–312.
41. Ilondo, M.M. et al. Receptor dimerization determines the effects of growth hormone in primary rat adipocytes and cultured human IM-9 lymphocytes. *Endocrinology*, 1994. 134: 2397–2403.
42. Silva, C.M. et al. Human growth hormone (GH) receptor is characterized as the 134-kilodalton tyrosine-phosphorylated protein activated by GH treatment in IM-9 cells. *Endocrinology*, 1993. 133: 2307–2312.
43. Silva, C.M., M.J. Weber, and M.O. Thorner. Stimulation of tyrosine phosphorylation in human cells by activation of the growth hormone receptor. *Endocrinology*, 1993. 132: 101–108.
44. Carter-Su, C. and L.S. Smit. Signaling via JAK tyrosine kinases: growth hormone receptor as a model system. *Recent Prog Horm Res*, 1998. 53: 61–83.
45. Billestrup, N. et al. Molecular mechanism of growth hormone signalling. *Endocr J*, 1998. 45: S41–S45.
46. Argetsinger, L.S. et al. Identification of JAK2 as a growth hormone receptor-associated tyrosine kinase. *Cell*, 1993. 74: 237–244.
47. Witthuhn, B.A. et al. JAK2 associates with the erythropoietin receptor and is tyrosine phosphorylated and activated following stimulation with erythropoietin. *Cell*, 1993. 74: 227–36.
48. Eriksson, H., M. Ridderstrale, and H. Tornqvist. Tyrosine phosphorylation of the growth hormone (GH) receptor and Janus tyrosine kinase is involved in the insulin-like actions of GH in primary rat adipocytes. *Endocrinology*, 1995. 136: 5093–5101.
49. Lichanska, A.M. and M.J. Waters. New insights into growth hormone receptor function and clinical implications. *Horm Res*, 2008. 69: 138–145.
50. Waters, M.J. et al. New insights into growth hormone action. *J Mol Endocrinol*, 2006. 36: 1–7.
51. Rowland, J.E. et al. In vivo analysis of growth hormone receptor signaling domains and their associated transcripts. *Mol Cell Biol,* 2005. 25: 66–77.
52. Starr, R. et al. A family of cytokine-inducible inhibitors of signalling. *Nature*, 1997. 387: 917–921.
53. Endo, T.A. et al. A new protein containing an SH2 domain that inhibits JAK kinases. *Nature*, 1997. 387: 921–924.

54. Greenhalgh, C.J. and W.S. Alexander. Suppressors of cytokine signalling and regulation of growth hormone action. *Growth Horm IGF Res*, 2004. 14: 200–206.
55. Ridderstrale, M. Signaling mechanism for the insulin-like effects of growth hormone: another example of a classical hormonal negative feedback. *Curr Drug Targets Immune Endocr Metabol Disord*, 2005. 5: 79–92.
56. Bluher, S., J. Kratzsch, and W. Kiess. Insulin-like growth factor I, growth hormone and insulin in white adipose tissue. *Best Pract Res Clin Endocrinol Metab*, 2005. 19: 577–587.
57. Gaskins, H.R. et al. Regulation of insulin-like growth factor-I ribonucleic acid expression, polypeptide secretion, and binding protein activity by growth hormone in porcine preadipocyte cultures. *Endocrinology*, 1990. 126: 622–630.
58. Chen, N.X., G.J. Hausman, and J.T. Wright. Influence of age and fetal hypophysectomy on porcine preadipocytes: insulin-like growth factor-I (IGF-I) response, receptor binding and IGF binding proteins secretion. *Growth Dev Aging*, 1995. 59: 193–206.
59. Wabitsch, M. et al. IGF-I and IGFBP-3 expression in cultured human preadipocytes and adipocytes. *Horm Metab Res,* 2000. 32: 555–559.
60. Tavakkol, A., F.A. Simmen, and R.C. Simmen. Porcine insulin-like growth factor-I (pIGF-I): complementary deoxyribonucleic acid cloning and uterine expression of messenger ribonucleic acid encoding evolutionarily conserved IGF-I peptides. *Mol Endocrinol,* 1988. 2: 674–681.
61. Kim, J.J. and D. Accili. Signalling through IGF-I and insulin receptors: where is the specificity? *Growth Horm IGF Res*, 2002. 12: 84–90.
62. Gerfault, V., I. Louveau, and J. Mourot. The effect of GH and IGF-I on preadipocytes from large white and Meishan pigs in primary culture. *Gen Comp Endocrinol*, 1999. 114: 396–404.
63. Landron, D. et al. Growth hormone binding to cultured preadipocytes from obese fa/fa rats increases during cell differentiation. *Horm Metab Res,* 1987. 19: 403–406.
64. Kern, P.A. et al. Insulin-like growth factor action and production in adipocytes and endothelial cells from human adipose tissue. *Diabetes*, 1989. 38: 710–717.
65. Bolinder, J. et al. Studies of acute effects of insulin-like growth factors I and II in human fat cells. *J Clin Endocrinol Metab*, 1987. 65: 732–737.
66. Richardson, R.L., G.J. Hausman, and J.T. Wright. In situ binding and immunocytochemistry of insulin-like growth factor I receptors in primary cultures of porcine adipose tissue stromal vascular cells treated with indomethacin. *J Anim Sci,* 1994. 72: 969–975.
67. Zizola, C.F. et al. Changes in IGF-I receptor and IGF-I mRNA during differentiation of 3T3-L1 preadipocytes. *Biochimie,* 2002. 84: 975–980.
68. Louveau, I. and F. Gondret. Regulation of development and metabolism of adipose tissue by growth hormone and the insulin-like growth factor system. *Domest Anim Endocrinol*, 2004. 27: 241–255.
69. Rosen, C.J. Mouse models for understanding the growth hormone insulin-like growth factor-I axis. *Horm Res*, 2007. 68: 2–4.
70. Kloting, N. et al. Autocrine IGF-1 action in adipocytes controls systemic IGF-1 concentrations and growth. *Diabetes*, 2008. 57: 2074–2082.
71. Redelman, D. et al. Neuroendocrine hormones such as growth hormone and prolactin are integral members of the immunological cytokine network. *Cell Immunol*, 2008. 252: 111–121.
72. Calabro, J. and E.T. Yeh. Obesity, inflammation, and vascular disease: the role of the adipose tissue as an endocrine organ. *Subcell Biochem*, 2007. 42: 63–91.
73. Schaffler, A. et al. Brain talks with fat: evidence for a hypothalamic–pituitary–adipose axis? *Neuropeptides*, 2005. 39: 363–367.

74. Otero, M. et al. Leptin, from fat to inflammation: old questions and new insights. *FEBS Lett*, 2005. 579: 295–301.
75. Casanueva, F.F. and C. Dieguez. Neuroendocrine regulation and actions of leptin. *Front Neuroendocrinol*, 1999. 20: 317–363.
76. Clement, K. et al. A mutation in the human leptin receptor gene causes obesity and pituitary dysfunction. *Nature*, 1998. 392: 398–401.
77. Strobel, A. et al. A leptin missense mutation associated with hypogonadism and morbid obesity. *Nat Genet*, 1998. 18: 213–215.
78. Berg, A.H. and P.E. Scherer. Adipose tissue, inflammation, and cardiovascular disease. *Circ Res*, 2005. 96: 939–949.

# 8 Glucocorticoids as Modulators of Adipose Inflammation

*Nicholas M. Morton*

## CONTENTS



## 8.1 ADIPOSE INFLAMMATION IN OBESITY

Obesity is associated with a chronic inflammation of the adipose tissue that exacerbates and may underlie insulin resistance and metabolic disease. This concept originated from the finding that the tumor necrosis factor-alpha (TNF-α) pro-inflammatory cytokine was elevated in the adipose tissues of obese mice and in the circulation of morbidly obese humans (Hotamisligil et al., 1993, Hotamisligil et al., 1995). Transgenic mice lacking TNF-α (Uysal et al., 1997) or elements of its downstream inflammatory signalling cascade confirmed that obesity and its associated insulin-resistance and diabetes were ameliorated when inflammatory pathways were impaired (Yuan et al., 2001, Hirosumi et al., 2002, Kaneto et al., 2004, Tuncman et al., 2006).

It subsequently became apparent that increased adipose tissue mass served as a source of many cytokines (as well as chemoattractants, prothrombotic, vasoactive, and angiomodulatory factors), some of which have been directly linked to tissue insulin-resistance, e.g., TNF-α and interleukin (IL)-6 (Rotter et al., 2003), monocyte chemoattractant protein (MCP)-1 (Sartipy and Loskutoff 2003 ), serum amyloid (SA)-A3 (Lin et al., 2001), macrophage inhibitory factor (MIF) (Atsumi et al., 2007), CXC motif chemokine ligand-14 (Nara et al., 2007), and many others. In addition, increased infiltration and changes in functions of cells of both the innate immune system such as macrophages (Xu et al., 2003, Weisberg et al., 2003, Zeyda et al., 2007), dendritic cells (Nguyen et al., 2007) and T cells of the adaptive immune system (Wu et al., 2007), particularly cytotoxic CD8+ T cells (Rausch et al., 2008) contribute to inflammatory processes in the adipose tissues of obese mice and humans.

Key to the early events of immune system changes in adipose tissue appears to be activation of Toll-like receptors (TLRs)-2 and 4 on both macrophages and adipocytes by the high circulating free fatty acids found in obesity. In effect the free fatty acids "hi-jack" these innate immune system bacterial lipotoxin recognition systems (Lin et al., 2000, Shi et al., 2006, Song et al., 2006, Nguyen et al., 2007), signalling through pro-inflammatory pathways, and echoing the shared evolutionary pathways of immunity and metabolism (Hotamisligil, 2006). Further, activation of cellular stress mechanisms (Wellen and Hotamisligil, 2005) within adipocytes appears to induce a hybrid necrotic/apoptotic-like state (Cinti et al., 2005, Murano et al., 2008), associated with localised hypoxia (Ye et al., 2007, Rausch et al., 2008) that may be secondary to impaired or delayed vascularization and angiogenesis in the rapidly expanding adipose tissue (Rupnick et al., 2002, Kim et al., 2007, Lijnen, 2008).

The infiltration of immune cells to these sites may therefore initially be a rescue-and-clear-operation that malfunctions, perhaps unable to keep pace with the poorly vascularized expansion of the fat cells, leading to hypoxia, cellular stress and ultimately, chronic inflammation. The resulting local increase and overspill of fatty acids and cytokines from the increased adipose tissue mass along with a reduction in anti-inflammatory cytokines (IL-10) and insulin sensitizing adipokines such as adiponectin (Berg et al., 2001, Yamauchi et al., 2001) causes systemic insulin resistance in muscle, liver, and locally in the adipose tissue itself. High systemic levels of these pro-inflammatory mediators may subsequently impair pancreatic β-cell insulin secretion (Donath et al., 2008) and contribute to atherosclerotic processes in the vessel walls (Lau et al., 2005).

## 8.2 GLUCOCORTICOIDS AND INFLAMMATION

Glucocorticoids are synthesized primarily in the adrenal glands in response to the release of adrenocorticotrophin from the hypothalamus (Goulding and Flowers, 2001, Munck et al., 1992). They have many functions, but most importantly with respect to this chapter, they modulate immune cell function and regulate nutrient metabolism directly at the adipocyte level. Glucocorticoids potently suppress inflammatory processes at high doses, such as when they are used therapeutically to treat inflammatory conditions such as rheumatoid arthritis (Gudbjornsson et al., 2002). Their principal effect is to dampen the signalling cascades that lie downstream of pro-inflammatory cytokines such as TNF-α, IL-1β, IL-6, and the innate immune system bacterial lipotoxin recognition Toll-like receptors (TLRs), namely the nuclear transcription factor (NF)-κB and activator protein (AP)-1 signalling cascades (Riechardt and Schutz, 1998) . Furthermore, glucocorticoids promote the resolution phase of inflammation by conferring phagocytic capacity to macrophages that ingest and process apoptotic leukocytes (neutrophils) at the sites of inflammation (Giles et al., 2001).

### 8.2.1 Glucocorticoids and Adipose Inflammation

The chronic low-grade inflammatory state in adipose tissue is distinct from the classical inflammatory processes associated with acute infection and tissue injury, despite their employment of the same cellular signalling machinery (Hotamisligil, 2006). It is important to note that obesity involves chronic modest elevations in tissue and systemic cytokine levels (e.g., plasma levels of IL-6 are elevated 3-fold in obesity versus 1000-fold in acute sepsis (Friedland et al., 1992 Hoene and Weigart, 2007) along with a similarly muted increase in tissue glucocorticoid levels, ~2- to 3-fold higher (Rask et al., 2001, Sandeep et al., 2005, Masuzaki et al., 2001) versus Cushing's syndrome that can involve up to a 50-fold higher midnight plasma cortisol in the most extreme cases with a loss of the diurnal nadir (Arnaldi et al., 2003, Papanicolaou et al., 1998).

At these lower "modulatory" levels, glucocorticoids can cause both release and inhibition of inflammatory mediators from adipocytes, suggesting the balance of glucocorticoid action is complex, dynamic, and dependent on the levels of active glucocorticoid (of both plasma and intracellular origin) and the influence of the local inflammatory or metabolic (e.g., insulin sensitivity) milieu. For example, dexamethasone (synthetic glucocorticoid) treatment of the clonal mouse adipocyte 3T3-L1 cell line increases expression and secretion of pro-inflammatory serum amyloid (SAA)-3 (Fasshauer et al., 2004) and prothrombotic plasminogen-activated inhibitor (PAI)-1 (Udden et al., 2002) but suppresses IL-6 (Fasshauer et al., 2003).

Glucocorticoids also reduce expression of the adiponectin insulin-sensitizing adipokine from 3T3 adipocytes (Fasshauer et al., 2003b). Adipose monocyte chemoattractant (MCP)-1 levels are also suppressed by glucocorticoids in adipocytes in vitro (Fasshauer et al., 2004b). In addition to affecting cytokine expression and secretion, glucocorticoids may activate cellular stress and survival mechanisms in adipocytes, such as tribbles homolog (TRB)-3 signaling (Yacoub et al., 2006), which is known to impair insulin signaling, at least in liver (Koo et al., 2004). Further, cellular stress pathways may be positively regulated by the glucocorticoid-inducible CCAAT/enhancer binding protein (C/EBP)-β, (Boruk et al., 1998, Arai et al., 2007, Sai et al., 2008), leading to cellular degradation and apoptotic processes that can spark subsequent macrophage recruitment (Cinti et al., 2005, Murano et al., 2008).

Cellular and endoplasmic reticulum stress mechanisms have been directly linked to insulin resistance and diabetes in obesity (Ozcan et al., 2004). Glucocorticoids may also directly impair glucose uptake (i.e., a form of insulin resistance) in adipocytes (Sakoda et al., 2000). Further, they can promote free fatty acid release from mature adipocytes through hormone-sensitive lipase (HSL)-mediated lipolysis (Slavin et al., 1994), contributing to a feed-forward vicious cycle of whole body insulin resistance (high free fatty acids impair skeletal muscle insulin sensitivity) and local TLR activation on macrophages and adipocytes (see above). Recent studies have identified a common mechanism between glucocorticoid signaling and signaling induced by saturated fatty acids whereby they converge on the pro-inflammatory cellular signaling system operating through ceramide (Holland et al., 2007).

Aside from their direct actions on adipokine release, cellular stress, and insulin sensitivity in mature adipocytes, glucocorticoids are likely to exacerbate the chronic inflammatory processes in obese adipose tissues by driving the accelerated differentiation of—and fat accumulation in—adipose stromal cell pre-adipocytes (Ailhaud et al., 1991, Gaillard et al., 1991, Wolf, 1999, Bjorntorp and Rosmond, 2000) and locally inhibiting the development of the necessary vasculature support of the expanding adipose tissue due to their potent angiostatic effects (Small et al., 2005).

Thus, glucocorticoids are implicated in modulating all the stages of adipose tissue dysfunction in obesity, from adipocyte differentiation and lipid accumulation, cellular stress, insulin resistance, pro-inflammatory cytokine release from adipocytes and stromal immunocytes, adipose vascularization, and potential macrophage phagoctyosis of dying adipocytes (Figure 8.1).

## 8.3 11β-HYDROXYSTEROID DEHYDROGENASE TYPE 1

11β-Hydroxysteroid dehydrogenase (11β-HSD), discovered over 50 years ago, catalyzes the interconversion of active glucocorticoids (cortisol in humans, corticosterone in rodents) and their inactive 11-keto forms (cortisone, 11-dehydrocorticosterone) within cells. 11β-HSD activity was initially purified from rat liver (Monder et al., 1993) and subsequently two 11β-HSD isozymes that represented distinct gene products were cloned (Tannin et al., 1991, Albiston et al., 1994). Type 1 11β-HSD has a low affinity for active cortisol and corticosterone and is a nicotine adenine dinucleotide phosphate (NADPH)-dependent enzyme. 11β-HSD1 is widely expressed in many tissues (Ricketts et al., 1998) most highly in liver (Tannin et al., 1991) and at lower levels in adipose tissue (Bujalska, et al., 1997, Rask et al., 2002) and in other organs and cells with high GR expression (Whorwood et al., 1991). In intact cells, 11β-HSD1 acts as an 11-ketoreductase, reactivating inert cortisone into cortisol (Seckl and Walker, 2001). This is dependent on the provision of co-factor by a co-localized endoplasmic reticulum enzyme known as hexose-6-phosphate dehydrogenase (H6PDH) (Atanasov et al., 2004, Lavery et al., 2006).

### 8.3.1 Substrate Levels for 11β-HSD1

In vivo, the main source of 11-ketosteroid is the dehydrogenation of cortisol (humans) and corticosterone (rodents) by 11β-HSD2, which occurs mainly in the

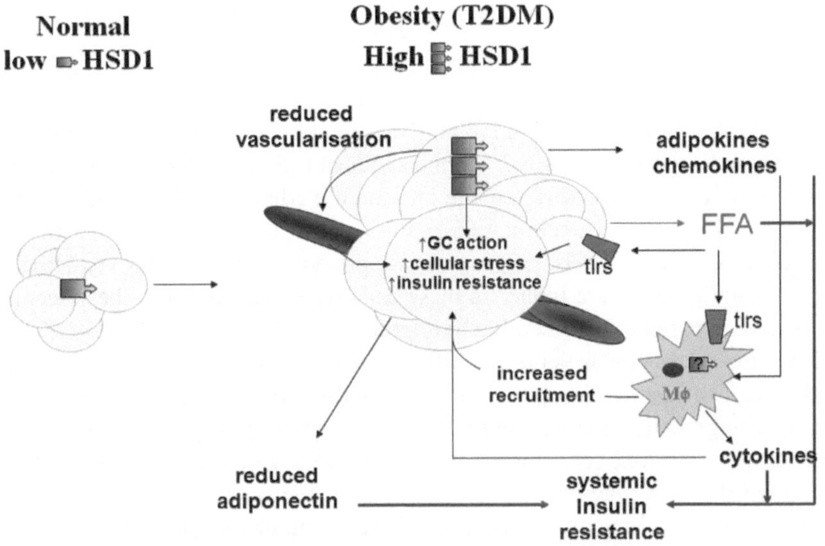


**FIGURE 8.1** Potential effects of increased adipose glucocorticoid action (11β-HSD1 activity) in obese adipose tissue. Increased glucocorticoid levels within the adipocytes lead to increased cellular stress and insulin resistance and subsequent free fatty acid (FFA) release that causes local activation of Toll-like receptors (tlrs) and increased pro-inflammatory signaling (and exacerbation of insulin resistance in adipocytes). Increased FFA levels also contribute to systemic insulin resistance. Direct action of glucocorticoids on adipokine and chemokine secretion leads to increased macrophage accumulation and activation, further increasing the adipocytokine and cytokine secretion from adipocytes and macrophages. Insulin-sensitizing adipokine levels are reduced. This increases local insulin resistance in adipocytes (e.g., TNF-α) and contributes to systemic insulin resistance. Spillover of glucocorticoids (peripheral fat in humans) may impair angiogenic vascular support, leading to hypoxia and cellular stress and promoting inflammatory effects in expanding adipose tissue. Mθ = macrophage. T2DM = type 2 (insulin-resistant) diabetes mellitus. GC = glucocorticoid.

kidney (Whitworth et al., 1989, Katz et al., 1999). Active cortisol circulates at ~100 nmol/l at the night time nadir and rises to peak ~500 nmol/l in the morning. However, cortisol is bound (>95%) to plasma proteins (Dunn et al., 1981) such as CBG (cortisol binding globulin). Thus, free cortisol levels are approximately 0.5 to 1 nmol/l at the nadir, whereas cortisone, which maintains day-long levels from 50 to 100 nmol/l in humans (Walker et al., 1992), is unbound and available for reactivation within tissues. Plasma concentrations of 11-dehydrocorticosterone are approximately 50 nmol/l, and 5 nmol/l in rats and mice, respectively (Kotelevtsev et al., 1997). Thus, circulating cortisone levels equal or exceed free cortisol levels for a significant part of the diurnal cycle. Studies in 11β-HSD1$^{-/-}$ mice (see below) underscored the significance of the enzyme to augment intracellular glucocorticoid action in vivo (Kotelevtsev et al., 1997) and showed that 11β-HSD1 is the sole enzyme capable of reactivating inert 11-ketosteroids to active glucocorticoids.

### 8.3.2 11β-HSD1 in Adipocytes

As described above, glucocorticoids play a key role in the regulation of adipose tissue metabolism and in the differentiation of pre-adipocytes into adipocytes. 11β-HSD1, but not 11β-HSD2, mRNA is expressed in rat (Napolitano et al., 1998) and in human white adipose tissue (Bujalska et al., 1997, Morton and Seckl, 2008). 11β-HSD1 levels increase as clonal mouse fibroblasts differentiate into mature adipocytes (3T3F442A and 3T3-L1 cells) and the reaction direction is 11-ketoreduction (Napolitano et al., 1998). Unlike the undifferentiated 3T3 cells, primary mouse pre-adipocytes express 11β-HSD1, and reductase activity levels are higher in these cells when isolated from the visceral adipose tissue compared to peripheral fat pre-adipocytes (Peixoto et al., 2008).

This suggests that an as yet unidentified post-transcriptional level effect exaggerates the cortisol reactivation potential of 11β-HSD1 in visceral fat pre-adipocytes, potentially augmenting the accumulation of visceral adiposity in obesity. However, human adipose stromal cells appear to lack H6PDH, and thus predominantly inactivate cortisol which may cause pre-adipocyte proliferation (Bujalska et al., 2002). This may result in increased visceral adipocyte numbers when those cells subsequently differentiate and begin to express H6PDH, thus activating 11β-HSD1 reductase activity and increasing active glucocorticoid levels.

### 8.3.3 Regulation of Adipocyte 11β-HSD1: Implications for Adipose Glucocorticoid Action

11β-HSD1 expression and activity are regulated by a wide array of factors including pro- and anti-inflammatory cytokines, adrenergic agonists, CRH and ACTH, and metabolic factors (Seckl et al., 2004, Morton and Seckl, 2008). Regulation of adipocyte 11β-HSD1 by the prominent metabolic hormones insulin and glucocorticoids is complex, with roughly equal numbers of conflicting results over a range of disparate experimental paradigms in humans and rodents that may reflect protocol-, species-, relative insulin sensitivity- and differentiation (pre-adipocytes versus adipocytes) -dependent effects that require further investigation (Morton and Seckl 2008).

The peroxisome proliferator-activated receptor (PPAR)-γ transcription factor that switches on the entire adipogenic program in pre-adipocytes (Tontonoz and Spiegelman, 2008), downregulates 11β-HSD1 in adipocytes in vitro and in mice in vivo when activated by synthetic ligands such as the anti-diabetic thiazolidinedione (Berger et al., 2001), perhaps providing a rationale for their adipose insulin-sensitizing effects. This suggests that PPARγ is necessary to switch on the adipocyte expression of 11β-HSD1, but that in mature adipocytes, PPARγ activation may increase adipocyte insulin sensitivity by suppressing intracellular glucocorticoid reactivation. The oxysterol liver X receptor (LXR) agonists also partially downregulate 11β-HSD1 in vitro and in vivo (Stulnig et al., 2002) which suggests a link between glucocorticoid and oxysterol metabolism. This was confirmed at the substrate level in adipocytes (Wamil et al., 2008) where oxysterols compete for 11β-HSD1 glucocorticoid reactivation. Adipose 11β-HSD1 may therefore contribute to the regulation of the adipose oxycholesterol pool. This has implications for increased fat in obesity by serving as a sump for atherogenic cholesterol metabolites, and suggests adipose tissue levels are linked to regulation of local glucocorticoid reactivation (Wamil et al., 2008).

Another master metabolic transcription factor known as C/EBPβ, is critical for adipocyte differentiation and binds to several sites on the 11β-HSD1 promoter. C/EBPβ increases 11β-HSD1 promoter activity whereas C/EBPβ acts as a dominant negative repressor of 11β-HSD1 transcription when added together with C/EBPβ (Williams et al., 2000). These data suggest the possibility that 11β-HSD1 regulation in adipocytes by insulin and glucocorticoids is indirectly mediated through changes in C/EBP-related proteins. More recently, C/EBPβ has been implicated as the major glucocorticoid-inducible regulator of the 11β-HSD1 gene and may be the dominant regulator in non-hepatic cells (Sai et al., 2008).

Indeed activation of C/EBPβ in response to cellular stresses such as the AMP kinase activator AICAR and C2 ceramide, which signal for cellular energy depletion and saturated fat-induced insulin-resistance, respectively, increased pre-adipocyte 11β-HSD1 levels (Arai et al., 2007). Since under many circumstances the regulation of 11β-HSD1 in liver and adipose tissue is discordant, indeed often reciprocal (Livingstone et al., 2000, Morton et al., 2005), as for other well characterized C/EBP-regulated genes such as PEPCK (Olswang et al., 2003), the relative abundance of the C/EBP transcription factors appears to be an important determinant of 11β-HSD1 expression and its hormonal responsiveness. Transcriptional control of adipocyte and liver 11β-HSD1 is believed to be mediated through the 11β-HSD1 gene P2 promoter, which gives it a distinct regulatory profile from the alternative, in particular cytokine-sensitive, P1 promoter used in lung, with implications for altered regulation within inflammatory contexts in that tissue (Bruley et al., 2007).

Despite a great deal of information about regulatory mechanisms, it is perhaps surprising that the genetic basis underlying increased adipose 11β-HSD1 in idiopathic obesity remains uncertain. Attempts to link polymorphisms in the 11β-HSD1 gene with obesity have not been successful (Draper et al., 2002, Caramelli et al., 2001), although specific polymorphisms in the 11β-HSD1 gene do associate with hypertension (Francks et al., 2004), insulin sensitivity and/or diabetes (Nair et al., 2004), and apolipoprotein levels (Robitaille et al., 2004). Studies of identical twins also support environmental causes for the association of increased adipose 11β-HSD1 expression and obesity (Kannisto et al., 2004).

In summary, a number of key regulatory transcription factors are implicated in the control of adipose 11β-HSD1. However, conclusive proof of the major transcriptional changes that accompany, and thus may underlie, the dysregulation of adipose glucocorticoid action in obesity, are as yet unknown.

## 8.4 11β-HSD1 IN OBESITY

Obese rodents with defects in the leptin receptor (Zucker Lepr*fa/fa* rats) or in the leptin gene (leptin-deficient Lep*ob/ob* mice) revealed that obesity is associated with increased whole adipose tissue 11β-HSD1 (Livingstone et al., 2000, Masuzaki et al., 2001, Morton et al., 2004b). However, this is not a universal finding and is likely strain- and adipose depot-dependent in rodents (Morton et al., 2005, Liu et al., 2005). For example, mice with obesity of polygenic origin (similar to the situation in idiopathic human obesity) have low adipose, but elevated liver 11β-HSD1 (Morton et al., 2005).

While this supports the evidence for reciprocal, likely hormone-mediated, tissue regulation of the enzyme (see above), it also suggests that tissue 11β-HSD1 levels must be determined in each model of obesity and the impact of altered glucocorticoid regeneration placed within the broader metabolic context of the organism (expression of 11β-HSD1 is at least 10-fold higher in liver, but effects on adipose tissue function, e.g., lipolysis and adipokine release, may have a relatively greater impact upon systemic insulin sensitivity). Liver 11β-HSD1 remains an important therapeutic target, but is beyond the scope of this chapter (see Morton and Seckl, 2008 for a review). The implications are clear for human obesity: potential therapeutic intervention targeting 11β-HSD1 should first be matched with the individual's 11β-HSD1 activity profile.

In humans, 11β-HSD1 mRNA and activity is increased in subcutaneous abdominal adipose tissue of obese subjects both in vivo and in vitro (Rask et al., 2001, 2002, Lindsay et al., 2003, Wake et al., 2003, Kannisto et al., 2004, Goedecke et al., 2006, Michailidou et al., 2007, Paulsen et al., 2007, Morton and Seckl, 2008). Further studies have directly confirmed increased 11β-HSD1 activity using tissue microdialysis in obese subcutaneous adipose tissue (Sandeep et al., 2005). Recent studies suggest that in some cases 11β-HSD1 mRNA levels are also increased in visceral omental adipose tissue of obese women and are strong predictors of fat cell size in this visceral depot (Michailidou et al., 2007, Paulsen et al., 2007). Since many adipokines are expressed at higher levels in visceral fat and this depot may be the most active site in terms of inflammatory processes, the impact that elevated visceral fat glucocorticoid action and production would have locally and on the portal (blood draining from visceral fat to the liver) supply of adipokines and fatty acids to the liver remains important to determine.

### 8.4.1 Visceral Fat 11β-HSD1: A Pronounced Contribution to Portal Insulin Resistance?

Higher portal corticosterone was reported in a model of adipose-specific 11β-HSD1 overexpression (see below), indicating the possibility of a paracrine effect of excess cortisol and/or corticosterone to modulate both adjacent inflammatory and immune cells in the adipose tissue and liver metabolic and inflammatory mechanisms (Masuzaki et al., 2001) through a "spillover" into the portal circulation (Morton and Seckl, 2008). Recent studies in humans indicate that splanchnic (central blood supply from visceral organs including liver, as opposed to peripheral blood) cortisol production was comparable to rates of adrenal cortisol secretion (Andrew et al., 2005), and estimated using a mathematical model that visceral adipose contributes somewhat more cortisol production than does the liver to this source of re-activated hormone.

This view was challenged, however, by similar studies in dogs (Basu et al., 2006) and more recently direct measurement of portal vein cortisol in humans indicated that visceral (fat) 11β-HSD1 is unlikely to contribute to portal cortisol levels (Stimson et al., 2008). A further complication is that far from being a source of cortisol, the visceral compartment (visceral fat, gut, pancreas, spleen) actually releases cortisone into the portal veins in obese humans (Basu et al., 2008). Future studies should determine whether this potentially important source of glucocorticoid is a factor in the viscerally obese. At least in rodent studies, the visceral compartment appears to be the depot where the largest increase in 11β-HSD1 expression (Livingstone et al., 2000, Masuzaki et al., 2001, Morton et al., 2004b) and, crucially, glucocorticoid "reactivation potential" (Peixoto et al., 2008) may occur. Whether glucocorticoid spillover occurs does not change the expected effects of increased visceral adipose 11β-HSD1 to affect adipocyte lipolysis, cellular stress, insulin resistance, and adipokine release. This is important both from the point of view of local tissue inflammatory cell function (macrophages) and drainage to the liver—the site of secondary, presumably detrimental, effects of free fatty acid and adipokine spillover.

## 8.5 TRANSGENIC MANIPULATION OF ADIPOSE 11β-HSD1

### 8.5.1 Adipose 11β-HSD1 Overexpressing Mice

Mice overexpressing 11β-HSD1 selectively in adipose tissue under the adipocyte fatty acid binding protein (aP2) promoter express two- to three-fold more 11β-HSD1 in all adipose depots, but not in other tissues. This modest elevation of 11β-HSD1 is similar to the levels reported in human obese adipose tissue. The transgene doubled corticosterone levels within adipose tissue and increased release of corticosterone into the portal circulation without altering systemic glucocorticoid levels. Transgenic aP2-11β-HSD1mice exhibited intra-abdominal obesity associated with higher levels of GR in visceral than in peripheral fat depots (Masuzaki et al., 2001) as well as high fat diet-exacerbated insulin-resistant diabetes and dyslipidemia (elevated free fatty acid and triglyceride levels).

Within adipocytes, aP2-11β-HSD1 mice showed changes concordant with decreased sensitivity to insulin and/or increased corticosterone levels, with decreased expression of insulin-sensitizing adiponectin (Yamauchi et al., 2002) and increased expression of the arch insulin-resistance TNF-α adipokine (Hotamisligil et al., 1993). Serum TNF-α levels were also elevated, indicating the importance of adipose-specific changes of whole body adipokine and insulin sensitivity. Interestingly, resistin, an adipokine that promotes insulin resistance in mice (Steppan et al., 2001), was reduced in aP2-HSD1 transgenic adipose tissue.

Notably, angiotensinogen mRNA, normally expressed at low levels in adipose tissue, was strikingly elevated in aP2-11β-HSD1 transgenic adipose tissue. Indeed elevated levels of this glucocorticoid-regulated transcript appear to drive the marked hypertension seen in aP2-11β-HSD1 mice (Masuzaki et al., 2003). Finally, the animals exhibited hyperphagia that may be partly related to leptin resistance at the hypothalamic level where this adipokine elicits its primary function as a satiety hormone (Halaas et al., 1995). The aP2-11β-HSD1 transgenic mouse, with its subtly elevated adipose glucocorticoid reactivation (similar to that found in human obese adipose tissue) closely modelled many major features of human metabolic syndrome, and argues that this adipose defect may serve as a primary driver of many features of this disease.

### 8.5.2 Global 11β-HSD1 Knockout (11β-HSD1$^{-/-}$) Mice

Global 11β-HSD1$^{-/-}$ mice demonstrated the importance of intracellular glucocorticoid re-amplification by the enzyme in glucocorticoid action (Kotelevtsev et al., 1997). The 11β-HSD1$^{-/-}$ mice resisted stress and high fat diet induced hyperglycemia (Kotelevtsev et al., 1997). When the transgenic line was back-crossed onto the obesity-prone C57Bl/6J mouse genetic background, they gained significantly less weight than controls, despite a relative hyperphagia when fed a high fat diet. This appeared due to an enhanced metabolic rate as inferred by increased core body temperature (Morton et al., 2004a). The 11β-HSD1$^{-/-}$ mice preferentially gained adipose tissue in the "metabolically safer" peripheral depots rather than in the "metabolically disadvantageous" visceral sites.

Although the explanation for this redistribution of fat, the inverse of the fat distribution pattern of the aP2-11β-HSD1 transgenic mice, is uncertain, these mice showed higher mRNA levels of the adipogenic transcription factor PPARγ receptor in their adipose tissues. Further, 11β-HSD1$^{-/-}$ mice showed a greater increase of adipose PPARγ receptors with high fat feeding than wild-type mice (Morton et al., 2004a). PPARγ ligands caused insulin sensitization and fat redistribution to the periphery and this may underpin the favorable morphology (Kelly et al., 1999, Sewter et al., 2002) seen when increased circulating free fatty acids act as endogenous ligands (Xu et al., 1999) for the PPARγ receptors. Insulin sensitization is evident at the adipocyte level, since isolated primary 11β-HSD1$^{-/-}$ adipocytes show increased basal and insulin-stimulated glucose uptake. Additionally, the 11β-HSD1$^{-/-}$ mice also show greater induction of uncoupling protein-2 in mesenteric adipose tissue than wild-type mice which may allow local calorie wastage rather than storage as fat (Morton et al., 2004a).

Uncoupling protein-2 is downregulated by glucocorticoids (Xu et al., 1999) and upregulated by PPARγ activation (Digby et al., 2000). The significance of this altered PPARγ expression is particularly important with respect to obesity. Notably, we have shown that 11β-HSD1$^{-/-}$ mice exhibit increased PPARγ expression and downstream genes linked to β-oxidation in liver. This observation was confirmed by others who saw increased liver β-oxidation after HSD1 inhibitor treatment (Berthiaume et al., 2007a). In adipocytes, liver, and macrophages, activation of PPAR signalling downregulates 11β-HSD1 expression (Berger et al., 2001, Mai et al., 2007, Hermanowski-Vosatka et al., 1999), suggesting an important feedback loop whereby PPAR receptor ligands (potentially high free fatty acids in obesity) may limit glucocorticoid action by reducing their intracellular re-activation.

With respect to inflammatory pathways, recent evidence suggests that macrophage-specific knockout of PPARγ causes an impairment in the switching of macrophages from a pro-inflammatory (M1-like) to an anti-inflammatory (M2-like) phenotype (Odergaard et al., 2007). Higher PPARγ levels in knockout macrophages may be expected to skew macrophage activation toward an anti-inflammatory phenotype in 11β-HSD1$^{-/-}$ mice, thus contributing to the protected metabolic phenotype of the mice (see below). Intriguingly, the induction of an M2-like anti-inflammatory profile includes increases in macrophage β-oxidation (Vats et al., 2006) via PPAR co-activator (PGC)-1, suggesting overlapping mechanisms operate in the liver and macrophages and possibly adipocytes. Thus we attributed higher PPARγ-responsive uncoupling protein (UCP)-2 expression in the 11β-HSD1$^{-/-}$ adipose to increased energy dissipation within the adipocytes (Morton et al., 2004a). This adipose calorie-burning effect has been confirmed by others using 11β-HSD1 inhibitors and showed that carnitine palmitoyl transferase (CPT)-1 activity increased specifically in rat visceral fat and was associated with reduced visceral fat cell size and fat depot mass (Berthiaume et al., 2007b).

The 11β-HSD1$^{-/-}$ mice also exhibit key changes in adipokine profile that are consistent with their insulin-sensitized and diabetes-resistant phenotype with high fat feeding. Adipose leptin mRNA and plasma leptin levels are reduced, particularly in peripheral adipose where 11β-HSD1 expression is highest in mice (perhaps a cause of the transiently increased food intake) (Morton et al., 2004a, Densmore et al., 2006). Along with its role in appetite and energy expenditure, leptin is a potent regulator of immune function with pro-inflammatory activities (Otero et al., 2006). Adipocyte resistin and TNF-α mRNAs were reduced whereas adiponectin was increased, again compatible with an adipose-mediated insulin-sensitized phenotype. Overall the 11β-HSD1$^{-/-}$ mice showed improved glucose tolerance, in part due to increased adipocyte insulin sensitivity, lower fasting free fatty acids—an indirect indicator of adipose insulin sensitization—and reduced intra-tissue glucocorticoid levels in the face of modest hypercorticosteronemia associated with a protective adipokine profile (Morton et al., 2004b). Future studies should determine the subcellular origins of this improved adipokine profile within adipose tissue with respect to the increased infiltration of leukocytes (see below).

### 8.5.3 Adipose 11β-HSD2 Overexpressing Mice: Model of Adipose-Specific Glucocorticoid Deficiency

To confirm the key importance of reduced adipose tissue glucocorticoid levels in the beneficial phenotype of the 11β-HSD1$^{-/-}$ mouse, a transgenic mouse ectopically expressing the glucocorticoid-inactivating 11β-HSD2 isozyme in adipose tissue was generated, again exploiting the aP2 promoter (Kershaw et al., 2005). Expression levels were similar to those in kidneys, where 11β-HSD2 performs its physiological role to exclude corticosterone from the high affinity mineralocorticoid receptor.

Echoing the 11β-HSD1$^{-/-}$ phenotype, the aP2-11β-HSD2 mice resisted weight gain on a high fat diet due to reduced fat mass accumulation and insulin sensitization associated with decreased leptin and resistin, but increased adiponectin, PPARγ, and uncoupling protein-2 expression in fat tissues. This model reinforces the concept that reducing levels of active corticosterone exclusively in adipose tissue engenders a favorable metabolic disease-resistant phenotype and improved adipokine profile. These effects were unexpectedly pronounced in subcutaneous adipose tissue, where the transgene was inexplicably expressed at higher levels than other tissues.

Thus, while the greatest disease risk is associated with increased visceral adiposity (Wajchenberg, 2000, Kissebah et al., 1982), the ap2-HSD2 mice demonstrate that peripheral fat glucocorticoid metabolism (where 11β-HSD1 is expressed at highest levels in mice (Morton et al., 2004b) has a large impact on metabolism and systemic insulin sensitivity. Recent studies addressing adipose cortisol production in humans strengthen the idea that subcutaneous fat 11β-HSD1 is elevated and contributes directly to local (Sandeep et al., 2005) and whole body (released into the circulation) cortisol production (Stimson et al., 2008). The precise roles of glucocorticoids, other hormones (such as insulin), transcription factors (such as PPARγ), and their downstream adipokines in each adipose depot require further investigation.

## 8.6 DOWNREGULATION OF 11β-HSD1 AS ADAPTATION TO CHRONIC FAT FEEDING

Given the association of high adipose 11β-HSD1 with increased (visceral) fat deposition, insulin resistance, and a detrimental adipokine profile, it was unexpected that normal mice markedly downregulated 11β-HSD1 mRNA and activity in fat in response to a high fat diet (Morton et al., 2004b). Similar findings were reported in rats (Drake et al., 2005). Strikingly, the metabolic disease-resistant A/J mouse strain exhibited lower basal levels of 11β-HSD1 mRNA and activity in visceral and peripheral fat depots. Moreover, A/J mice downregulated adipose 11β-HSD1 more markedly upon high fat feeding than the metabolic disease-prone C57Bl/6J strain (Morton et al., 2004b). Thus the A/J strain almost completely "switched off" adipose 11β-HSD1 upon high fat feeding, in effect becoming like 11β-HSD1$^{-/-}$ and aP2-HSD2 mice in line with their maintained insulin sensitivity.

This suggested that downregulation of adipose 11β-HSD1 may be an adaptive mechanism, more pronounced in metabolic disease-resistance, that counteracts the adverse metabolic consequences of a chronic high fat diet. In similar experiments in rats, high fat diet-mediated downregulation of adipose 11β-HSD1 was transient and reversed after several months as weight increased and insulin resistance developed, suggesting that this metabolic adaptation was sustained with worsening obesity in rats (Drake et al., 2005)

## 8.7 11β-HSD1 AND MACROPHAGES

As noted above, adipose tissue becomes infiltrated with pro-inflammatory macrophages during obesity. From studies investigating the effects of glucocorticoids (Giles et al., 2001) on macrophage function in classical inflammatory models, the potential expression of 11β-HSD1 within adipose macrophages becomes highly relevant to the adipose inflammation issue. The 11β-HSD1 enzyme is induced upon macrophage activation by stimuli such as lipopolysaccharide (LPS) in human monocytes and mouse macrophages (Thieringer et al., 2001, Gilmour et al., 2006, Ishii et al., 2006). The enzyme is also increased by pro-inflammatory cytokines in fibroblasts (Hardy et al., 2006, Hardy et al., 2008) and smooth muscle cells (Cai et al., 2000) and by cytokines such as TNF-α and IL-1β where it may modulate inflammatory processes such as rheumatoid arthritis and atherosclerosis. 11β-HSD1 augments the process of glucocorticoid-induced macrophage phagocytosis (Giles et al., 2001), thus promoting resolution of inflammatory processes (Gilmour et al., 2006).

The latter concept seems clear within the context of classical inflammatory paradigms such as LPS-mediated activation of macrophages (Thieringer et al., 2001, Gilmour et al., 2006, Ishii et al., 2006, Chapman et al., 2007). However, when 11β-HSD1$^{-/-}$ mice are challenged with LPS-mediated endotoxemia, they exhibit exaggerated systemic and (splenic) macrophage inflammatory responses in part through the TGF-β-induced SHIP-1, NF-κB, and MAPK intracellular signalling cascade (Zhang and Daynes, 2007). As shown earlier, this was accompanied by increased secretion of cytokines, particularly the glucocorticoid-suppressible IL-6 (Gilmour et al., 2006) from the 11β-HSD1$^{-/-}$ macrophages (Zhang and Daynes, 2007). The authors attributed this macrophage hyper-responsiveness to the subtly higher in vivo glucocorticoid levels driving an altered myeloid differentiation process (control and 11β-HSD1$^{-/-}$ macrophages exhibited similar LPS responses when differentiated in vitro). In this case, the subtly higher glucocorticoid levels in 11β-HSD1$^{-/-}$ plasma appeared to be the driving force, with glucocorticoids priming a pro-inflammatory response (Zhang and Daynes, 2007).

Others observed direct pro-inflammatory effects of 11β-HSD1 inhibition in a macrophage (J774.1) cell line in vitro (Ishii et al., 2006), further adding to the complex role of 11β-HSD1 in the regulation of inflammatory responses. In any case, these data together suggest that regulation of 11β-HSD1 within adipose tissue macrophages may be a key determinant of inflammatory processes within fat during obesity. Are there parallels between obese adipose tissue macrophage and classical macrophage activation? The LPS-TLR systems are purported to be activated in adipocytes and macrophages through high free fatty acid levels in obesity (Lin et al., 2000, Shi et al., 2007, Song et al., 2007, Nguyen et al., 2007). It may be expected then that obesity would lead to increased activation and expression of adipose macrophage 11β-HSD1.

However, as described above, two issues remain unclear. The first is whether the more muted chronic inflammatory response within adipose tissue during obesity leads to the same effect on adipose tissue macrophages. Second, it is apparent that the downstream effects of altered macrophage 11β-HSD1 levels on inflammatory responses are not obviously predictable, that is, pro- and anti-inflammatory effects have been described with 11β-HSD1 manipulation in macrophages, and glucocorticoids play a dual role in the secretion of cytokines, depending on the prevailing levels. To complicate matters further, adipose tissue macrophages may constitute a distinct subpopulation that has both M2 anti-inflammatory and M1 pro-inflammatory characteristics (Zeyda et al., 2007).

In order to address these questions, we investigated adipose tissue macrophage 11β-HSD1 expression and the effects of 11β-HSD1 deficiency on the pro-inflammatory phenotype during dietary obesity in mice. Preliminary data suggest that adipose tissue macrophage 11β-HSD1 levels are in fact very low and that, like adipocyte 11β-HSD1 (Morton et al., 2004b), adipose tissue macrophage 11β-HSD1 is downregulated during dietary obesity in mice (Battle et al., 2008, unpublished data). Further studies are under way to fully characterize the phenotype of 11β-HSD1$^{-/-}$ adipose tissue macrophages, but indicate an anti-inflammatory phenotype including reduced infiltration of adipose macrophages, consistent with the lower adipose TNF-α (Morton et al., 2004a) that likely derives from the adipose macrophages and metabolic protection seen in the model (Battle et al., 2008, unpublished observations). This would be consistent with an increased PPARγ expression-mediated switch (see above) toward an anti-inflammatory, metabolically protective, adipose macrophage phenotype.

## 8.8 OTHER SITES OF 11β-HSD1 EXPRESSION PERTINENT TO ADIPOSE INFLAMMATION

Glucocorticoids are angiostatic, and 11β-HSD1$^{-/-}$ mice show improved recovery after myocardial infarction in vivo due to increased angiogenesis as well as in vitro in angiogenesis assays using aortic rings from the model (Small et al., 2005). Angiogenesis is a key determinant of adipose tissue expansion during obesity and inhibition of this process prevents obesity (Rupnick et al., 2002, Brakenhielm et al., 2004, Kim et al., 2007, Lijnen et al., 2008). If this effect translates to angiogenic processes in adipose tissue, 11β-HSD1 deficiency would potentially promote a more robust angiogenic response to the rapid expansion of adipose tissue with obesity in vivo, improving vascularization and reducing downstream inflammatory consequences. Both 11β-HSD1 and 11β-HSD2 appear to be expressed in human endothelial cells and modulate inducible nitric oxide synthase (iNOS) activity (Liu et al., 2008). The activity in rodents is less clear but suggests that expression of 11β-HSD isoforms may be both species and anatomical (perhaps fat depot) site-specific (Hadoke et al., 2006).

Along with the recent discovery that T cells of the adaptive immune system are involved in coordinating the early and chronic inflammatory events in adipose tissue during obesity (Wu et al., 2007, Rausch et al., 2008), it is also of note that T lymphocytes expressed 11β-HSD1, and this increased with aging and as cells became polarized into Th1 and Th2 subtypes (Zhang et al., 2005). B cells also express 11β-HSD1 (Zhang et al., 2005) although their role in adipose inflammation is unclear. Dendritic (bone marrow-derived) cells are also implicated in the infiltration and inflammation of adipose of obesity (Nguyen et al., 2007) and this cell type, at least as derived after extensive culture and a positive selection process in vitro, also expressed 11β-HSD1 (Zhang et al., 2005).

In summary, a number of other key inflammatory and vascular cells within adipose tissues have the potential to express 11β-HSD1 and this may impact adipose inflammation in obesity. However, as predicted above for macrophages (see above), this remains to be conclusively demonstrated in cells derived from fat where functional subpopulations of these cell types exist.

## 8.9 11β-HSD1 INHIBITORS AS THERAPEUTICS

To date, pharmacological intervention with 11β-HSD1 inhibition in vivo has shown anti-inflammatory effects, such as lower MCP-1 levels in dyslipidemic apoE null mice (Hermanowski-Vosatka, et al., 2005). The beneficial metabolic and anti-inflammatory effects bode well for using the enzyme as a target to treat obesity and diabetes (Hughes et al., 2008). Lower adipose TNF-α and resistin and higher adiponectin and adipose PPARγ in 11β-HSD1$^{-/-}$ animals all agree with an overall anti-inflammatory, metabolic protective effect within the context of the chronic low grade inflammation of obesity. However, due to the complex regulation of adipokines by glucocorticoids, insulin, and other cytokines, the full effects of 11β-HSD1 inhibition in adipose tissue, particularly in humans, remain to be completely defined.

Another matter to be determined is whether glucocorticoid spillover from adipocytes expressing high levels of 11β-HSD1 in obesity modulates the inflammatory responses of neighboring macrophages and immune cells in a paracrine manner (e.g., through TLRs) or whether the infiltrating immunocytes activate an intracrine 11β-HSD1 response to modulate their function during the progression of obesity, as occurs in the classical inflammatory paradigms described above. This seems less likely from preliminary studies in our laboratory insofar as macrophages are concerned (Battle et al., unpublished observations).

## 8.10 CONCLUSIONS

It may be speculated that the loss of glucocorticoid reactivation by adipose (mainly adipocyte) 11β-HSD1 reduces adipose inflammation, at least by direct effects, to improve insulin sensitivity, reduce cellular stress, and favorably alter adipokine secretion in adipocyte and fat distribution in vivo. Reduced glucocorticoid spillover (likely only in subcutaneous fat in humans) may potentially produce beneficial effects on adipose macrophages, other immunocytes, and local vasculature. There may be potentially beneficial effects on other adipose cell types: reduced pre-adipocyte differentiation and reduced macrophage and immunocyte infiltration. Some of these effects may be mediated by altered expression of 11β-HSD1 within those cell types in an intracrine manner.

The inherent assumption here is that the subtly elevated glucocorticoid levels in adipose tissue in obesity are largely detrimental and pro-inflammatory. This moderates the paradoxical notion that glucocorticoids are purely anti-inflammatory, and suggests that in fact no loss of such a protective anti-inflammatory glucocorticoid effect occurs with 11β-HSD1 inhibition (those effects are specific to pharmacological dose-mediated dampening of the inflammatory cascade). Loss of a pro-resolution effect mediated by 11β-HSD1 in macrophages (Gilmour et al., 2006) appears insufficiently severe to appreciably exacerbate this chronic condition in adipose tissue.

## REFERENCES


Ailhaud G et al. 1991. Growth and differentiation of regional adipose tissue: molecular and hormonal mechanisms. *Int J Obes. Suppl* 2: 87– 90.

Albiston, AL et al. 1994. Cloning and tissue distribution of the human 11β-hydroxysteroid dehydrogenase type 2 enzyme. *Mol Cell Endocrinol* 105: R11–R17.

Andrew R et al. 2005. The contribution of visceral adipose tissue to splanchnic cortisol production in healthy humans. *Diabetes* 54: 1364–1370.

Arai N et al. 2007. Ceramide and adenosine 5' monophosphate- activated protein kinases are two novel regulators of 11β- hydroxysteroid dehydrogenase type 1 expression and activity in cultured preadipocytes. *Endocrinology* 148: 5268–5277.

Arnaldi G et al. 2003. Diagnosis and complications of Cushing's syndrome: a consensus statement. *J Clin Endocrinol Metab* 88: 5593–5602.

Atanasov AG et al. 2004. Hexose-6-phosphate dehydrogenase determines the reaction direction of 11βa- hydroxysteroid dehydrogenase type 1 as an oxoreductase. *FEBS Lett* 571: 129–133.

Atsumi T et al. 2007. The pro-inflammatory cytokine macrophage migration inhibitory factor regulates glucose metabolism during systemic inflammation. *J Immunol* 179: 5399–5406.

Basu R et al. 2008. The liver is the site of splanchnic cortisol production in obese nondiabetic humans. *Diabetes* (epub ahead of print).

Basu R et al. RA 2006. Splanchnic cortisol production in dogs occurs primarily in the liver: evidence for substantial hepatic specific 11β-hydroxysteroid dehydrogenase type 1 activity. *Diabetes* 55: 3013–3019.

Battle J et al. et al. 2008. The role of 11β-hydroxysteroid dehydrogenase type 1 in adipose inflammation in obesity. Abstract presented at Third Symposium of Centre for Cardiovascular Sciences, Edinburgh.

Berg AH et al. 2001. The adipocyte-secreted protein Acrp30 enhances hepatic insulin action. *Nat Med* 7: 947–953.

Berger J et al. 2001. Peroxisome proliferator-activated receptor-γ ligands inhibit adipocyte 11β-hydroxysteroid dehydrogenase type 1 expression and activity. *J Biol Chem* 276: 12629–12635.

Berthiaume M et al. 2007b. Depot-specific modulation of rat intra-abdominal adipose tissue lipid metabolism by pharmacological inhibition of 11β hydroxysteroid dehydrogenase type 1. *Endocrinology* 148: 2391–2397.

Berthiaume M et al. 2007a. 11β-HSD1 inhibition improves triglyceridemia through reduced liver VLDL secretion and partitions lipids toward oxidative tissues. *Am J Physiol Endocrinol Metab* 293: E1045–E1052.

Björntorp P and Rosmond R. 2000 Obesity and cortisol. *Nutrition* 16: 924–936.

Boruk M, Savory JG, and Hache RJ. 1998. AF2-dependent potentiation of CCAAT enhancer binding protein β-mediated transcriptional activation by glucocorticoid receptor. *Mol Endocrinol* 12: 1749–1763.

Bråkenhielm E et al. 2004. Angiogenesis inhibitor, TNP-470, prevents diet-induced and genetic obesity in mice. *Circ Res* 94: 1579–1588.

Bruley C et al. 2007. A novel promoter for the 11β-hydroxysteroid dehydrogenase type 1 gene is active in lung and is C/EBPβ-independent. *Endocrinology* 147: 2879–2885.

Bujalska I, Kumar S, and Stewart PM. 1997. Does central obesity reflect Cushing's disease of the omentum? *Lancet* 349: 1210–1213.

Bujalska, IJ et al. 2002 A switch in dehydrogenase to reductase activity of 11β-hydroxysteroid dehydrogenase type 1 upon differentiation of human omental adipose stromal cells. *J Clin Endocrinol Metab* 87: 1205–1210.

Cai TQ et al. 2001. Induction of 11β-hydroxysteroid dehydrogenase type 1 but not -2 in human aortic smooth muscle cells by inflammatory stimuli. *J Steroid Biochem Mol Biol* 77: 117–122.

Caramelli E et al. 2001. Lack of mutations of type 1 11β-hydroxysteroid dehydrogenase gene in patients with abdominal obesity. *Endocr Res* 27: 47–61.

Chapman KE et al. 2008. The role and regulation of 11β-hydroxysteroid dehydrogenase type 1 in the inflammatory response. *Mol Cell Endocrinol* (epub ahead of print).

Cinti S et al. 2005. Adipocyte death defines macrophage localization and function in adipose tissue of obese mice and humans. *J Lipid Res* 46: 2347–2355.

Densmore VS et al. 2006. 11β-hydroxysteroid dehydrogenase type 1 induction in the arcuate nucleus by high-fat feeding: a novel constraint to hyperphagia? *Endocrinology* 147 4486–4495.

Digby, JE et al. 2000 Depot-related and thiazolidinedione-responsive expression of uncoupling protein 2 (UCP2) in human adipocytes. *Int J Obes* 24: 585–592.

Donath MY et al. 2008. Cytokines and β cell biology: from concept to clinical translation. *Endocr Rev* 29: 334–350.

Drake, AJ et al. 2005. Reduced adipose glucocorticoid reactivation and increased hepatic glucocorticoid clearance as an early adaptation to high-fat feeding in Wistar rats. *Endocrinol* 146: 913–919.

Draper, N et al. 2002. Association studies between microsatellite markers within the gene encoding human 11β-hydroxysteroid dehydrogenase type 1 and body mass index, waist to hip ratio, and glucocorticoid metabolism. *J Clin Endocrinol Metab* 87: 4984–4990.

Caramelli, E et al. 2001. Lack of mutations of type 1 11β-hydroxysteroid dehydrogenase gene in patients with abdominal obesity. *Endocrine Research* 27: 47–61.

Dunn, JF, Nisula, BC, and Rodbard, D. 1981. Transport of steroid hormones: binding of 21 endogenous steroids to both testosterone-binding globulin and corticosteroid-binding globulin in human plasma. *J Clin Endocrinol Metab* 53: 58–68.

Engeli, S et al. 2004. Regulation of 11β-HSD genes in human adipose tissue: Influence of central obesity and weight loss. *Obesity Res* 12: 9–17.

Fasshauer M et al. 2004a. Serum amyloid A3 expression is stimulated by dexamethasone and interleukin-6 in 3T3-L1 adipocytes. *J Endocrinol* 183: 561–567.

Fasshauer M et al. 2004b. Monocyte chemoattractant protein 1 expression is stimulated by growth hormone and interleukin-6 in 3T3-L1 adipocytes. *Biochem Biophys Res Commun* 317: 598–604.

Fasshauer M et al. 2003a. Interleukin (IL)-6 mRNA expression is stimulated by insulin, isoproterenol, tumour necrosis factor α, growth hormone, and IL-6 in 3T3-L1 adipocytes. *Horm Metab Res* 35: 147–152.

Fasshauer M et al. 2003b. Adiponectin gene expression and secretion is inhibited by interleukin-6 in 3T3-L1 adipocytes. *Biochem Biophys Res Commun* 301: 1045–1050.

Franks PW, Knowler WC, and Nair S, 2004. Interaction between an 11β-HSD1 gene variant and birth era modifies the risk of hypertension in Pima Indians. *Hypertension* 44: 681–688.

Friedland JS et al. 1992. Prolonged elevation of interleukin-8 and interleukin-6 concentrations in plasma and of leukocyte interleukin-8 mRNA levels during septicemic and localized *Pseudomonas pseudomallei* infection. *Infect Immun* 60: 2402–2408.

Gaillard D et al. 1991. Control of terminal differentiation of adipose precursor cells by glucocorticoids. *J Lipid Res* 1991 32: 569–579.

Giles KM et al. 2001. Glucocorticoid augmentation of macrophage capacity for phagocytosis of apoptotic cells is associated with reduced p130Cas expression, loss of paxillin/pyk2 phosphorylation, and high levels of active Rac. *J Immunol* 167: 976–986.

Gilmour JS et al. 2006. Local amplification of glucocorticoids by 11β-hydroxysteroid dehydrogenase type 1 promotes macrophage phagocytosis of apoptotic leukocytes. *J Immunol* 176: 7605–7611.

Goedecke JH et al. 2006. Glucocorticoid metabolism within superficial subcutaneous rather than visceral adipose tissue is associated with features of the metabolic syndrome in South African women. *Clin Endocrinol* 65: 81– 87.

Goulding NJ and Flowers RJ. 2001. Glucocortioids: Milestones in Drug Therapy Series. Birkhauser Verlag: Basel.

Gudbjornsson B, Juliusson UI, and Gudjonsson FV. 2002. Prevalence of long-term steroid treatment and the frequency of decision making to prevent steroid-induced osteoporosis in daily clinical practice. *Ann Rheum Dis* 61: 32–36.

Hadoke PW et al. 2006. Intra-vascular glucocorticoid metabolism as a modulator of vascular structure and function. *Cell Mol Life Sci* 63: 565–578.

Halaas JL et al. 1995. Weight-reducing effects of the plasma protein encoded by the obese gene. *Science* 269: 543–546.

Hardy R et al. 2008. Local and systemic glucocorticoid metabolism in inflammatory arthritis. *Ann Rheum Dis* 67: 1204–1210.

Hardy RS et al. 2006. Differential expression, function and response to inflammatory stimuli of 11β-hydroxysteroid dehydrogenase type 1 in human fibroblasts: a mechanism for tissue-specific regulation of inflammation. *Arthritis Res Ther* 8:R108.

Hermanowski-Vosatka A et al. 2005. 11β-HSD1 inhibition ameliorates metabolic syndrome and prevents progression of atherosclerosis in mice. *J Exp Med* 202: 517–527.

Hermanowski-Vosatka A et al. 2000. PPARα agonists reduce 11β-hydroxysteroid dehydrogenase type 1 in the liver. *Biochem Biophys Res Commun* 279: 330–336.

Hirosumi J et al. 2002. Central role for JNK in obesity and insulin resistance. *Nature* 420: 333–336.

Hoene M and Weigert C. 2007. The role of interleukin-6 in insulin resistance, body fat distribution and energy balance. *Obes Rev* 9: 20–29.

Holland WL et al. 2007. Inhibition of ceramide synthesis ameliorates glucocorticoid-, saturated fat-, and obesity-induced insulin resistance. *Cell Metab* 5: 167–179.

Hotamisligil GS et al. 1995. Increased adipose tissue expression of tumor necrosis factor-α in human obesity and insulin resistance. *J Clin Invest* 95: 2409–2415.

Hotamisligil GS. 2006. Inflammation and metabolic disorders. *Nature* 444: 860–867.

Hotamisligil G, Shargill N, and Spiegelman BM. 1993. Adipose expression of tumor necrosis factor-α: direct role in obesity-linked insulin resistance. *Science* 259: 87– 91.

Hughes KA, Webster SP, and Walker BR. 2008. 11β-Hydroxysteroid dehydrogenase type 1 (11β-HSD1) inhibitors in type 2 diabetes mellitus and obesity. *Expert Opin Investig Drugs* 17: 481–496.

Ishii T et al. 2007. Augmentation of 11β-hydroxysteroid dehydrogenase type 1 in LPS-activated J774.1 macrophages: role of 11β-HSD1 in pro-inflammatory properties in macrophages. *FEBS Lett* 581: 349–354.

Kaneto H et al. 2004. Possible novel therapy for diabetes with cell-permeable JNK inhibitory peptide. *Nat Med* 10: 1128–1132.

Kannisto K et al. 2004. Overexpression of 11β-hydroxysteroid dehydrogenase-1 in adipose tissue is associated with acquired obesity and features of insulin resistance: studies in young adult monozygotic twins. *J Clin Endocrinol Metab* 89: 4414–4421.

Katz J et al. 1999. An in vivo study of the cortisol–cortisone shuttle in subcutaneous abdominal adipose tissue. *Clin Endocrinol* 50: 63–68.

Kelly IE et al. 1999. Effects of a thiazolidinedione compound on body fat and fat distribution of patients with type 2 diabetes. *Diabetes Care* 22: 288–293.

Kershaw E et al. 2005. Adipocyte-specific glucocorticoid inactivation protects against diet-induced obesity. *Diabetes* 54: 1023–1031.

Kim YM et al. 2007. Assessment of the anti-obesity effects of the TNP-470 analog, CKD-732. *J Mol Endocrinol* 38: 455–465.

Kissebah AH et al. 1982. Relation of body fat distribution to metabolic complications of obesity. *J Clin Endocrinol Metab* 54, 254–260.

Koo SH et al. 2004. PGC-1 promotes insulin resistance in liver through PPARα-dependent induction of TRB-3. *Nat Med* 10: 530–534.

Kotelevtsev Y et al. 1997. 11ß-hydroxysteroid dehydrogenase type 1 knockout mice show attenuated glucocorticoid inducible responses and resist hyperglycaemia on obesity or stress. *Proc Natl Acad Sci USA* 94: 14924–14929.

Lau DC et al. 2005. Adipokines: molecular links between obesity and atheroslcerosis. *Am J Physiol Heart Circ Physiol* 288: H2031–H2041.

Lavery GG et al. 2006. Hexose-6-phosphate dehydrogenase knockout mice lack 11β-hydroxysteroid dehydrogenase type 1-mediated glucocorticoid generation. *J Biol Chem* 281:5144–5145.

Lijnen HR. 2008. Angiogenesis and obesity. *Cardiovasc Res* 78: 286–293.

Lin Y et al. 2000. The lipopolysaccharide-activated Toll-like receptor (TLR)-4 induces synthesis of the closely related receptor TLR-2 in adipocytes. *J Biol Chem* 275: 24255–24263.

Lin Y et al. 2001. Hyperglycemia-induced production of acute phase reactants in adipose tissue. *J Biol Chem* 276: 42077–42083.

Lindsay RS et al. 2003. Subcutaneous adipose 11β-hydroxysteroid dehydrogenase type 1 activity and messenger ribonucleic acid levels are associated with adiposity and insulinemia in Pima Indians and Caucasians. *J Clin Endocrinol Metab* 88: 2738–2744.

Liu Y et al. 2008. Glucocorticoid response elements and 11β-hydroxysteroid dehydrogenases in the regulation of endothelial nitric oxide synthase expression. *Cardiovasc Res* (epub ahead of print).

Liu YJ et al. 2005. Increased glucocorticoid receptor and 11β-hydroxysteroid dehydrogenase type 1 expression in hepatocytes may contribute to the phenotype of type 2 diabetes in db/db mice. *Diabetes* 54: 32–40.

Livingstone DEW et al. 2000. Understanding the role of glucocorticoids in obesity: tissue-specific alterations of corticosterone metabolism in obese Zucker rats. *Endocrinology* 141: 560–563.

Mai K et al. 2007. Rosiglitazone decreases 11β-hydroxysteroid dehydrogenase type 1 in subcutaneous adipose tissue. *Clin Endocrinol* 67: 419–425.

Masuzaki H et al. 2001: A transgenic model of visceral obesity and the metabolic syndrome. *Science* 294: 2166–2170.

Masuzaki H et al. 2003. Transgenic amplification of glucocorticoid action in adipose tissue causes high blood pressure in mice. *J Clin Invest* 112: 83–90.

Michailidou Z et al. 2007. Omental fat 11β-HSD1, but not GR, is correlated with fat cell size independently of obesity. *Obesity* 15: 1155–1163.

Monder C and White PC. 1993. 11β-hydroxysteroid dehydrogenase. *Vit Horm* 47: 187–271.

Morton NM et al. 2005. A polygenic model of the metabolic syndrome with reduced circulating and intra-adipose glucocorticoid action. *Diabetes* 54: 3371–3378.

Morton NM and Seckl JR. 2008. 11β-hydroxysteroid dehydrogenase type 1 and obesity. *Front Horm Res* 36: 146–164.

Morton N et al. 2004a. Novel adipose tissue-mediated resistance to diet induced visceral obesity in 11β-hydroxysteroid dehydrogenase type 1-deficient mice. *Diabetes* 53: 931–938.

Morton N, Ramage L, and Seckl J. 2004b. Downregulation of adipose 11β-hydroxysteroid dehydrogenase type 1 by high fat feeding in mice: a potential adaptive mechanism counteracting metabolic disease. *Endocrinology* 145: 2707–2712.

Munck A and Náray-Fejes-Tóth A. 1992. The ups and downs of glucocorticoid physiology: permissive and suppressive effects revisited. *Mol Cell Endocrinol* 90: C1–C4.

Murano I et al. 2008. Dead adipocytes, detected as crown-like structures, are prevalent in visceral fat depots of genetically obese mice. *J Lipid Res* 49: 1562–1568.

Nair S et al. 2004. The 11β hydroxysteroid dehydrogenase type 1 gene: genetic polymorphisms are associated with type 2 diabetes in Pima indians independently of obesity and expression in adipocyte and muscle. *Diabetologia* 47: 1088–1095.

Napolitano A et al. 1998. 11β-hydroxysteroid dehydrogenase type 1 in adipocytes: expression is differentiation-dependent and hormonally regulated. *J Steroid Biochem Molec Biol* 64: 251–260.

Nara N et al. 2007. Disruption of CXC motif chemokine ligand 14 in mice ameliorates obesity-induced insulin resistance. *J Biol Chem* 282: 30794–30803.

Nguyen MT et al. 2007. A subpopulation of macrophages infiltrates hypertrophic adipose tissue and is activated by free fatty acids via Toll-like receptors 2 and 4 and JNK-dependent pathways. *J Biol Chem* 282: 35279–35292.

Odegaard JI et al. 2007. Macrophage-specific PPARγ controls alternative activation and improves insulin resistance. *Nature* 447: 1116–1120.

Olswang Y et al. 2003. Glucocorticoids repress transcription of phosphoenolpyruvate carboxykinase (GTP) gene in adipocytes by inhibiting its C/EBP-mediated activation. *J Biol Chem* 278: 12929–12936.

Otero M et al. 2006. Towards a pro-inflammatory and immunomodulatory emerging role of leptin. *Rheumatology* 45: 944–950.

Ozcan U et al. 2004. Endoplasmic reticulum stress links obesity, insulin action, and type 2 diabetes. *Science* 306: 457–461.

Papanicolaou DA et al. 1998. A single midnight serum cortisol measurement distinguishes Cushing's syndrome from pseudo-Cushing states. *J Clin Endocrinol Metab* 83: 1163–1167.

Paulsen SK et al. 2007. 11β-HSD type 1: impact of gender, obesity and fat localisation. *Obesity* 15: 1954– 1960.

Peixoto RA et al. 2008. Pre-adipocyte 11β-hydroxysteroid dehydrogenase type 1 is a ketoreductase and contributes to diet-induced visceral obesity in vivo. *Endocrinology* 149: 1861–1868.

Rask E et al. 2002. Tissue-specific changes in peripheral cortisol metabolism in obese women: increased adipose 11β-hydroxysteroid dehydrogenase type 1 activity. *J Clin Endocrinol Metab* 87: 3330–3336.

Rask E et al. 2001. Tissue-specific dysregulation of cortisol metabolism in human obesity. *J Clin Endocrinol Metab* 86: 1418–1421.

Rausch ME et al. 2008. Obesity in C57BL/6J mice is characterized by adipose tissue hypoxia and cytotoxic T cell infiltration. *Int J Obes* 32: 451–463.

Reichardt HM and Schütz G. 1998. Glucocorticoid signalling: multiple variations of a common theme. *Mol Cell Endocrinol* 146: 1–6.

Ricketts ML et al. 1998. Immunohistochemical localization of type 1 11β-hydroxysteroid dehydrogenase in human tissues. *J Clin Endocrinol Metab* 83: 1325–1335.

Robitaille J et al. 2004. Molecular screening of the 11β-HSD1 gene in men characterized by the metabolic syndrome. *Obesity Res* 12: 1570–1575.

Rotter V, Nagaev I, and Smith U. 2003. Interleukin-6 (IL-6) induces insulin resistance in 3T3-L1 adipocytes and is, like IL-8 and tumor necrosis factor-α, overexpressed in human fat cells from insulin-resistant subjects. *J Biol Chem* 278: 45777–45784.

Rupnick MA et al. 2002. Adipose tissue mass can be regulated through the vasculature. *Proc Natl Acad Sci USA* 99: 10730–10735.

Sai S et al. 2008. Glucocorticoid regulation of the promoter of 11β-hydroxysteroid dehydrogenase type 1 is indirect and requires CCAAT/enhancer-binding protein β. *Mol Endocrinol* 22: 2049–2060.

Sakoda H et al. 2000. Dexamethasone-induced insulin resistance in 3T3-L1 adipocytes is due to inhibition of glucose transport rather than insulin signal transduction. *Diabetes* 49: 1700–1708.

Sandeep TC et al. 2005. Increased in vivo regeneration of cortisol in adipose tissue in human obesity and effects of the 11β-hydroxysteroid dehydrogenase type 1 inhibitor carbenoxolone. *Diabetes* 54: 872– 879.

Sartipy P and Loskutoff DJ. 2003. Monocyte chemo-attractant protein 1 in obesity and insulin resistance. *Proc Natl Acad Sci USA* 100: 7265–7270.

Schweizer RA et al. 2004. Rapid hepatic metabolism of 7-ketocholesterol by 11β-hydroxysteroid dehydrogenase type 1: species-specific differences between the rat, human, and hamster enzyme. *J Biol Chem* 279: 18415–18424.

Seckl JR and Walker BR. 2001. 11β-Hydroxysteroid dehydrogenase type 1: a tissue-specific amplifier of glucocorticoid action. *Endocrinology* 142: 1371–1376.

Seckl J, et al. 2004. Glucocorticoids and 11β-hydroxysteroid dehydrogenase in adipose tissue. *Rec Prog Horm Res* 59: 359–393.

Sewter CP et al. 2002. Regional differences in the response of human pre-adipocytes to PPARγ and RXRα agonists. *Diabetes* 51: 718–723.

Shi H et al. 2006. TLR4 links innate immunity and fatty acid-induced insulin resistance. *J Clin Invest* 116: 3015–3025.

Slavin BG, Ong JM, and Kern PA. 1994. Hormonal regulation of hormone-sensitive lipase activity and mRNA levels in isolated rat adipocytes. *J Lipid Res* 35, 1535–1541.

Small GR et al. 2005. Preventing local regeneration of glucocorticoids by 11β-hydroxysteroid dehydrogenase type 1 enhances angiogenesis. *Proc Natl Acad Sci USA* 102: 12165–12170.

Song MJ et al. 2006. Activation of Toll-like receptor 4 is associated with insulin resistance in adipocytes. *Biochem Biophys Res Commun* 346: 739–745.

Steppan C et al. 2001. The hormone resistin links obesity to diabetes. *Nature* 409: 307–312.

Stimson RH et al. 2008 Cortisol release from adipose tissue by 11β-hydroxysteroid dehydrogenase type 1 in humans. *Diabetes* (epub ahead of print).

Stulnig TM et al. 2002. Liver X receptors downregulate 11β-hydroxysteroid dehydrogenase type 1 expression and activity. *Diabetes* 51: 2426–2433.

Tannin GM et al. 1991. Cloning and sequencing of the human cDNA for corticosteroid 11β-dehydrogenase: the human gene for 11β-hydroxysteroid dehydrogenase: structure, tissue distribution, and chromosomal localization. *J Biol Chem* 266: 16653–16658.

Thieringer R et al. 2001. 11β-Hydroxysteroid dehydrogenase type 1 is induced in human monocytes upon differentiation to macrophages. *J Immunol* 167: 30–35.

Tomlinson JW et al. 2002. Expression of 11β-hydroxysteroid dehydrogenase type 1 in adipose tissue is not increased in human obesity. *J Clin Endocrinol Metab* 87: 5630–5635.

Tontonoz P and Spiegelman BM. 2008. Fat and beyond: the diverse biology of PPARγ. *Annu Rev Biochem* 77: 289–312.

Tuncman G et al. 2006. Functional in vivo interactions between JNK1 and JNK2 isoforms in obesity and insulin resistance. *Proc Natl Acad Sci USA* 103: 10741–10746.

Uddén J, Eriksson P, and Hoffstedt J. 2002.Glucocorticoid-regulated adipose tissue secretion of PAI-1, but not IL-6, TNF-α or leptin in vivo. *Horm Metab Res* 34: 698–702.

Udden J, Folkesson R, and Hoffstedt J. 2001. Downregulation of uncoupling protein 2 mRNA in women treated with glucocorticoids. *Int J Obes* 25: 1615–1618.

Uysal KT et al. 1997. Protection from obesity-induced insulin resistance in mice lacking TNF-α function. *Nature* 389: 610–614.

Vats D et al. 2006. Oxidative metabolism and PGC-1β attenuate macrophage-mediated inflammation. *Cell Metab* 4: 13–24.

Wajchenberg BL. 2000. Subcutaneous and visceral adipose tissue: their relation to the metabolic syndrome. *Endocr Rev* 21: 697–738.

Wake DJ et al. 2003. Local and systemic impact of transcriptional upregulation of 11β-hydroxysteroid dehydrogenase type 1 in adipose tissue in human obesity. *J Clin Endocrinol Metab* 88: 3983–3988.

Walker BR et al. 1992. Mineralocorticoid excess and inhibition of 11β-hydroxysteroid dehydrogenase in patients with ectopic ACTH syndrome. *Clin Endocrinology* 37: 483–492.

Wamil M et al. 2008. 7-Oxysterols modulate glucocorticoid activity in adipocytes through competition for 11β-hydroxysteroid dehydrogenase. *Endocrinology* 149: 5909–5918.

Weisberg SP et al. 2003. Obesity is associated with macrophage accumulation in adipose tissue. *J Clin Invest* 112: 1796–1808.

Wellen KE and Hotamisligil GS. 2005. Inflammation, stress, and diabetes. *J Clin Invest* 115: 1111–1119.

Whitworth JA et al. 1989. The kidney is the major site of cortisone production in man. *Clin Endocrinol* 31: 255–361.

Whorwood CB et al. 1991. Tissue localization of 11β-hydroxysteroid dehydrogenase and its relationship to the glucocorticoid receptor. *J Steroid Biochem Molec Biol* 41: 21–28.

Williams LJS et al. 2000. C/EBPβ regulates hepatic transcription of 11ß-hydroxysteroid dehydrogenase type 1: a novel mechanism for cross-talk between the C/EBP and glucocorticoid signalling pathways. *J Biol Chem* 275: 30232–30239.

Wolf G. 1999. The molecular mechanism of the stimulation of adipocyte differentiation by a glucocorticoid. *Nutr Rev* 57: 324–326.

Wu H et al. 2007. T cell accumulation and regulated on activation, normal T cell expressed and secreted upregulation in adipose tissue in obesity. *Circulation* 115: 1029–1038.

Xu H et al. 2003. Chronic inflammation in fat plays a crucial role in the development of obesity-related insulin resistance. *J Clin Invest* 112: 1821–1830.

Xu HE et al. 1999. Molecular recognition of fatty acids by peroxisome proliferator-activated receptors. *Mol Cell* 3: 397–403.

Yacoub Wasef SZ et al. 2006. Glucose, dexamethasone, and the unfolded protein response regulate TRB3 mRNA expression in 3T3-L1 adipocytes and L6 myotubes. *Am J Physiol Endocrinol Metab* 291: E1274–E1280.

Yamauchi T et al. 2002. Adiponectin stimulates glucose utilization and fatty acid oxidation by activating AMP-activated protein kinase. *Nat Med* 8: 1288–1295.

Yamauchi T et al. 2001. The fat-derived hormone adiponectin reverses insulin resistance associated with both lipoatrophy and obesity. *Nat Med* 7: 941–946.

Ye J et al. 2007. Hypoxia is a potential risk factor for chronic inflammation and adiponectin reduction in adipose tissue of ob/ob and dietary obese mice. *Am J Physiol Endocrinol Metab* 293: E1118–E1128.

Yuan M et al. 2001. Reversal of obesity- and diet-induced insulin resistance with salicylates or targeted disruption of IKKβ. *Science* 293: 1673–1677. Erratum: 2002. *Science* 295: 277.

Zeyda M et al. 2007. Human adipose tissue macrophages are of an anti-inflammatory phenotype but capable of excessive pro-inflammatory mediator production. *Int J Obes* 31: 1420–1428.

Zhang TY and Daynes RA. 2007. Macrophages from 11β-hydroxysteroid dehydrogenase type 1-deficient mice exhibit an increased sensitivity to lipopolysaccharide stimulation due to TGF-β-mediated upregulation of SHIP1 expression. *J Immunol* 179: 6325–6335.

Zhang TY, Ding X, and Daynes RA. 2005. The expression of 11β-hydroxysteroid dehydrogenase type I by lymphocytes provides a novel means for intracrine regulation of glucocorticoid activities. *J Immunol* 174: 879–889.

# 9 Prostaglandins as Mediators of Adipose Inflammation

*Martha Lappas*

## CONTENTS



## 9.1 INTRODUCTION

Inflammation caused by obesity, especially central obesity, is thought to be the underlying basis of a significant number of diseases including cardiovascular disease, metabolic syndrome, hypertension, diabetes, hyperlipidemia, and cancer [Trayhurn, 2005]. The importance of adipose tissue in such epidemic diseases led pharmaceutical companies to target adipocyte metabolism in their searches for drugs for treating or reducing the risk of these conditions.

Adipose tissue, in addition to adipocytes and pre-adipocytes, contains stromal vascular cells that include pericytes, fibroblasts, monocytes, macrophages, and cells of the endothelium (endothelial and vascular smooth muscle cells). It has long been known that macrophages are bone-marrow derived and provide an important link between obesity and related inflammatory disorders. However, recent studies have shown that the adipocytes also demonstrate significant intrinsic inflammatory properties. For example, adipocytes contribute almost one third of the IL-6 concentration in the circulations of patients who are obese [Fantuzzi, 2005]. Further, it has been demonstrated that adipocytes are responsive to infectious and inflammatory signals [Lappas et al., 2004], with a downstream activation of multiple inflammatory signalling cascades [Lin et al., 2000; Rajala and Sherer, 2003] to induce expression and secretion of several adipokines, triacylglycerol, free albumin-bound non-esterified fatty acids, and inflammatory proteins [Kershaw and Flier, 2004; Trayhurn and Wood, 2004; Pittas et al., 2004; Frayn et al., 2005].

The metabolically active molecules released by adipose tissue may have effects on distant target tissues (e.g., liver, skeletal muscle, pancreas) and/or local paracrine effects in adipose tissues. For example, adipose tissue expresses the Toll-like lipopolysaccharide (LPS) receptor TLR4, and when stimulated with endotoxin, these receptors activate the nuclear factor-κB (NF-κB) signal transduction pathways [Berg et al., 2004]. In turn, these pathways induce the expression of inflammatory mediators such as IL-6, TNF-α, and prostaglandins [Fain et al., 2004; Lappas et al., 2004; Lappas et al., 2005a] that make significant contributions to systemic inflammation.

In this chapter, a brief overview of prostaglandins and the enzymes involved in their formation will be provided. A review of the current literature on the role of prostaglandins as modulators of inflammation, adipocyte differentiation, and lipolysis will also be presented. Studies that examined the association between adipose tissue derived prostaglandins and their relationship to inflammatory disorders will then be discussed.

## 9.2 PROSTAGLANDIN BIOSYNTHESIS

Prostaglandins, found in almost every tissue in humans and animals, are formed from polyunsaturated fatty acids, rapidly metabolized, and diverse in their effects. They can act in an autocrine fashion or endocrine/paracrine fashion, where they participate in physiological processes such as inflammation, immune response, kidney function, bone metabolism, ovulation, and adipocyte differentiation and function [reviewed in Dubois et al., 1998].

Prostaglandins are potent and ubiquitous lipid mediators derived enzymatically from the actions of multiple pathways involving both phospholipase (PLA2) and cyclooxygenase (COX) isozymes. Through the activity of one or more PLA2 enzymes, non-esterified arachidonic acid is released from membrane phospholipids such as phosphatidylinositol and phosphatidylethanolamine. The substrate arachidonic acid can then be metabolized through at least three different pathways: (1) the cyclooxygenase pathway leading to the formation of prostaglandins including prostacyclin (PGI2) and thromboxanes (TXs); (2) the lipoxygenase pathway, leading to the synthesis of leukotrienes (LTs) and hydroxyeicosatetraenoic acids (HETEs); and

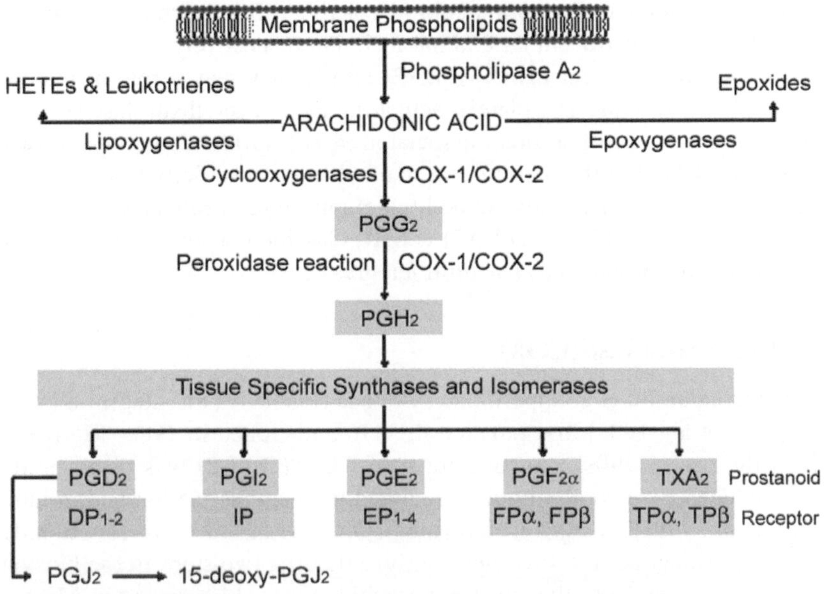


**FIGURE 9.1** Prostaglandin biosynthesis pathways.

(3) the epoxygenase pathway, leading to the formation of epoxides (see Figure 9.1). Cyclopentenone prostaglandins designated prostaglandin A2 (PGA2), PGA1, and PGJ2 are formed by dehydration within the cyclopentenone ring of PGE2, PGE1, and PGD2, respectively [reviewed in Straus and Glass, 2001].

Prostaglandins of the A and J series contain a cyclopentenone ring structure characterized by the presence of a chemically reactive α, β-unsaturated carbonyl. Cyclopentenone (A2/J2) isoprostanes (IsoPs) are prostaglandin (PG)-like compounds generated in vivo from the free radical-induced peroxidation of arachidonic acid. Studies definitively show that cyclopentenone IsoPs are formed in large amounts in vivo, and this is in marked contrast to cyclopentenone prostaglandins, for which little evidence exists that they are endogenously produced [Milne et al., 2005]. Recent studies have demonstrated that cyclopentenone IsoPs inhibit the inflammatory response [Musiek et al., 2005] and thus may serve as brakes to prevent excessive damage due to inflammation.

### 9.2.1 Phospholipase A2 (PLA2) Isozymes

PLA2 represents a ubiquitous family of esterases that hydrolyze the *sn*-2 acyl ester bonds of 1,2 diacyl-*sn*-3 glycerophospholipids, thereby liberating equimolar amounts of 1-acyl lysophosphatide and free fatty acid [Flower and Blackwell, 1976]. Mammalian cells contain several structurally different PLA2 enzymes that exhibit distinct localization, function, and mechanisms of regulation [reviewed in Lappas and Rice, 2004; Schaloske and Dennis, 2006]. The secretory PLA2 (sPLA2) and cytosolic PLA2 (cPLA2) enzymes display different biophysical characteristics and biochemical requirements for optimal enzyme activity that are consistent with their sites of action, that is extracellular and intracellular, respectively.

The calcium-independent PLA2 (iPLA2) family may play a role in membrane phospholipid remodelling. The platelet activating factor acetylhydrolase (PAF-AH), otherwise known as the lipoprotein-associated PLA2 (lp-PLA2) family, consists of a large group of PLA2s that exhibit unusual substrate specificity toward PAF and oxidized phospholipids. Recently, an additional family has been added; it is categorized as the lysosomal PLA2 (LPLA2) cohort, classified according to its catalytic mechanism and functional and structural features.

### 9.2.2 Cyclooxygenase (COX)

COX or prostaglandin H endoperoxide synthase (PGHS) is an integral membrane protein [Otto et al., 1993] first purified in 1976 and cloned in 1988 [Merlie et al., 1988]. In the early 1990s, a second form of COX (COX-2) was discovered and could be induced by several stimuli associated with cell activation and inflammation [Xie et al., 1991]. A third isoenzyme, COX-3, is of unknown function. COX enzymes are bifunctional in that they catalyze the first two steps in the biosynthesis of prostaglandins from the substrate arachidonic acid [reviewed in Marnett et al., 1999].

The cyclooxygenase site oxidizes arachidonic acid to the unstable intermediate hydroperoxy endoperoxide (PGG2). This process involves the incorporation of two molecules of oxygen. The hydroperoxidase site of COX reduces the 15-hydroperoxyl group of PGG2 to the hydroxyl endoperoxide (PGH2) [Ohki, 1979].

### 9.2.3 Tissue-Specific Synthases

PGH2 is transformed by a range of enzymatic and non-enzymatic mechanisms into the primary prostanoids: PGE2, PGF2α, PGD2, PGI2, and TxA2. The enzymes for coupling PGH2 synthesis to downstream metabolism include two types of PGD synthase (PGDS) (lipocalin-type, L-PGDS and hematopoietic type, H-PGDS), along with microsomal PGE synthase (mPGES)-1 and -2, cytosolic PGE synthase (cPGES), prostacyclin synthase, PGF synthase, and thromboxane synthase.

## 9.3 MECHANISMS OF PROSTAGLANDIN ACTION

### 9.3.1 G-Protein Coupled Transmembrane Receptors

After prostaglandins are formed, they are secreted from cells via a carrier-mediated process. E, D, I, and F series prostaglandins act via G-protein coupled transmembrane receptors [Tsuboi et al., 2002]. Receptors for prostaglandins can be divided into five main types: EP (including the four sub-sets of EP1, EP2, EP3, and EP4), FP, DP, IP, and TP. The first letter corresponds to the type of prostaglandin that is a major ligand, i.e., EP for PGE2 and FP for PGF2α, DP for PGD2, IP for PGI2, and TP for TXA2.

Eicosanoid receptor expression is hormonally regulated. EP1 and EP3 receptors initiate smooth muscle contraction through mechanisms that include calcium mobilization and inhibition of cyclic AMP (cAMP). EP2 and EP4 receptors act through increased cAMP formation and relax smooth muscle. FP receptors result in elevated intracellular free calcium and smooth muscle contraction, whereas IP receptors increase cAMP, resulting in relaxation.

### 9.3.2 Nuclear Protein Receptors and Transcription Factors

The cyclopentenone prostaglandins interact with other specific cellular targets including nuclear protein receptors and transcription factors via covalent modification of specific cysteine residues in the DNA binding sites of these proteins. For instance, 15d-PGJ2 is a high affinity ligand for the peroxisome proliferator-activated receptor-γ (PPARγ) and is predominantly found in adipose tissue [Kliewer et al., 1995]. Other activities of the cyclopentenone prostaglandins are mediated by the reactive α,β-unsaturated carbonyl group located in the cyclopentenone ring. For example, 15d-PGJ2 attenuates the activation of the transcription factor NF-κB by preventing the phosphorylation of its inhibitor protein [Rossi et al., 1997].

## 9.4 PROSTAGLANDIN PRODUCTION BY ADIPOSE TISSUE

It has long been known that prostanoids are released by rat adipose tissue [Shaw and Ramwell, 1968], although it was thought that they were made by the non-fat cells of rat adipose tissue [Parker et al., 1989; Chatzipanteli et al., 1992]. However, subsequent studies have clearly shown that freshly isolated rat and human adipocytes both secrete prostaglandins, especially PGE2 and PGI2 [Richelsen, 1987; Richelsen, 1992; Richelsen et al., 1992; Fain et al., 2001b; Fain et al., 2002], with visceral adipose tissue releasing more PGE2 than subcutaneous adipose tissue [Fain et al., 2004]. Cells isolated after digestion of subcutaneous and/or visceral adipose tissue do produce less PGE2 or PGI2 than undigested tissue debris [Fain et al., 2002; Fain et al., 2004], but there is significant upregulation over time (between 4 and 48 hours) in PGE2 release in adipocytes but not in adipose tissue [Fain et al., 2004] which is also associated with an increase in COX-2 immunoreactive protein expression [Fain et al., 2002]. PGE2 is catalyzed by 15-hydroxyprostaglandin dehydrogenase (PGDH) to generate 15-keto-PGE2, which is, in turn, further catabolized by 15-oxoprostaglandin-$\Delta^{13}$-reductase (PGR-2) to 13,14-dihydro-15-keto-PGE2 [Chou, et al., 2007].

Because high activities of both PGDH and PGR-2 have been detected in adipose tissue [Anggard et al., 1971], PGE2 catabolism is thought to be highly active in adipocytes. Indeed, PGE2 is the most abundant prostaglandin produced in the 3T3-L1 pre-adipocyte cell line [Hyman et al., 1982], pre-adipocytes [Fain et al., 2004], adipocytes [Bell-Parikh et al., 2003], and adipose tissue [Richelsen, 1992; Fain et al., 2004]. The secretion of 15d-PGJ2 by 3T3-L1 adipocytes [Bell-Parikh et al., 2003] and PGF2α by pre-adipocytes [Yu et al., 1995] has also been reported.

All the key enzymes in PGD and PGJ2 synthesis are expressed in both pre-adipocytes and subcutaneous and omental human adipose tissue, with COX-1 and PGDS more highly expressed in omental adipose tissue [Jowsey et al., 2003; Quinkler et al., 2006]. High levels of mPGES-1 are expressed in adipose tissue and adipocytes, with epididymal (visceral) levels of mPGES-1 being higher than inguinal (subcutaneous, abdominal) adipose tissue. Furthermore, protein expression of mPGES-1 is also detected in undifferentiated and differentiated mouse 3T3-L1 adipocytes and in human primary subcutaneous pre-adipocytes at all stages of adipogenesis. Adipose tissue also expresses low levels of mPGES-2 and cPGES [Hétu et al., 2007]. A recent study identified and characterized a membrane-associated intracellular calcium-dependent, adipose-specific PLA2 named adipose-specific PLA2 (AdPLA) [Duncan et al., 2008].

## 9.5 MULTIFACETED ROLES OF PROSTAGLANDINS

In addition to the regulation of whole body energy homeostasis, adipocytes are known to fulfill important endocrine functions and may serve as the link between obesity and inflammation. For that reason, determining the mechanisms that underlie adipocyte differentiation and function has become an area of intense investigation.

As discussed above, mature adipocytes and cultured pre-adipocytes produce significant amounts of prostaglandins, and several lines of evidence indicate that these prostaglandins may play an important physiological role in adipose tissue metabolism and development. In the following sections, the regulation of adipokines by prostaglandins, which may be important in the inflammatory response in adipose tissue, and the role of prostaglandins in adipocyte differentiation and lipolysis will be discussed.

### 9.5.1 Regulation of Adipokines by Prostaglandins in Adipose Tissue

Prostaglandins may be important components of the inflammatory response within adipose tissue, particularly in obesity, by potentially influencing the production of key inflammation-related adipokines. PGD2 and the J2 series prostaglandins induce reductions in adiponectin and leptin mRNA expression and release. In contrast, PGD2 induced a marked stimulation of IL-6 and MCP-1 mRNA expression and release [Peeraully et al., 2006], and treatment with the natural PPARγ ligand PGJ2 reduced TNF-α expression levels in retroperitoneal and mesenteric white adipose tissues of obese rats [Okuno et al., 1998].

On the other hand, exogenous PGE2 can stimulate leptin release by mouse adipose tissue when the basal formation of PGE2 is blocked by dexamethasone [Fain et al., 2000], and arachidonic acid or PGE2 stimulates leptin release by subcutaneous adipose tissue explants from obese humans. The stimulatory effect of arachidonic acid on leptin formation was blocked by NS-398, a COX-2 inhibitor [Fain et al., 2001b]. Nerve growth factor (NGF), an inflammatory response protein secreted by adipocytes, is dramatically upregulated by prostaglandin PGD2 and by the J series prostaglandins, PGJ2 and delta12-PGJ2 [Bulló, et al., 2005].

### 9.5.2 Role of Prostaglandins in Adipocyte Differentiation and Maturation

Adipocyte differentiation is a complex and staged process [Fajas et al., 1998]. The differentiation of pre-adipocyte fibroblasts to adipocytes is a crucial process to many disease states including obesity, diabetes, cardiovascular, and autoimmune diseases. Indeed, obesity is due to the hypertrophy of adipocytes (abnormal increase in adipose tissue mass due to the excessive accumulation of triglycerides in adipocytes) and to the recruitment of new adipocytes from precursor cells, two processes largely dependent on regulation of adipocyte differentiation. Manufactured and secreted by pre-adipocytes and adipocytes, prostaglandins are complex regulators of adipocyte differentiation, mediated by cell surface and nuclear receptors. While the exact function of each prostaglandin in the adipocyte is not completely understood, reports indicate that prostaglandins exert both positive and negative influences over pre-adipocyte conversion to triglyceride-storing adipocytes.

PGI2, and its stable analogue carbacyclin (cPGI2), exclusively affect adipose precursor cells where they behave as adipogenic–hyperplastic effectors, leading to an increase in the number of adipocytes from dormant pre-adipocytes. PGI2 promotes and/or amplifies terminal differentiation of cultured pre-adipocytes by inducing pre-adipocyte intracellular increases of both cAMP and free calcium [Vassaux et al., 1992a; Vassaux et al., 1993; Darimont et al., 1994]. Moreover, PGI2 increases the expression of both C/EBP$\beta$ and C/EBP$\delta$ in pre-adipose cells [Aubert et al., 2000].

The paracrine adipogenic effect of PGI2 has also been reported to be controlled by angiotensin II. Ob1771 adipose cells challenged with angiotensin II produce PGI2 which can then induce pre-adipose cells to differentiate into adipose cells [Darimont et al., 1994]. In contrast, PGI2 is unable to trigger any signal transduction in differentiated adipocytes [Vassaux et al., 1992a] consistent with the decrease in the expression of the cell surface IP receptor. In addition to the biological effects of PGI2 mediated by its cell-surface receptor, its ability to promote differentiation may also be mediated by PPAR$\gamma$ [Brun et al., 1996].

A greater degree of adipocyte differentiation was observed in cultures from obese rats (compared to lean rats) and was associated with lower basal PGE2 synthesis [Gaskins et al., 1989]. PGE2 has been reported to either activate [Shillabeer et al., 1998] or inhibit [Lepak and Serrero, 1993] early adipogenesis. However, more recent studies have shown that PGE2 stimulates proliferation of 3T3-L1 cells induced to differentiate, suggesting that PGE2 positively influences adipocyte differentiation through triggering the clonal expansion phase [Fajas et al., 2003]. Further recent studies have shown that 15-keto-PGE2, which is catabolized from PGE2, enhances adipogenesis of 3T3-L1 cells via PPAR$\gamma$ [Chou et al., 2007].

Other prostaglandins that influence fat cell development include PGF2$\alpha$ which inhibits differentiation of various pre-adipose cell lines including 3T3-L1, 3T3-F442A, Ob1771, and rat primary pre-adipocyte cells [Catalioto et al., 1991; Miller et al., 1996; Casimir et al., 1996; Gaillard et al., 1989]. One potential mechanism by which PGF2$\alpha$ inhibits adipogenesis is through stimulated phosphorylation of PPAR$\gamma$ by activated MAPK, leading to abrogation of its transactivating activity [Hu et al., 1996; Reginato et al., 1998]. A different mode of action for PGF2$\alpha$ has also been proposed for primary rat pre-adipocytes. In both undifferentiated and differentiated cells, PGF2α stimulates mRNA expression and production of TGF-β, which is also an inhibitor of adipocyte differentiation [Lepak and Serreno, 1995]. Stimulation of FP prostanoid receptors is associated with the inhibition of differentiation in 3T3-L1 cells and primary rat pre-adipocytes [Casimir et al., 1996; Serreno and Lepak, 1997].

Controversy surrounds the role of PGJ2 and its metabolites in adipocyte differentiation. Early studies showed that the PGD2 metabolite 15d-PGJ2 promotes terminal adipocyte differentiation by increasing the activity of the adipogenic transcription factor PPARγ [Forman et al., 1995; Adams et al., 1997; Kliewer et al., 1995; Smith et al., 2002], and this is associated with increase in COX-2 expression [Inuzuka et al., 1999]. However, more recent studies show that although 15d-PGJ2 is formed by 3T3-L1 cells, it is unaltered during adipocyte differentiation, and suppression of its formation by COX inhibition fails to influence differentiation [Bell-Parikh et al., 2003]. Furthermore, although exogenous added 15d-PGJ2 can drive adipocyte maturation by ligating PPARγ, the concentrations required exceed those of the endogenous prostaglandin released into the medium and the intracellular levels. Other studies have shown that PGA2, PGD2, PGJ2, and TXB2 have no effects on the adipocyte differentiation process in 3T3-L1 [Fajas et al., 2003].

The generation of prostaglandins contributing to pre-adipocyte differentiation is mediated in part by PLA2 and COX enzymes and tissue-specific synthases. Epididymal fat pad weight, the sizes of adipocytes, and serum levels of PGE2 are reduced in group IVA PLA2-deficient mice compared to the wild-type mice suggesting the involvement of the enzyme in the storage of lipids in the adipose tissue [Ii et al., 2008]. A very recent study demonstrated that AdPLA, which is highly expressed in white adipose tissue, is induced during pre-adipocyte differentiation to adipocytes [Duncan et al., 2008]. Collectively, this suggests that PLA2 contributes to the storage of lipids in mature adipocytes, namely hypertrophy of the cells, probably through the generation of prostaglandins including PGE2.

COX-2 plays an important role in early adipocyte differentiation, regulating entry into the cell cycle, whereas its role in terminal differentiation is dispensable [Fajas et al., 2003]. Others have shown that COX-1 and COX-2 negatively modulate adipose cell differentiation, and inhibition of COX-2 alleviated TNF-α-dependent inhibition of adipocyte differentiation [Yan et al., 2003]. In addition, arachidonic acid-dependent inhibition of adipocyte differentiation was associated with sustained COX-2 expression that was reversed by a selective COX inhibitor [Petersen et al., 2003].

Upon differentiation of 3T3-L1 cells, there is a downregulation in COX-1 and COX-2 mRNA and protein expression, whereby COX-1 and COX-2 became intracellularly more diffuse upon differentiation [Xie et al., 2006]. This is associated with a concomitant decrease in PGE2, 6-keto PGF1α, and PGD2. However, the role of COX in adipocyte differentiation has been a matter of dispute [MacDougald and Lane, 1995]. Perhaps the finding that some of the commonly used COX inhibitors are also PPARγ agonists may explain, at least in part, why conflicting results were obtained [Lehmann et al., 1997].

Conflicting data concern the expression of mPGES-1 during differentiation, with an increase and nuclear redistribution of mPGES-1 protein expression (compared with actin) during differentiation of mouse 3T3-L1 adipocytes reported [Xie et al., 2006]. Others found no changes in protein expression of pre-adipocytes and mature adipocytes [Hétu and Riendeau, 2007]. The expression of PGR-2 [Chou et al., 2007] and L-PGDS [Xie et al., 2006] is upregulated in the late phase of 3T3-L1 adipocyte differentiation. Mice with a knockout of the gene encoding for PGDS exhibited adipocyte hypertrophy independent of diet [Ragolia et al., 2005]. This is in keeping with the observation that the expression of PGDS is almost 20 times higher in omental than subcutaneous adipocytes and that omental adipocytes are smaller than subcutaneous adipocytes [Quinkler et al., 2006].

Adipogenesis by prostaglandins can be controlled by a number of inflammatory mediators including NF-κB and adiponectin. Indeed, these inflammatory signalling pathways are differentially regulated during adipogenesis. For example, NF-κB expression and activity are elevated in fully differentiated adipocytes [Berg et al., 2004]. The induction of NF-κB suggests an immunomodulatory switch and activation of inflammatory functions including PLA2, COX-2, and PGE2 formation [Lappas et al., 2005a]. Adiponectin, which is present within normal bone marrow, can inhibit fat cell formation by marrow-derived pre-adipocytes via induction of COX-2 gene expression and secretion of PGE2 [Yokota et al., 2002]. Indeed we have previously shown that adiponectin stimulates PGE2 release from human adipose tissue [Lappas et al., 2005b].

### 9.5.3 Role of Prostaglandins in Adipose Tissue Lipolysis

Lipolysis, the breakdown of fat stored in fat cells, is an important pathophysiological factor behind insulin resistance and the associated metabolic abnormalities observed in obese subjects. During this process, free fatty acids are released into the bloodstream and circulate throughout the body. The excessive accumulation of triglycerides in adipocytes is associated with an abnormal increase in adipose tissue mass [Kershaw and Flier, 2004]. Very early studies suggested that prostaglandins in part regulated lipolysis in adipose tissue [Shaw and Ramwell, 1968; Illiano and Cuatrecasas, 1971]. More recent studies have been able to unravel the complex and often opposing roles of prostaglandins in adipose tissue lipolysis. For example, it has been suggested that PGI2 and PGE2 have separate target cells in adipose tissue and appear to act in concert rather than in an opposite manner, controlling both hyperplastic and hypertrophic development [Vassaux et al., 1992b].

In mature adipocytes, PGE2, via interaction with its specific receptor, suppresses the production of cAMP (which induces lipolysis through the activation of hormone-sensitive lipase), thus inhibiting lipolysis and contributing to the maintenance of high intracellular triacylglycerol content [Vassaux et al., 1992b; Chatzipanteli et al., 1992; Mater et al., 1998]. Similarly, PGE2 inhibits basal lipolysis over 24 hour incubation of mouse adipose tissue in primary culture [Fain et al., 2000]. Low concentrations of PGE1 and PGE2 inhibit glycerol production in adipose tissue and counteract the stimulation of glycerol release induced by catecholamines and glucagon [Steinberg et al., 1964]. This is due to interference with the activation of tissue lipase usually produced by exposure of adipose tissue to these hormones.

## 9.6 PROSTAGLANDINS, ADIPOSE TISSUE, INFLAMMATION, AND DISEASE

### 9.6.1 Obesity and Related Metabolic Disorders

Adipose tissue in obesity undergoes changes in cell size that alter its normal physiological function. This altered adipocyte function is associated with chronic and systemic inflammatory responses characterized by abnormal adipokine production (e.g., leptin and adiponectin), increased synthesis of acute phase reactants (e.g., C-reactive protein), and the activation of pro-inflammatory signalling pathways. Adipose tissue is highly integrated into overall physiological regulation through cross-talk with other organs and multiple metabolic systems. Indeed, obesity predisposes individuals to increased risks of developing many diseases including atherosclerosis, type 2 diabetes, cardiovascular disease, insulin resistance, and metabolic syndrome. In the ensuing sections, a review of the current literature on the role of prostaglandins in obesity and associated metabolic disorders will be presented.

The levels of PGE2 and mPGES-1 protein are reduced in epididymal and inguinal adipose tissue in C57Bl/6 mice subjected to a high fat diet for 12 weeks compared to their lean counterparts [Hétu and Riendeau, 2007]. Expression of mPGES-2 and cPGES in epididymal adipose tissue was significantly elevated in obese tissues, possibly reflecting a partial compensation for the decrease of mPGES-1. The mPGES-2 and cPGES expression levels, however, remained relatively constant in inguinal adipose tissue. These data indicate that mPGES-1 plays a role in the regulation of PGE2 synthesis in the adipose tissue, and that its downregulation may be involved in the alterations of lipolysis and adipogenesis associated with obesity.

Similarly, adipocytes isolated from human subcutaneous and visceral white adipose tissues of morbidly obese individuals with BMI levels of 45 have a tendency to release lower amounts of PGE2 compared with their release by fat from individuals with BMIs of 32 (but this was not significant) [Fain et al., 2004]. Another observation was an overall decrease release rate of PGE2 by adipocytes isolated from obese Zucker rats (*fa/fa*) [Gaskins et al., 1989]. Moreover, caloric restriction-induced weight loss decreased the expression of EP3 in adipose tissues of obese subjects and led to the concomitant increased expression of molecules with anti-inflammatory properties [Clément et al., 2004]. However, others noted no significant differences in prostaglandins of the E-type and the F-type due to nutritional status or body build in fed-and-starved lean, normal, and obese women [Curtis-Prior et al., 1979].

A few studies reported an association with increased BMI and sPLA2 levels. In Pima Indians, a population with a very high prevalence of obesity and insulin resistance, sPLA2 is positively correlated with percent body fat, BMI, fasting plasma insulin concentration, and 2-hour glucose levels in non-diabetic individuals [Weyer et al., 2002]. Plasma sPLA2-IIA levels were positively correlated with obesity indices including visceral adipose tissue, with levels about 30% higher among men characterized by a higher accumulation of visceral adipose tissue [Paradis et al., 2006]. Plasma level of sPLA2-IIA correlated with that of C-reactive protein and served as an independent risk factor and predictor of coronary heart disease [Kugiyama et al., 1999].

Overweight and obese school children are also positively associated with Lp-PLA2 concentrations [Nagel et al., 2007]. In keeping with this, mice with PLA2 null mutations exhibit reduced adipose tissue and increased insulin sensitivity [Huggins et al., 2002]. In subcutaneous adipose tissue, COX-1 mRNA expression increased with BMI [Quinkler et al., 2006]. Heterozygous COX-2 mice develop obesity which is not secondary to a defect in leptin release by adipose tissue. It has been shown that homozygous COX-2$^{-/-}$ mice have decreased body weight and basal adipose tissue release of PGE2 or 6-keto-PGF1α compared to heterozygous COX-2$^{+/-}$ mice. NS-398, a specific COX-2 inhibitor, inhibited leptin release by adipose tissue from control, COX-1$^{-/-}$ and COX-2$^{+/-}$ mice, but had no effect on leptin release by adipose tissue from COX-2$^{-/-}$ mice [Fain et al., 2001a]. Two German study populations showed that the His variant of the *COX-2* Arg298His polymorphism is associated with reduced risk of type 2 diabetes [Nitz et al., 2007] and this may in part mediated by lowered BMI [Lindner et al., 2007].

Recent studies have shown that the fat cells surrounding coronary arteries may play a central and previously unrecognized role in the development of cardiovascular disease through a direct role in the pathogenesis of the atherosclerosis. Epicardial adipocytes have been shown to produce inflammatory mediators including substantial COX-2 activity. PGE2, the major COX-2 metabolite, has been shown to play an important part in angiogenesis and inflammation [Stoll et al., 2006].

Gestational diabetes mellitus (GDM) is considered a prediabetic state [Pendergrass et al., 1995], and the pathophysiologies of both conditions are clearly related. Based on the role of sPLA2 and COX-2 in the etiology of insulin resistance, preliminary data from our laboratory demonstrate increased mRNA expression of sPLA2-IIA and COX-2 in subcutaneous adipose tissue from women with GDM when compared to normal pregnant women (Figure 9.2). However, we noted no difference in the mRNA expression of cPLA2 and COX-1. Thus, prostaglandins may also play a role in the inflammatory response associated with GDM.

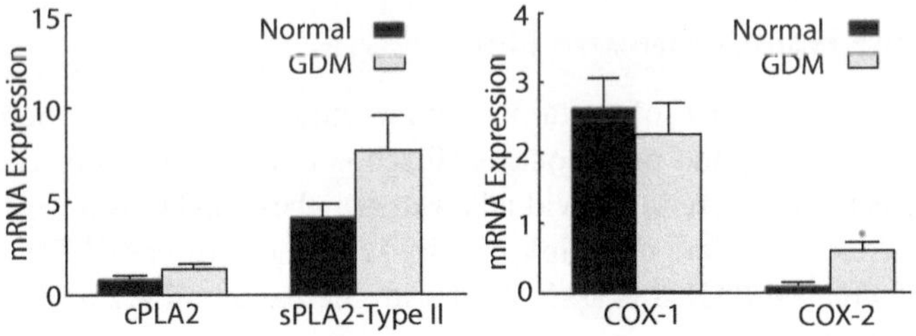


**FIGURE 9.2** Phospholipase A2 (PLA2) and cyclooxygenase (COX) expression in normal pregnant women and women with gestational diabetes mellitus (GDM). PLA2 and COX mRNA expression levels relative to β-actin in subcutaneous adipose tissue derived from normal pregnant women and pregnant women with gestational diabetes. Each bar represents the mean expression of PLA2 or COX relative to β-actin mRNA. * = $p < 0.05$ versus normal adipose tissue (Student's t-test).

### 9.6.2 Cancer

Epidemiological studies indicate that obesity is a significant risk factor for the development of cancer, although the exact mechanisms have not yet been identified. Recent years have seen increased interest in the role of prostaglandins in cancer. Upregulation of the inducible isoform COX-2 has been identified in many human cancers including human breast carcinoma in which COX-2 overexpression has also been detected in approximately 40% of cases. Furthermore, epidemiologic studies show a protective effect of COX inhibitory drugs with respect to both colon and breast cancer [Williams et al., 1999].

The synthesis of estrogens by cytochrome P450 aromatase (aromatase), a product of the CYP19 gene, is an important characteristic of breast carcinogenesis. Disproportionately high aromatase expression and activity in undifferentiated breast adipose fibroblasts adjacent to malignant epithelial cells likely contributes to cancer development and progression. In human breast adipose tissue, PGE2 increases estrogen biosynthesis, via the enzyme aromatase, thus possibly leading to increased risk of developing breast cancer.

Malignant breast epithelial cells secrete PGE2, which, via EP2 and EP4, stimulates aromatase expression in a cAMP- and PKC-dependent manner in adjacent breast adipose fibroblasts, leading to increased local concentrations of estrogen in addition to reducing BRCA1 expression (a tumor suppressor gene) [Chen et al., 2007; Subbaramaiah et al., 2008]. Prostaglandin-mediated estrogen overproduction may be an important site-specific consequence of COX-2 upregulation in breast cancer. In support of this, a positive correlation exists between aromatase and COX expression in human breast cancer specimens [Brueggemeier et al., 1999], and increased aromatase mRNA and activity and reduced amounts of BRCA1 are present in a mouse model overexpressing COX-2 in the mammary gland [Subbaramaiah et al., 2008]. On the other hand, 15d-PGJ2 inhibits aromatase expression in breast adipose mesenchymal cells [Winnett et al., 2003]. Thus, PGE2 and 15d-PGJ2 may have reciprocal roles to play in the regulation of aromatase expression in the breast.

### 9.6.3 Other Immune-Mediated Diseases

In humans, the prevalence of certain immune-mediated diseases such as asthma [Mannino et al., 2006] and periodontitis [Pischon et al., 2007] is increased or disease activity is more severe in individuals who are obese and thus it is proposed that adipocyte-secreted proteins may link obesity with these diseases. For example, in asthmatics, increased BMI was associated with increased sPLA2 activity [Misso et al., 2008].

Recent evidence suggests that prostaglandins released from adipose tissue may play a role in aging. Visceral adipose tissues of old mice exhibited higher mRNA expression COX-2 compared with tissues of young mice. Further, adipocytes, but not stromal vascular cells, from visceral adipose tissues of old mice produced more PGE2 than those from young mice [Wu et al., 2007]. This may explain why the incidence of type 2 diabetes increases with age.

In Graves' disease, the orbit of the eye can become severely inflamed and infiltrated with T lymphocytes as part of the autoimmune process. Inflammation and adipogenesis are two parallel processes associated with increased activity in severe Graves' ophthalmopathy. The orbital fibroblasts convert to fat-like cells causing the eye to protrude, which is disfiguring and can lead to blindness. It has been shown that human peripheral blood T cells express COX-2 and produce PGD2 and members of the PGJ family of prostaglandins including 15d-PGJ2. These products convert pre-adipocyte orbital fibroblasts to adipocytes *in vitro* and this adipogenic differentiation can be blocked by COX-2 selective inhibitors [Feldon et al., 2006].

## 9.7 CONCLUSIONS

An increase in adipocyte size, as a result of lipid accumulation, and increased adipocyte numbers, as a result of accelerated adipocyte differentiation and maturation are major characteristics of obesity. Dysregulated adipogenesis leads to an increase in the secretion of pro-inflammatory products from adipose tissue, thus providing a link between subclinical inflammation to obesity and associated pathologies. The levels of arachidonic acid, which is derived from dietary essential fatty acids, are high in obesity and diabetic states, and serve as precursors for prostaglandins that act via cell surface G-protein coupled receptors and nuclear protein receptors. Prostaglandins play a number of roles in adipose tissues, including modulators of inflammation, adipocyte differentiation, and lipolysis. Understanding phospholipid metabolism in adipose tissue is critical and may open new opportunities for the treatment of obesity and its related metabolic diseases.

## ACKNOWLEDGMENTS

The author is the recipient of an R.D. Wright Fellowship (Grant 454777) from the National Health and Medical Research Council (NHMRC) of Australia. Our work in the area has largely been funded by the NHMRC (Project Grant 454310), Diabetes Australia Research Trust, and the Medical Research Foundation for Women and Babies. Michelle Ann Colomiere of the Department of Obstetrics and Gynaecology, University of Melbourne, Mercy Hospital for Women is acknowledged for graciously supplying the data for Figure 9.2.

## REFERENCES

Adams M et al. Activators of peroxisome proliferator-activated receptor γ have depot-specific effects in human preadipocyte differentiation. *J Clin Invest* 100: 3149–3153, 1997.

Anggard E, Larsson C, and Samuelsson B. The distribution of 15-hydroxy prostaglandin dehydrogenase and prostaglandin δ13-reductase in tissues of the swine. *Acta Physiol Scand* 81: 396–404, 1971.

Aubert J et al. Prostacyclin IP receptor upregulates the early expression of C/EBPβ and C/EBPδ in preadipose cells. *Mol Cell Endocrinol* 160: 149–156, 2000.

Bell-Parikh LC et al. Biosynthesis of 15-deoxy-δ12,14-PGJ2 and the ligation of PPARγ. *J Clin Invest* 112: 945–955, 2003.

Berg AH et al. Adipocyte differentiation induces dynamic changes in NF-κB expression and activity. *Am J Physiol Endocrinol Metab* 287: E1178–E1188, 2004.

Brueggemeier RW et al. Correlation of aromatase and cyclooxygenase gene expression in human breast cancer specimens. *Cancer Lett* 140: 27–35, 1999.

Brun RP et al. Differential activation of adipogenesis by multiple PPAR isoforms. *Genes Dev* 10: 974–984, 1996.

Bulló MB, Peeraully MR, and Trayhurn P. Stimulation of NGF expression and secretion in 3T3-L1 adipocytes by prostaglandins PGD2, PGJ2 and γ12-PGJ2. *Am J Physiol Endocrinol Metab* 289: E62–E67, 2005.

Casimir DA, Miller CW, and Ntambi JM. Preadipocyte differentiation blocked by prostaglandin stimulation of prostanoid FP2 receptor in murine 3T3–L1 cells. *Differentiation* 60: 203–210, 1996.

Catalioto RM et al. Autocrine control of adipose cell differentiation by prostacyclin and PGF2α. *Biochim Biophys Acta* 1091: 364–369, 1991.

Chatzipanteli K, Rudolph S, and Axelrod L. Coordinate control of lipolysis by prostaglandin E2 and prostacyclin in rat adipose tissue. *Diabetes* 41: 927–935, 1992.

Chen D et al. Prostaglandin E2 induces breast cancer-related aromatase promoters via activation of p38 and c-jun NH2-terminal kinase in adipose fibroblasts. *Cancer Res* 67: 8914–8922, 2007.

Chou W-L et al. Identification of a novel prostaglandin reductase reveals the involvement of prostaglandin E2 catabolism in regulation of peroxisome proliferator-activated receptor activation. *J Biol Chem* 282: 18162–18172, 2007.

Clément K et al. Weight loss regulates inflammation-related genes in white adipose tissue of obese subjects. *FASEB J* 18: 1657–1669, 2004.

Curtis-Prior PB et al. Plasma prostaglandin levels in fed and starved lean, normal and obese women. *J Cell Mol Life Sci* 35: 911–912, 1979.

Darimont C et al. Differentiation of preadipose cells: paracrine role of prostacyclin upon stimulation of adipose cells by angiotensin-II. *Endocrinology* 135: 2030–2036, 1994.

Dubois RN et al. Cyclooxygenase in biology and disease. *FASEB J* 12: 1063–1073, 1998.

Duncan RE et al. Identification and functional characterization of adipose-specific phospholipase A2 (AsPLA). *J Biol Chem* 283: 25428–25436, 2008.

Fain JN et al. Regulation of leptin release and lipolysis by PGE2 in rat adipose tissue. *Prostagl Other Lipid Mediat* 62: 343–350, 2000.

Fain JN, Ballou LR, and Bahouth SW. Obesity is induced in mice heterozygous for cyclooxygenase-2. *Prostagl Other Lipid Mediat* 65: 199–209, 2001a.

Fain JN et al. Stimulation of leptin release by arachidonic acid and prostaglandin E2 in adipose tissue from obese humans. *Metabolism* 50: 921–928, 2001b.

Fain JN et al. Comparison of PGE2, prostacyclin and leptin release by human adipocytes versus explants of adipose tissue in primary culture. *Prostagl Leukot Essent Fatty Acids* 67: 467–473, 2002.

Fain JN et al. Comparison of the release of adipokines by adipose tissue, adipose tissue matrix, and adipocytes from visceral and subcutaneous abdominal adipose tissues of obese humans. *Endocrinology* 145: 2273–2282, 2004.

Fajas LJ, Fruchart C, and Auwerx J. Transcriptional control of adipogenesis. *Curr Opin Cell Biol* 10: 165–173, 1998.

Fajas L et al. Selective cyclo-oxygenase-2 inhibitors impair adipocyte differentiation through inhibition of the clonal expansion phase. *J Lipid Res* 44: 1652–1659, 2003.

Fantuzzi G. Adipose tissue, adipokines, and inflammation. *J Allergy Clin Immunol* 115: 911–919, 2005.

Feldon SE et al. Activated human T lymphocytes express cyclooxygenase-2 and produce proadipogenic prostaglandins that drive human orbital fibroblast differentiation to adipocytes. *Am J Pathol* 169: 1183–1193, 2006.

Flower RJ and Blackwell GJ. The importance of phospholipase-A2 in prostaglandin biosynthesis. *Biochem Pharmacol* 25: 285–291, 1976.

Forman BM et al. 15-deoxy-δ12,14-prostaglandin J2 is a ligand for the adipocyte determination factor PPAR γ. *Cell* 83: 803–812, 1995.

Frayn KN, Fielding BA, and Karperef F. Adipose tissue fatty acid metabolism and cardiovascular disease. *Curr Opin Lipidol* 16: 409–415, 2005.

Gaillard D et al. Requirement and role of arachidonic acid in the differentiation of preadipose cells. *Biochem J* 257: 389–397, 1989.

Gaskins HR et al. Evidence for abnormal prostaglandin synthesis in obese Zucker rat adipose cell cultures. *J Nutr* 119: 458–462, 1989.

Hétu P-O and Riendeau D. Down-regulation of microsomal prostaglandin E2 synthase-1 in adipose tissue by high-fat feeding. *Obesity* 15: 60–68, 2007.

Hu E, Kim JB, Sarraf P, and Spiegelman BM. Inhibition of adipogenesis through MAP kinase-mediated phosphorylation of PPAR-γ. *Science* 274: 2100–2103, 1996.

Huggins KW, Boileau AC, and Hui DY. Protection against diet-induced obesity and obesity-related insulin resistance in group 1B PLA2-deficient mice. *Am J Physiol Endocrinol Metab* 283: E994–E1001, 2002.

Hyman BT and Stoll LL. Spector arachidonic acid: prostaglandin production by 3T3-L1 cells in culture. *Biochim Biophys Acta* 713: 375–385, 1982.

Ii H et al. Group IVA phospholipase A2 is associated with the storage of lipids in adipose tissue and liver. *Prostagl Other Lipid Mediat* 86: 12–17, 2008.

Illiano G and Cuatrecasas P. Endogenous prostaglandins modulate lipolytic processes in adipose tissue. *Nature New Biol* 234: 72–74, 1971.

Inuzuka H et al. Differential regulation of immediate early gene expression in preadipocyte cells through multiple signaling pathways. *Biochem Biophys Res Commun* 265: 664–668, 1999.

Jowsey IR et al. Prostaglandin D2 synthase enzymes and PPARγ are co-expressed in mouse 3T3-L1 adipocytes and human tissues. *Prostagl Other Lipid Mediat* 70: 267–284, 2003.

Kershaw EE and Flier JS. Adipose tissue as an endocrine organ. *J Clin Endocrinol Metab* 89: 2548–2556, 2004.

Kliewer SA et al. A prostaglandin J2 metabolite binds peroxisome proliferator-activated receptor γ and promotes adipocyte differentiation. *Cell* 83: 813–819, 1995.

Kugiyama K et al. Circulating levels of secretory type II phospholipase A2 predict coronary events in patients with coronary artery disease. *Circulation* 100: 1280–1284, 1999.

Lappas M and Rice GE. Phospholipase A2 isozymes in pregnancy and parturition. *Prostagl Leukot Essent Fatty Acids* 70: 87–100, 2004.

Lappas M, Permezel M, and Rice GE. Release of pro-inflammatory cytokines and 8-isoprostane from placenta, adipose tissue and skeletal muscle from normal pregnant women and women with gestational diabetes mellitus. *J Clin Endocrinol Metab* 89: 5627–5633, 2004.

Lappas M, Yee K, Permezel M, and Rice GE. Sulphasalazine and BAY 11-7082 interfere with the NF-κB and IKK-β pathway to regulate the release of pro-inflammatory cytokines from human adipose tissue and skeletal muscle, in vitro. *Endocrinology* 146: 1491–1497, 2005a.

Lappas M, Permezel M, and Rice GE. Leptin and adiponectin stimulate the release of proinflammatory cytokines and prostaglandins from human placenta and maternal adipose tissue via NF-κB, PPAR-γ and ERK 1/2. *Endocrinology* 146: 3334–3342, 2005b.

Lehmann JM et al. Peroxisome proliferator-activated receptors α and γ are activated by indomethacin and other non-steroidal anti-inflammatory drugs. *J Biol Chem* 272: 3406–3410, 1997.

Lepak NM and Serrero G. Inhibition of adipose differentiation by 9α, 11β-prostaglandin F2α. *Prostaglandins* 46: 511–517, 1993.

Lepak NM and Serreno G. Prostaglandin F2α stimulates transforming growth factor-α expression in adipocyte precursors. *Endocrinology* 136: 3222–3229, 1995.

Lin Y et al. The lipopolysaccharide-activated Toll-like receptor (TLR)-4 induces synthesis of the closely related receptor TLR-2 in adipocytes. *J Biol Chem* 275: 24255–24263, 2000.

Lindner I et al. Prostaglandin E synthase 2 (PTGES2) Arg298His polymorphism and parameters of the metabolic syndrome. *Mol Nutr Food Res* 51: 1447–1451, 2007.

MacDougald OA and Lane MD. Transcriptional regulation of gene expression during adipocyte differentiation. *Annu Rev Biochem* 64: 345–373, 1995.

Mannino DM et al. Boys with high body masses have an increased risk of developing asthma: findings from the National Longitudinal Survey of Youth. *Int J Obesity* 30: 6–13, 2006.

Marnett LJ et al. Arachidonic acid oxygenation by COX-1 and COX-2. Mechanisms of catalysis and inhibition. *J Biol Chem* 274: 22903–22906, 1999.

Mater MK et al. Arachidonic acid inhibits lipogenic gene expression in 3T3-L1 adipocytes through a prostanoid pathway. *J Lipid Res* 39: 1327–1334, 1998.

Merlie JP et al. Isolation and characterisation of the complementary DNA for sheep seminal vesicle prostaglandin endoperoxide synthase (cyclooxygenase). *J Biol Chem* 263: 3550–33553, 1988.

Miller CW, Casimir DA, and Ntambi JM. The mechanism of inhibition of 3T3-L1 preadipocyte differentiation by prostaglandin F2α. *Endocrinology* 137: 5641–5650, 1996.

Milne GL, Musiek ES, and Morrow JD. The cyclopentenone (A2/J2) isoprostanes: unique, highly reactive products of arachidonate peroxidation. *Antiox Redox Signal* 7: 210–220, 2005.

Misso NLA et al. Plasma phospholipase A2 activity in patients with asthma: association with body mass index and cholesterol concentration. *Thorax* 63: 21–26, 2008.

Musiek ES et al. Cyclopentenone isoprostanes inhibit the inflammatory response in macrophages. *J Biol Chem* 280: 35562–35570, 2005.

Nagel G et al. Prevalence and cluster of cardiometabolic biomarkers in overweight and obese schoolchildren: results from a large survey in southwest Germany. *Clin Chem* 54: 317–325, 2008.

Nitz I et al. Association of prostaglandin E synthase 2 (PTGES2) Arg298His polymorphism with type 2 diabetes in two German study populations. *J Clin Endocrinol Metab* 92: 3183–3188, 2007.

Ohki S et al. Prostaglandin hydroperoxidase, an integral part of prostaglandin endoperoxide synthetase from bovine vesicular gland microsomes. *J Biol Chem* 254: 829–836, 1979.

Okuno A et al. Troglitazone increases the number of small adipocytes without the change of white adipose tissue mass in obese Zucker rats. *J Clin Invest* 101: 1354–1361, 1998.

Otto JC, DeWitt DL, and Smith WL. N-glycosylation of prostaglandin endoperoxide synthases-1 and -2 and their orientations in the endoplasmic reticulum. *J Biol Chem* 268: 18234–18242, 1993.

Paradis M-E et al. Visceral adipose tissue accumulation, secretory phospholipase A2-IIA and atherogenecity of LDL. *Int J Obesity* 30: 1615–1622, 2006.

Parker J, Lane J, and Axelrod L. Cooperation of adipocytes and endothelial cells required for catecholamine stimulation of PGI2 production by rat adipose tissue. *Diabetes* 38: 1123–1132, 1989.

Peeraully M et al. Prostaglandin D2 and J2-series (PGJ2, Δ12-PGJ2) prostaglandins stimulate IL-6 and MCP-1, but inhibit leptin, expression and secretion by 3T3-L1 adipocytes. *Pflügers Arch* 453: 177–187, 2006.

Pendergrass M, Fazioni E, and De Fronzo R. Non-insulin-dependent diabetes mellitus and gestational diabetes mellitus: same disease, another name? *Diabetes Rev* 3: 566–583, 1995.

Petersen RK et al. Arachidonic acid-dependent inhibition of adipocyte differentiation requires PKA activity and is associated with sustained expression of cyclooxygenases. *J Lipid Res* 44: 2320–2330, 2003.

Pischon N et al. Obesity, inflammation, and periodontal disease. *Dent Res* 86: 400–409, 2007.

Pittas AG, Joseph NA, and Greenberg AS. Adipocytokines and insulin resistance. *J Clin Endocrinol Metab* 89: 447–452, 2004.

Quinkler M et al. Depot-specific prostaglandin synthesis in human adipose tissue: A novel possible mechanism of adipogenesis. *Gene* 380: 137–143, 2006.

Ragolia L et al. Accelerated glucose intolerance, nephropathy, and atherosclerosis in prostaglandin D2 synthase knock-out mice. *J Biol Chem* 280: 29946–29955, 2005.

Rajala MW and Sherer PE. The adipocyte: at the crossroads of energy homeostasis, inflammation, and atherosclerosis. *Endocrinology* 144: 3765–3773, 2003.

Reginato MJ et al. Prostaglandins promote and block adipogenesis through opposing effects on peroxisome proliferator-activated receptor γ. *J Biol Chem* 273: 1855–1858, 1998.

Richelsen B. Factors regulating the production of prostaglandin E2 and prostacyclin (prostaglandin I2) in rat and human adipocytes. *Biochem J* 247: 389–394, 1987.

Richelsen B. Release and effects of prostaglandins in adipose tissue. *Prostagl Leukot Essent Fatty Acids* 47: 171–182, 1992.

Richelsen B, Børglum JD, and Sørensen SS. Biosynthetic capacity and regulatory aspects of prostaglandin E2 formation in adipocytes. *Mol Cell Endocrinol* 85: 73–81, 1992.

Rossi A, Elia G, and Santoro MG. Inhibition of nuclear factor κB by prostaglandin A1: an effect associated with heat shock transcription factor activation. *Proc Natl Acad Sci USA* 94: 746–750, 1997.

Schaloske RH and Dennis EA. The phospholipase A2 superfamily and its group numbering system. *Biochim Biophys Acta* 1761: 1246–1259, 2006.

Serreno G and Lepak NM. Prostaglandin F2α receptor (FP receptor) agonists are potent adipose differentiation inhibitors for primary culture of adipocyte precursors in defined medium. *Biochem Biophys Res Commun* 233: 200–202, 1997.

Shaw JE and Ramwell PW. Release of prostaglandin from rat epididymal fat pad on nervous and hormonal stiumulation. *J Biol Chem* 243: 1498–1503, 1968.

Shillabeer G et al. Arachidonic acid metabolites of the lipoxygenase as well as the cyclooxygenase pathway may be involved in regulating preadipocyte differentiation. *Metabolism* 47: 461–466, 1998.

Smith TJ et al. Orbital fibroblast heterogeneity may determine the clinical presentation of thyroid-associated ophthalmopathy. J Clin Endocrinol Metab 87: 385–392, 2002.

Steinberg D et al. Effects of the prostaglandins on hormone-induced mobilization of free fatty acids. *J Clin Invest* 43: 1533–1540, 1964.

Stoll LL et al. Isolation and characterization of human epicardial adipocytes: potential role in vascular inflammation. *FASEB J* 20: A1074–A1074, 2006.

Straus DS and Glass CK. Cyclopentenone prostaglandins: New insights on biological activities and cellular targets. *Med Res Rev* 21: 185–210, 2001.

Subbaramaiah K et al. EP2 and EP4 receptors regulate aromatase expression in human adipocytes and breast cancer cells. *J Biol Chem* 283: 3433–3444, 2008.

Trayhurn P. Adipose tissue in obesity: an inflammatory issue. *Endocrinology* 146: 1003–1005, 2005.

Trayhurn P and Wood IS. Adipokines: inflammation and the pleiotropic role of white adipose tissue. *Br J Nutr* 92: 347–355, 2004.

Tsuboi K, Sugimoto Y, and Ichikawa A. Prostanoid receptor subtypes. *Prostagl Other Lipid Mediat* 68-69: 535–556, 2002.

Vassaux G et al. Prostacyclin is a specific effector of adipose cell differentiation. Its dual role as a cAMP- and $Ca^{2+}$-elevating agent. *J Biol Chem* 267: 11092–11097, 1992a.

Vassaux G et al. Differential response of preadipocytes and adipocytes to prostacyclin and prostaglandin E2: physiological implications. *Endocrinol* 131: 2393–2398, 1992b.
Vassaux G et al. Inhibition of prostacyclin-induced $Ca^{2+}$ mobilization by phorbol esters in Ob1771 preadipocytes. *Prostaglandins* 46: 441–451, 1993.
Weyer C et al. Humoral markers of inflammation and endothelial dysfunction in relation to adiposity and in vivo insulin action in Pima Indians. *Atherosclerosis* 161: 233–242, 2002.
Williams CS, Mann M, and DuBois RN. The role of cyclooxygenases in inflammation, cancer, and development. *Oncogene* 18: 7908–7916, 1999.
Winnett G, van Hagen D, and Schrey M. Prostaglandin J2 metabolites inhibit aromatase activity by redox-sensitive mechanisms: potential implications for breast cancer therapy. *Int J Cancer* 103: 600–605, 2003.
Wu D et al. Aging up-regulates expression of inflammatory mediators in mouse adipose tissue. *J Immunol* 179: 4829–4839, 2007.
Xie W et al. Expression of a mitogen-responsive gene encoding prostaglandin synthase is regulated by mRNA splicing. *Proc Natl Acad Sci USA* 88: 2692–2696, 1991.
Xie Y et al. Differentiation-dependent regulation of the cyclooxygenase cascade during adipogenesis suggests a complex role for prostaglandins. *Diabetes Obes Metab* 8: 83–93, 2006.
Yan H et al. Role of cyclooxygenases COX-1 and COX-2 in modulating adipogenesis in 3T3-L1 cells. *J Lipid Res* 44: 424–429, 2003.
Yokota T et al. Paracrine regulation of fat cell formation in bone marrow cultures via adiponectin and prostaglandins. *J Clin Invest* 15: 109:1303–1310, 2002.
Yu K, Bayona W, and Kallen CB, et al. Differential activation of peroxisome proliferator-activated receptors by eicosanoids. *J Biol Chem* 270: 23975–23983, 1995.

# 10 Inflammatory Actions of Adiponectin, Leptin, and Resistin

*Danielle M. Stringer, Vanessa C. DeClercq, Ryan W.T. Hunt, Maria S. Baranowski, Carla G. Taylor, and Peter Zahradka*

**CONTENTS**

## 10.1 INTRODUCTION

Obesity is associated with a chronic inflammatory state that contributes to the development of various metabolic and cardiovascular diseases. Adipose tissue has the ability to produce and secrete a variety of bioactive molecules such as adipokines that can regulate the inflammatory responses of several obesity-related complications [1]. Many of these pro-inflammatory adipokines such as leptin [2], resistin [3], C-reactive protein (CRP) [4], tumor necrosis factor-α (TNF-α), and interleukin-6 (IL-6) [5] are increased in obesity and may promote obesity-linked metabolic and cardiovascular diseases.

In contrast to most other adipokines that are upregulated in obesity, adiponectin is substantially reduced [6] and has been shown to have anti-inflammatory properties and play a protective role in the development of several obesity-related diseases [7,8]. This chapter will discuss the role of adipose tissue-derived molecules, specifically adiponectin, leptin, and resistin, in regulating inflammatory pathways that contribute to chronic disease.

## 10.2 OBESITY, ADIPOSE TISSUE DYSFUNCTION, AND ADIPOKINES

Obesity is defined as a body mass index (BMI) equal to or greater than 30 $kg/m^2$. Development of obesity requires a coordinated increase in fat cell size (adipocyte hypertrophy) and number (adipocyte hyperplasia) [1,9]. In terms of metabolic homeostasis, adipocytes play a key role in energy metabolism and triglyceride storage. As adipocytes enlarge, they become increasingly resistant to the antilypolytic effects of insulin, resulting in chronically elevated levels of circulating free fatty acids (FFAs) and FFA-induced impairments in insulin sensitivity. When the lipid storage capacity of enlarged adipocytes is exceeded, lipid storage spills over into other tissues such as liver, muscle, and pancreas, further increasing insulin resistance and lipotoxic effects [10].

The metabolic disturbances of obesity are accompanied by altered adipokine production which is directly related to adipocyte size [1]. Large dysfunctional adipocytes produce increased amounts of leptin, a pro-inflammatory adipokine, and reduced amounts of adiponectin, an anti-inflammatory adipokine, compared to smaller functional adipocytes. This imbalance of more pro-inflammatory and fewer anti-inflammatory adipokines in combination with chronic activation of the innate immune system contributes to the chronic low-grade inflammation in obesity. Although leptin and adiponectin are synthesized and released from the adipocytes in adipose tissue [11,12], other adipokines can be produced and released by other cells in adipose tissue or by a combination of adipose tissue cell types. For example, resistin is produced by macrophages infiltrating human adipose tissue [13], while TNF-α is produced by both adipocytes and infiltrating macrophages [14].

The number of infiltrated macrophages is positively correlated with BMI and adipocyte size in human abdominal subcutaneous adipose tissue. Surgically induced weight loss decreases the number of infiltrating macrophages, changes the inflammatory gene profile, and increases anti-inflammatory IL-10 in the remaining macrophages [15]. Interestingly, some nutritional and therapeutic treatments that reduce adipocyte cell size can improve the anti-inflammatory and pro-inflammatory adipokine profile without reductions in adipose mass or body weight [16]. Therefore, adipocyte size appears to be a key regulator of adipokine production and adipose tissue function. In obesity, the altered adipokine profile produced by large dysfunctional adipocytes exerts local effects in adipose tissue and systemic effects in other organs, and thus contributes to the pathogenesis of several chronic diseases including cardiovascular disease and diabetes mellitus via inflammatory pathways [14].

## 10.3 ADIPONECTIN

The human adiponectin gene is located on chromosome 3q27 and encodes a 224-amino acid sequence containing three distinct domains: the N-terminal hypervariable region, the collagen-like domain, and the C-terminal globular, complement C1q-like domain [6,17]. Four different research groups independently identified adiponectin. Scherer et al. (1995) discovered that adiponectin was an abundant serum protein in mice that was also produced by 3T3-L1 adipocytes. They named this protein adipocyte complement-related protein of 30 kDa (ACRP30) [18].

This group was also the first to identify several oligomeric forms of adiponectin [18]. Hu et al. (1996) confirmed that adiponectin, then called AdipoQ, was expressed in adipose tissue; however, they also noted increased levels in differentiated adipocytes and observed that adiponectin levels were decreased in obese humans and mice [12]. About the same time, Maeda et al. (1996) cloned the human adiponectin gene, referring to it as the adipose most abundant gene transcript 1, while Nakano et al. (1996) purified human adiponectin from plasma, naming it gelatin-binding protein of 28 kDa [19,20].

In the circulation, adiponectin can be found as three different oligomeric complexes: a trimer (low molecular weight), hexamer (medium molecular weight), and oligomer (high molecular weight) [6,21–25]. In addition to the full-length protein, a proteolytic cleavage fragment of adiponectin called globular adiponectin was also detected in the circulation [26]. Adiponectin trimers are formed through hydrophobic interactions among globular domains, whereas the formation of high molecular weight (HMW) complexes involves disulfide bonds between trimer molecules. Recent work provides insight into the regulation of HMW adiponectin formation. Wang et al. (2006) demonstrated that substitution of arginine with lysine disrupts the hydroxylation and glycosylation of lysine and inhibits the assembly of HMW adiponectin [27]. Thus, the formation of the HMW oligomeric complex is apparently regulated by hydroxylation and glycosylation of the lysine residues.

The different forms of adiponectin have different physiological effects in diseases involving inflammatory pathways. For example, Bobbert et al. (2005) showed that weight loss leads to an increase in total adiponectin and changes in the ratio of high to low molecular weight adiponectin [28]. Specifically, a positive correlation between body weight and low molecular weight adiponectin and an inverse correlation between body weight and high molecular weight adiponectin were observed. Furthermore, the ratio of HMW adiponectin to total adiponectin is decreased in patients with type 2 diabetes mellitus (T2DM) [27]. Nevertheless, this association remains controversial, since Polak et al. (2008) recently reported that high molecular weight adiponectin does not correlate with body weight [29].

### 10.3.1 Adiponectin Receptors

Two receptors have been identified for adiponectin. The adiponectin type 1 receptor (AdipoR1) is predominantly expressed in skeletal muscle, whereas the adiponectin type 2 receptor (AdipoR2) is mainly expressed in the liver [30]. AdipoR1 has a high affinity for the globular form. In contrast, both the globular and full-length adiponectin bind to AdipoR2 [30].

### 10.3.2 Adiponectin and Obesity

Epidemiological studies have shown a negative association between adiponectin levels and BMI [31]. Conversely, positive associations among circulating TNF-α, IL-6, and BMI have been documented [32]. As well, weight reduction has been shown to decrease circulating TNF-α and IL-6 levels [33] while increasing adiponectin levels [34]. Expression of adiponectin in adipocytes is suppressed with pro-inflammatory cytokines. For example, 3T3-L1 adipocytes treated with TNF-α exhibited reduced adiponectin expression [35]. Similar results were obtained with primary human adipocytes, showing that TNF-α treatment reduced expression and secretion of adiponectin [36]. Comparable reductions in adiponectin expression and secretion are observed following IL-6 treatment of 3T3-L1 adipocytes [37].

### 10.3.3 Adiponectin and Inflammatory Disease States

Adiponectin possesses anti-atherogenic [26,38,39] and anti-diabetic [40,41] properties. In non-obese individuals, adiponectin is normally found in the circulation at concentrations ranging from 2 to 20 μg/mL; however, obese individuals have significantly reduced plasma adiponectin levels [23,31]. An inverse correlation was observed among plasma adiponectin concentration, BMI [31] and waist circumference [42] in both sexes. In addition, the amount of adiponectin secreted by the adipose tissue is decreased with increased fat accumulation [43]. The strong inverse association between obesity and levels of adiponectin may prove to be a key link in obesity-related inflammatory diseases. By altering components of the inflammatory response, it is postulated that adiponectin may inhibit the progression of several chronic diseases.

Low-grade chronic inflammation is known to link obesity with related co-morbidities such as atherosclerosis, insulin resistance, and endothelial dysfunction. Several studies examined the role of inflammatory molecules and the risks of metabolic complications and cardiovascular disease. For example, CRP levels positively correlate with BMI as well as waist circumference and the risk of developing the metabolic syndrome [4]. Interestingly, a negative association was found between plasma CRP levels and plasma adiponectin levels in both sexes [44,45]. Additional adipokines such as IL-6 and TNF-α are also negatively correlated with adiponectin [5]. These anti-inflammatory properties of adiponectin may play a beneficial role in vascular and metabolic disorders.

### 10.3.4 Insulin Resistance

Increasing epidemiological and experimental studies demonstrated a protective role of adiponectin in the development of insulin resistance and diabetes. In both men and women with T2DM, plasma adiponectin concentrations are lower than in non-diabetic individuals [46]. Similarly, a negative association between plasma adiponectin concentrations and the development of insulin resistance and T2DM was documented in several clinical studies [47–50].

Diabetic mice treated with adiponectin display improved glucose tolerance [51]. Nawrocki et al. (2006) showed that adiponectin knockout mice fed a high fat diet developed insulin resistance [52]. Studies examining several types of tissues such as skeletal muscle, liver [53], and adipocytes [54] suggest that adiponectin may exert beneficial effects on insulin resistance through activation of AMP-activated protein kinase (AMPK). For example, adiponectin knockout mice given adenovirus expressing adiponectin show improved insulin sensitivity and increased AMPK activation [55]. Maeda et al. (2002) also reported improved insulin sensitivity with adenoviral delivery of adiponectin in adiponectin knockout mice; however, they also observed improved TNF-α levels [40], suggesting that the beneficial effect of adiponectin on insulin resistance may also work by suppressing the production of inflammatory cytokines.

### 10.3.5 Vascular Protection

Many studies demonstrated the beneficial action of adiponectin on endothelial homeostasis. A positive correlation between circulating adiponectin levels and arterial vasodilation was shown in humans [56]. Epidemiological studies have shown that plasma adiponectin levels negatively correlate with blood pressure in men with essential hypertension [57]. As well, adiponectin is significantly associated with systolic blood pressure in normotensive adults [58]. Recently, Devaraj et al. (2008) demonstrated that adiponectin treatment reduces mRNA and protein levels of CRP in human aortic endothelial cells, suggesting the link of obesity and vascular complications may be due to an increase in CRP production and aggravation of a pro-inflammatory state secondary to low levels of adiponectin [59].

Ohashi et al. (2006) have shown that adiponectin knockout mice develop diet-induced hypertension on high salt diets; when these mice are treated with an adiponectin adenovirus, the elevated blood pressure values are reversed [60]. Prostacyclin synthase (PGIS) and endothelial nitric oxide synthase (eNOS) are important enzymes involved in vasodilatation and are key determinants of endothelial function and angiogenesis. Investigation of possible mechanisms of adiponectin-related modulation of blood pressure regulation revealed that expression of PGIS and eNOS is downregulated in the aortae of adiponectin knockout mice, whereas treatment with adenovirus expressing adiponectin increases both PGIS and eNOS levels [60]. Adiponectin has also been shown to activate AMPK, leading to the phosphorylation and activation of eNOS[61].

Adiponectin also protects against atherosclerosis. In the absence of adiponectin, increased vascular smooth muscle cell proliferation and neointimal formation were observed [62]. In addition, when apoE knockout mice, normally prone to developing atherosclerosis, are treated with an adenovirus expressing adiponectin, they exhibit significant reductions in lesion size [38].

Adhesion molecules play a critical role in the inflammatory process in the vasculature and the progression of atherosclerosis and their expression has been linked to adiponectin [63]. For instance, apoE knockout mice treated with adenovirus expressing adiponectin showed reductions in the expression of vascular cell adhesion molecule-1 (VCAM-1), class A macrophage scavenger receptors (SR-As), and TNF-α in the aortic sinus [38]. In vitro, treatment of human aortic endothelial cells (HAECs) with adiponectin inhibited TNF-α-induced expression of adhesion molecules such as VCAM-1, E-selectin,, and intercellular adhesion molecule 1 (ICAM-1) [39]. Therefore, adiponectin shows potential in modifying the inflammatory process that underlies the development and progression of atherosclerosis.

### 10.3.6 Non-Alcoholic Fatty Liver Disease (NAFLD)

Inflammation due to obesity is also a risk factor for NAFLD, a condition that can ultimately lead to liver cirrhosis. Adiponectin is significantly lower in individuals with NAFLD [64]. Interestingly, serum adiponectin levels were significantly higher in obese children without liver abnormalities than in obese children with fatty infiltration or liver function abnormalities [65]. Similar data from adults showed significantly lower serum adiponectin levels in NAFLD patients with elevated liver enzymes than in healthy adults or NAFLD patients with normal liver enzymes [66]. Ma et el. (2008) examined liver biopsies from morbidly obese individuals with NAFLD or steatosis for the expression of adiponectin [67]. Adiponectin expression was decreased in the livers of those with NAFLD and immunohistochemistry revealed a negative correlation between adiponectin and the severity of inflammation [67].

Fetuin-A, a liver protein that is upregulated in hepatic steatosis, was recently shown to induce adipose tissue TNF-α and IL-6 expression while reducing adiponectin expression [68]. Additionally fetuin-A in humans was positively associated with plasma CRP levels and negatively associated with plasma adiponectin levels [68]. These data suggest an anti-inflammatory role for adiponectin in liver inflammation and hepatic steatosis.

### 10.3.7 Therapeutic Potential

Several studies have shown that thiazolidinediones (TZDs) exhibit anti-inflammatory properties and the ability to reduce plasma levels of TNF-α [69–71] and CRP [69,71]. Interestingly, treatment with TZDs resulted in increased plasma adiponectin levels in humans [35,72,73]. Ikeda et al. (2007) have shown that after only three days of treatment with TZDs, they noted significant increases in total and HMW adiponectin levels [74]. TZDs are known ligands of the transcription factor peroxisome proliferator-activated receptor (PPAR)-γ that, upon activation, can suppress inflammation and development of atherosclerosis [75]. It is still unclear whether the anti-inflammatory properties of TZDs are linked to their ability to increase adiponectin levels; nevertheless, identification of pharmaceuticals or alternatively nutraceuticals that target adiponectin may represent an innovative direction for developing novel anti-inflammatory agents.

## 10.4 LEPTIN

The product of the obese (*ob*) gene is the 16-kDa leptin hormone. Originally discovered as an unidentified satiety factor that decreases appetite and increases energy expenditure, leptin is not known to play a role in immunity, insulin signaling, reproduction, and inflammation. Leptin levels correlate positively with total body fat mass, BMI, and adipocyte size [76]. Spontaneous mutation of the *ob* gene in mice leads to the morbidly obese *ob/ob* mouse that is unable to produce leptin. Administration of leptin to *ob/ob* mice and humans that possess this rare mutation leads to reduced food intake and obesity and improved insulin sensitivity [77]. Reduced production of inflammatory cytokines such as IL-6 is also evident after administration of leptin when serum concentrations are low [78].

### 10.4.1 Synthesis and Secretion

Secreted primarily from white adipose tissue (WAT), leptin is also secreted at lower levels by gastric epithelial cells, intestine, brain, skeletal muscle, and placenta [2]. Subcutaneous WAT appears to contribute more to increased circulating levels of leptin than visceral WAT [79]. Leptin circulates in its free form in serum, but it can also be bound to a protein (ObRe) in plasma. Leptin levels are low in lean individuals (~3 ng/mL), and free leptin is roughly equivalent to bound leptin in lean subjects. However, free leptin may be 25 times higher than leptin bound to ObRe in the plasma of obese subjects [80]. In people lacking functional leptin receptors, circulating leptin levels can be as high as 600 ng/mL [81]. The effects of leptin on appetite regulation are well characterized [11,77,82].

### 10.4.2 Leptin Receptor

ObRb is expressed primarily in the hypothalamus but is also found in membranes of adipocytes, macrophages, pancreatic β-cells, skeletal muscle, adrenal, and liver cells. ObRb is a member of the class I cytokine receptor family and may participate in cross-talk with other cytokines such as IL-6 [83]. Due to alternative splicing, five other leptin receptors (ObRa–f) can be synthesized and these isoforms are present in differing amounts in a variety of cell types. ObRb, a transmembrane receptor, is the only one capable of intracellular signaling. ObRa, ObRc, and ObRd possess shortened cytoplasmic domains whose functions remain to be fully elucidated. ObRe is the soluble circulating leptin binding protein found in plasma.

The gene for the ObRb is located on the *db* locus on mouse chromosome 4. Spontaneous mutation of this gene in C57BL/6J mice produces the *db/db* mouse. This strain is characterized by obesity, hyperglycemia, hyperinsulinemia, and dyslipidemia, which makes these mice suitable models for severe T2DM [84]. In vivo, *db/db* mice also show impaired glucose tolerance and insulin resistance. In rats, spontaneous mutation of the *fa* (*fat*) gene leads to a dysfunctional ObRb. The *fa/fa* Zucker rat is morbidly obese and insulin-resistant and serves as a model for metabolic syndrome.

### 10.4.3 Leptin Regulation

Expression of leptin is mainly stimulated by food intake and energy use, but may also be influenced by other hormones. Insulin produced in response to food intake increases leptin levels, which in turn exerts a feedback effect on the brain that results in suppression of food intake. Hyperinsulinemia also leads to increased expression of leptin in humans; however, when a hyperinsulinemic clamp is employed, leptin levels do not appear to increase [85]. Leptin levels are higher in women than men; this is partially explained by the fact that testosterone inhibits leptin expression while estrogen leads to an increase in leptin mRNA [86].

Activation of the leptin receptor by leptin activates JAK2, which phosphorylates numerous tyrosine residues found on the cytoplasmic domain of the ObRb. The phosphorylation of Tyr1138 [87] enables docking of STAT3 and its subsequent phosphorylation by JAK2. Upon phosphorylation, STAT3 dimerizes and is transported to the nucleus where it modulates expression of various neuropeptide genes involved in regulating energy balance and appetite [88]. STAT3 also induces the expression of suppressor of cytokine signaling (SOCS3) which binds to phosphorylated Tyr985 and blocks JAK2. The resultant decrease in JAK2 activity represses leptin receptor signaling. When leptin was administered peripherally for 48 hours, mRNA coding for SOCS3 increased four-fold in *ob/ob* mice [89]. However, increased SOCS3 level may be one of the main mechanisms responsible for leptin resistance; SOCS3 mRNA and protein levels are significantly increased in mice that have lowered leptin sensitivity from diet-induced obesity. Overexpression of SOCS3 also leads to suppressed signaling of ObRb, while knocking out the SOCS3 gene leads to increased signaling associated with ObRb [87].

### 10.4.4 Leptin, Obesity, and Inflammation

In the obese state, adipose tissue is subject to chronic low grade inflammation and macrophage infiltration. Increased levels of leptin due to obesity lead to elevated expression of various inflammatory factors. Conversely, an inflammatory state can lead to elevated leptin secretion, thus limiting appetite. Leptin may not be the main cause of problems associated with obesity such as T2DM or cardiovascular disease; however, its dysregulation along with other important adipokines such as adiponectin and resistin will exacerbate the conditions associated with increased obesity [90].

### 10.4.5 Stimulation of Leptin Secretion by Inflammatory Mediators

The production of leptin in vitro is increased by numerous factors including glucocorticoids [91], TNF-α, glucose, insulin, IL-1, and estrogens [83]. One of the first inflammatory cytokines found to be associated with obesity was TNF-α. Similar findings for many other inflammatory cytokines followed shortly, and leptin was added to this list in 1994 [92]. Increased levels of TNF-α have been shown to induce expression of leptin both in vitro and in vivo. In the case of obesity, leptin levels are already elevated and cells have typically become resistant, which means they are no longer sensitive to further stimulation.

### 10.4.6 Stimulation of Inflammatory Factors

It has also been suggested that leptin is a major factor in the low grade inflammatory state associated with obesity [78]. Leptin has been shown to induce numerous inflammatory products and factors in a state of increased adiposity and in cases of illness accompanied by a lack of appetite. In the brain, leptin may act not only as a regulator of appetite but also as an inflammatory signal through its ability to induce COX-2 via interleukin-1β [93]. Inhibition of COX-2 by pharmacological agents and gene knockout leads to a preserved appetite, with no changes in leptin levels, during LPS-induced inflammation [94,95].

IL-$\beta^{-/-}$ mice also showed no increases in leptin after intraperitoneal injection of LPS or turpentine [96]. In mice, acute inflammatory conditions stimulated by LPS also lead to elevated leptin levels along with increased gene expression [97]. In the same study, administration of TNF and IL-1 also led to elevated levels of serum leptin. It has been demonstrated in human primary adipocytes in the presence of dexamethasone (inhibitor of upregulation of COX-2), that arachidonic acid and prostaglandin E2 can stimulate the release of leptin and leptin mRNA accumulation [98]. However, when pegylated recombinant leptin is administered in conjunction with weight loss, it can attenuate the production of several inflammatory factors such as TNF-α, although CRP was markedly increased [78]. This may benefit patients who are dieting, as leptin levels usually fail to promote food intake due to fasting and decreased adipose tissue [11]. It has recently been shown that a 10% weight loss can be maintained if leptin levels are not allowed to fall as they normally would following a decrease in body weight [2].

### 10.4.7 Leptin and Inflammatory Disease States

Elevated levels of leptin are associated with obesity, insulin resistance, platelet aggregation, vascular smooth muscle migration and proliferation, and increased oxidative stress and inflammation.

#### 10.4.7.1 Atherosclerosis

CRP, a marker for inflammation and indicator of increased cardiovascular risk, is increased in the presence of elevated leptin levels in coronary artery endothelial cells and is associated with the development of atherosclerosis [99]. Induction of ObRb by leptin leads to increased reactive oxygen species production and phosphorylation of MAPK, which may lead to increased CRP expression [100]. Conversely, decreased leptin levels associated with weight loss lead to lowered levels of CRP. The leptin receptor has also been identified in human atherosclerotic lesions [101] and elevated leptin levels induce platelet aggregation, even in healthy individuals [102].

#### 10.4.7.2 Diabetes Mellitus

In parallel with insulin resistance, elevated leptin levels correlate positively with the inhibition of glucose uptake [103]. It is hypothesized that leptin may decrease insulin secretion from pancreatic β-cells by regulating the amounts of triglycerides and free fatty acids present in these cells [104]. Furthermore, increased levels of cytokines such as TNF-α [105] and leptin have been implicated in the development of T2DM through exacerbation of inflammatory symptoms and altered cell signaling mechanisms [103].

Leptin can modulate insulin signaling in cells that express leptin receptors through several mechanisms, including (1) a reduction in insulin-dependent tyrosine phosphorylation of IRS-1 by decreasing IRS mobility, and (2) increased activity of IRS-1-associated PI3K. A potential reason for the relationship of insulin resistance and leptin resistance is the cross-talk between ObRb and PI3K [88]. Activation of PI3K by receptor-bound leptin likely involves JAK2 activation and subsequent phosphorylation of IRS proteins. The effect of leptin on PI3K is likely tissue-dependent as well as PI3K isoform-specific [106]. Leptin may modulate insulin signaling pathways in obese individuals and therefore play a role in T2DM; however, further research is required to fully elucidate the role of leptin in insulin resistance and T2DM.

#### 10.4.7.3 Psoriasis

Increased serum leptin levels have been correlated with increased severity of psoriasis [107,108] but this is not conclusive because other studies show that serum levels of leptin in those suffering from psoriasis do not differ from levels in healthy patients when matched with individuals of similar BMI measurements [109]. IL-6 and IL-1β are both elevated in patients suffering from psoriasis and leptin can induce both TNF-α and IL-1β production by blood monocytes [109]. The severity of psoriasis and its outcomes appear related to the degree of obesity along with enhancement by leptin and other adipokines, such as resistin, of numerous inflammatory processes that aggravate symptoms of psoriasis.

### 10.4.8 Potential Therapies

Administration of exogenous leptin to obese patients has proven to provide little benefit because such patients are usually already resistant to the effects of leptin and have highly elevated serum leptin levels [110]. Leptin has recently been shown to improve glucose levels in type 1 diabetic mice despite a lack of insulin [111]. In cases of leptin depletion such as lipodystrophy, administration of leptin is able to improve glycemic control and reduce triglycerides [112], thereby relieving FFA-mediated insulin resistance and lipid accumulation in organs [113].

## 10.5 RESISTIN

Resistin is the product of the RSTN gene located on mouse chromosome 8 and human chromosome 19. The RSTN gene encodes a 114-amino acid polypeptide with a 20-amino acid signal sequence secreted as a 94-amino acid polypeptide of 12.5 kDa [114]. Resistin belongs to the resistin-like molecule (RELM) family of cysteine-rich proteins that also includes RELMα, RELMβ, and RELMγ. The discovery of resistin is credited to three research groups [115–117]; however, the protein was named resistin by Lazar's group due to the observation that this novel protein promoted insulin resistance. Like adiponectin, resistin is known to circulate in two distinct forms: a high molecular weight hexamer and a more biologically active trimer form [118]. Although originally thought to be expressed exclusively in white adipose tissue, it is now known that resistin mRNA can be found in many tissues including hypothalamus, adrenal gland, spleen, skeletal muscle, pancreas, monocytes and macrophages, and the gastrointestinal tract [119].

### 10.5.1 Regulation

Resistin expression is modulated by a variety of endocrine factors. In rodent adipocytes, resistin expression is induced by corticosteroids, prolactin, testosterone, and growth hormone. Insulin, epinephrine, and somatotrophin produce inhibitory effects [120]. Resistin gene expression is induced by C/EBPα [121] and repressed by PPARγ [122] through direct binding of these transcription factors to the resistin promoter. Gender differences in resistin expression have been demonstrated in rodents, with white adipose tissue of male rats expressing more resistin than that of female rats [123,124]; however, mouse studies show the opposite gender-specific expression of resistin [125]. In humans, expression appears higher in women than in men [126–128].

### 10.5.2 Resistin and Obesity

Numerous reports from humans and animal models revealed a positive relationship between adiposity and serum resistin concentrations [129–136]; however, other studies found no relationship between them [126,127,137–139]. The apparent discrepancy between animal and human studies regarding the relationship between obesity and resistin may be due in part to the different sources of resistin. Adipose tissue can be thought of as a connective tissue framework upon which epithelial tissues rest and within which muscle and nervous tissue are embedded. Within this network exist adipocytes along with large numbers of non-adipocyte cells including fibroblasts, mast cells, macrophages, leukocytes, and other cells involved in the inflammatory response. While it has been reported that adipocytes are the sole sources of resistin in mice [117], studies of humans suggest that adipocytes express negligible amounts [122]. On the other hand, resistin mRNA was readily detectable in circulating mononuclear cells of humans [140]. Furthermore, while resistin was released by explants of human adipose tissue, it originated from the non-fat cells present in tissues and not from adipocytes [13]. Regardless of the source, resistin expression is most-likely upregulated in obese states in both humans and animals.

### 10.5.3 Resistin and Insulin Resistance

Obesity is a risk factor for development of insulin resistance. A widely accepted theory for the pathogenesis of insulin resistance secondary to obesity implicates the negative regulation of insulin signaling by adipokines. Resistin was originally named for its apparent ability to induce insulin resistance in mice [117]. Since that initial observation, other research groups documented a positive association between resistin and insulin resistance in both animals [134,141] and humans [3,127,142,143]. It is important to note, however, that several animal and human studies have failed to show a relationship, even an inverse relationship, between serum and plasma resistin and measures of insulin resistance in animals and humans with insulin resistance or T2DM [126,131,144–149].

### 10.5.4 Interactions of Resistin and Inflammatory Molecules

While much attention has been devoted to the increased risk of diabetes, cancer, and cardiovascular disease associated with obesity, mounting evidence indicates that other pathological states are also exacerbated in obese patients and many of these complications may result from the low-grade inflammation associated with obesity. Although resistin was originally implicated in the pathogenesis of insulin resistance, recent evidence shows that it may also be involved in inflammatory processes, providing a potential link between insulin resistance and inflammation. Bokarewa et al. (2005) demonstrated a definitive role of resistin in inflammation [150]. Treatment of isolated peripheral blood mononuclear cells (PBMCs) and leukocytes with resistin resulted in the production and secretion of several pro-inflammatory cytokines including TNF-α, IL-6, IL-1β, and, interestingly, resistin.

Treatment of these cells with resistin resulted in elevated cytosolic to nuclear translocation of NF-κB. Pre-incubation with parthenolide, a selective NF-κB inhibitor, ablated the resistin-stimulated production of pro-inflammatory cytokines, suggesting that resistin exerts its inflammatory actions through NF-κB signaling. Induction of cytokine production by resistin is not limited to peripheral cells. Overexpression of resistin in differentiated 3T3-L1 adipocytes results in increased production of TNF-α, IL-6, and monocyte chemotactic protein-1 (MCP-1) and, interestingly, decreased production of the anti-inflammatory IL-10 cytokine [151]. This study was the first to demonstrate that resistin can act in an autocrine manner to promote inflammation.

Pro-inflammatory agents, such as TNF-α, IL-6, IL-1β, and LPS have been shown to increase expression of resistin in a number of systems including PMBCs, white adipose tissue, rodent white blood cells, and 3T3-L1 adipocytes [152–154]. However, other reports demonstrate an opposing effect or no effect in that treatment with pro-inflammatory agents such as TNF-α and LPS downregulated or failed to elevate resistin expression in 3T3-L1 adipocytes and FVB mice [155,156]. In addition to cell culture and animal studies, the relationship of resistin and inflammation has been explored in human studies. A positive correlation between CRP, an acute phase reactant involved in the inflammatory process, and circulating resistin concentrations was found in several pathological conditions [157–159], even after adjustments for sex and body mass index [3,148], suggesting that the pro-inflammatory effects of resistin may be independent of overall obesity.

### 10.5.5 Resistin and Inflammatory Disease States

#### 10.5.5.1 Atherosclerosis

Emerging evidence implicates resistin in the initial stages of atherosclerosis via activation of vascular endothelial cells. Treatment of endothelial cells with human recombinant resistin induces the release of endothelin-1, increases the expression of the adhesion molecules VCAM-2, ICAM-1, and the MCP-1 chemokine, and potentiates CD40 ligand-induced endothelial cell activation [160,161]. In addition to promoting the atherosclerotic inflammatory process, resistin also stimulates smooth muscle cell proliferation through both extracellular signal-regulated kinase and Akt signaling [162-164]. In vivo studies found that resistin protein is present in both murine and human atherosclerotic lesions, with mRNA levels increasing proportionally to the degree of atherosclerotic development [162].

In human aortic aneurysms, areas of macrophage infiltration are associated with resistin-positive immunohistochemical staining, while normal arteries and veins do not display such a pattern [164]. Results from population-based studies have also shown a relationship between resistin and coronary artery calcification, a quantitative index of atherosclerosis [165]. These data implicate resistin in the development of atherosclerosis.

#### 10.5.5.2 Rheumatoid Arthritis (RA)

Through the production of adipokines, adipose tissue contributes to inflammatory and degenerative processes underlying common joint diseases. Resistin, in particular, has been implicated in the inflammatory process underlying RA, the most common form of chronic inflammatory joint disease leading to cartilage and bone destruction. Resistin is present in synovial fluid and tissue from patients with RA, and co-localizes with B-lymphocytes, macrophages and plasma cells [166]. Synovial fluid resistin concentration is positively correlated with CRP, erythrocyte sedimentation rate, and disease activity score in patients with RA [166,167]. Furthermore, synovial fluid from patients with RA contains significantly higher resistin levels compared to patients with non-inflammatory joint diseases [150,166]. The use of serum resistin as a marker for RA is controversial.

While one study has shown higher serum resistin levels in patients with RA compared to healthy controls and a positive correlation with CRP and erythrocyte sedimentation rate [168], another documents no relationship between blood resistin and duration of RA, circulating CRP, or white blood cell counts [150]. Interestingly, intra-articular injection of recombinant resistin into the knee joints of healthy mice resulted in leukocyte infiltration, hypertrophy of synovial lining, and panus formation—all signs of arthritis [150]. Development of resistin-induced arthritis was independent of gender and strain of mouse. Taken together, these results suggest a role for resistin in the pathogenesis of RA.

## 10.6 CONCLUSIONS

Accumulating evidence suggests that adipose tissue plays an important role in regulating the production and secretion of many bioactive molecules. In obesity, this regulation shifts toward a pro-inflammatory state, contributing to the development of obesity-related diseases. Key adipokines such as adiponectin, leptin, and resistin may alter the development of obesity-related complications by modulating inflammatory pathways involved in metabolic and cardiovascular diseases. These adipokines have the ability to act on multiple inflammatory pathways, enzymes, adhesion molecules, and transcription factors. Thus, targeting adipose derived molecules may be effective for preventing or treating inflammatory-related diseases.

## ACKNOWLEDGMENTS

The Natural Sciences and Engineering Research Council (NSERC) of Canada has generously supported our research program. DMS, VCD, and MSB are recipients of NSERC Canada graduate scholarships.

## REFERENCES

1. DeClercq V, Taylor C, and Zahradka P. Adipose tissue: the link between obesity and cardiovascular disease. *Cardiovasc Hematol Disord Drug Targets* 2008; 8: 228–237.
2. Ahima RS. Revisiting leptin's role in obesity and weight loss. *J Clin Invest* 2008; 118: 2380–2383.
3. McTernan PG et al. Resistin and type 2 diabetes: regulation of resistin expression by insulin and rosiglitazone and the effects of recombinant resistin on lipid and glucose metabolism in human differentiated adipocytes. *J Clin Endocrinol Metab* 2003; 88: 6098–6106.
4. Florez H et al. C-reactive protein is elevated in obese patients with the metabolic syndrome. *Diabetes Res Clin Pract* 2006; 71: 92–100.
5. Bruun JM et al. Regulation of adiponectin by adipose tissue-derived cytokines: in vivo and in vitro investigations in humans. *Am J Physiol Endocrinol Metab* 2003; 285: E527–E533.
6. Beltowski J, Jamroz-Wisniewska A, and Widomska S. Adiponectin and its role in cardiovascular diseases. *Cardiovasc Hematol Disord Drug Targets* 2008; 8: 7–46.
7. Ouchi N et al. Reciprocal association of C-reactive protein with adiponectin in blood stream and adipose tissue. *Circulation* 2003; 107: 671–674.
8. Takemura Y, Walsh K, and Ouchi N. Adiponectin and cardiovascular inflammatory responses. *Curr Atheroscler Rep* 2007; 9: 238–243.
9. Avram MM, Avram AS, and James WD. Subcutaneous fat in normal and diseased states 3. Adipogenesis: from stem cell to fat cell. *J Am Acad Dermatol* 2007; 56: 472–492.

10. DeFronzo RA. Dysfunctional fat cells, lipotoxicity and type 2 diabetes. *Int J Clin Pract* 2004; 143: 9–21.
11. Ahima RS. Adipose tissue as an endocrine organ. *Obesity* 2006; 14: 242S–249S.
12. Hu E, Liang P, and Spiegelman BM. AdipoQ is a novel adipose-specific gene dysregulated in obesity. *J Biol Chem* 1996; 271: 10697–10703.
13. Fain JN et al. Resistin release by human adipose tissue explants in primary culture. *Biochem Biophys Res Commun* 2003; 300: 674–678.
14. Bastard JP et al. Recent advances in the relationship between obesity, inflammation, and insulin resistance. *Eur Cytokine Netw* 2006; 17: 4–12.
15. Cancello R et al. Reduction of macrophage infiltration and chemoattractant gene expression changes in white adipose tissue of morbidly obese subjects after surgery-induced weight loss. *Diabetes* 2005; 54: 2277–2286.
16. Noto A et al. Dietary conjugated linoleic acid decreases adipocyte size and favorably modifies adipokine status and insulin sensitivity in obese, insulin-resistant rats. *Metabolism* 2007; 56: 1601–1611.
17. Kissebah AH et al. Quantitative trait loci on chromosomes 3 and 17 influence phenotypes of the metabolic syndrome. *Proc Natl Acad Sci USA* 2000; 97: 14478–14483.
18. Scherer PE et al. A novel serum protein similar to C1q, produced exclusively in adipocytes. *J Biol Chem* 1995; 270: 26746–26749.
19. Maeda K et al. cDNA cloning and expression of a novel adipose specific collagen-like factor, apM1 (AdiPose Most abundant gene transcript 1). *Biochem Biophys Res Commun* 1996; 221: 286–289.
20. Nakano Y et al. Isolation and characterization of GBP28, a novel gelatin-binding protein purified from human plasma. *J Biochem* 1996; 120: 803–812.
21. Kobayashi H et al. Selective suppression of endothelial cell apoptosis by the high molecular weight form of adiponectin. *Circ Res* 2004; 94: e27–e31.
22. Pajvani UB et al. Complex distribution, not absolute amount of adiponectin, correlates with thiazolidinedione-mediated improvement in insulin sensitivity. *J Biol Chem* 2004; 279: 12152–12162.
23. Tsao TS, Lodish HF, and Fruebis J. ACRP30, a new hormone controlling fat and glucose metabolism. *Eur J Pharmacol* 2002; 440: 213–221.
24. Tsao TS et al. Role of disulfide bonds in Acrp30/adiponectin structure and signaling specificity: different oligomers activate different signal transduction pathways. *J Biol Chem* 2003; 278: 50810–50817.
25. Xu A et al. Testosterone selectively reduces the high molecular weight form of adiponectin by inhibiting its secretion from adipocytes. *J Biol Chem* 2005; 280: 18073–18080.
26. Ouchi N et al. Adipocyte-derived plasma protein, adiponectin, suppresses lipid accumulation and class A scavenger receptor expression in human monocyte-derived macrophages. *Circulation* 2001; 103: 1057–1063.
27. Wang Y et al. Post-translational modifications of the four conserved lysine residues within the collagenous domain of adiponectin are required for the formation of its high molecular weight oligomeric complex. *J Biol Chem* 2006; 281: 16391–16400.
28. Bobbert T et al. Changes of adiponectin oligomer composition by moderate weight reduction. *Diabetes* 2005; 54: 2712–2719.
29. Polak J et al. Total adiponectin and adiponectin multimeric complexes in relation to weight loss-induced improvements in insulin sensitivity in obese women: NUGENOB study. *Eur J Endocrinol* 2008; 158: 533–541.
30. Yamauchi T et al. Cloning of adiponectin receptors that mediate antidiabetic metabolic effects. *Nature* 2003; 423: 762–769.
31. Arita Y et al. Paradoxical decrease of an adipose-specific protein, adiponectin, in obesity. *Biochem Biophys Res Commun* 1999; 257: 79–83.

32. Yudkin JS et al. C-reactive protein in healthy subjects: associations with obesity, insulin resistance, and endothelial dysfunction *Arterioscler Thromb Vasc Biol* 1999; 19: 972–978.
33. Ziccardi P et al. Reduction of inflammatory cytokine concentrations and improvement of endothelial functions in obese women after weight loss over one year. *Circulation* 2002; 105: 804–809.
34. Yang WS et al. Weight reduction increases plasma levels of an adipose-derived anti-inflammatory protein, adiponectin. *J Clin Endocrinol Metab* 2001; 86: 3815–3819.
35. Maeda N et al. PPARgamma ligands increase expression and plasma concentrations of adiponectin, an adipose-derived protein. *Diabetes* 2001; 50: 2094–2099.
36. Kappes A and Loffler G. Influences of ionomycin, dibutyryl-cycloAMP and tumour necrosis factor-α on intracellular amount and secretion of apM1 in differentiating primary human preadipocytes. *Horm Metab Res* 2000; 32: 548–554.
37. Fasshauer M et al. Adiponectin gene expression and secretion is inhibited by interleukin-6 in 3T3-L1 adipocytes. *Biochem Biophys Res Commun* 2003; 301: 1045–1050.
38. Okamoto Y et al. Adiponectin reduces atherosclerosis in apolipoprotein E-deficient mice. *Circulation* 2002; 106: 2767–2770.
39. Ouchi N et al. Novel modulator for endothelial adhesion molecules: adipocyte-derived plasma protein adiponectin. *Circulation* 1999; 100: 2473–2476.
40. Maeda N et al. Diet-induced insulin resistance in mice lacking adiponectin/ACRP30. *Nat Med* 2002; 8: 731–737.
41. Weyer C et al. Hypoadiponectinemia in obesity and type 2 diabetes: close association with insulin resistance and hyperinsulinemia. *J Clin Endocrinol Metab* 2001; 86: 1930–1935.
42. Ryo M et al. Adiponectin as a biomarker of the metabolic syndrome. *Circ J* 2004; 68: 975–981.
43. Hoffstedt J et al. Adipose tissue adiponectin production and adiponectin serum concentration in human obesity and insulin resistance. *J Clin Endocrinol Metab* 2004; 89: 1391–1396.
44. Engeli S et al. Association between adiponectin and mediators of inflammation in obese women. *Diabetes* 2003; 52: 942–947.
45. Ouchi N et al. Association of hypoadiponectinemia with impaired vasoreactivity. *Hypertension* 2003; 42: 231–234.
46. Hotta K et al. Plasma concentrations of a novel, adipose-specific protein, adiponectin, in type 2 diabetic patients. *Arterioscler Thromb Vasc Biol* 2000; 20: 1595–1599.
47. Choi KM et al. Serum adiponectin concentrations predict developments of type 2 diabetes and the metabolic syndrome in elderly Koreans. *Clin Endocrinol* 2004; 61: 75–80.
48. Snehalatha C et al. Plasma adiponectin is an independent predictor of type 2 diabetes in Asian indians. *Diabetes Care* 2003; 26: 3226–3229.
49. Spranger J et al. Adiponectin and protection against type 2 diabetes mellitus. *Lancet* 2003; 361: 226–228.
50. Yamamoto Y et al. Adiponectin, an adipocyte-derived protein, predicts future insulin resistance: two-year follow-up study in Japanese population. *J Clin Endocrinol Metab* 2004; 89: 87–90.
51. Berg AH et al. The adipocyte-secreted protein Acrp30 enhances hepatic insulin action. *Nat Med* 2001; 7: 947–953.
52. Nawrocki AR et al. Mice lacking adiponectin show decreased hepatic insulin sensitivity and reduced responsiveness to peroxisome proliferator-activated receptor γ agonists. *J Biol Chem* 2006; 281: 2654–2660.
53. Yamauchi T et al. Adiponectin stimulates glucose utilization and fatty-acid oxidation by activating AMP-activated protein kinase. *Nat Med* 2002; 8: 1288–1295.

54. Wu X et al. Involvement of AMP-activated protein kinase in glucose uptake stimulated by the globular domain of adiponectin in primary rat adipocytes. *Diabetes* 2003; 52: 1355–1363.
55. Satoh H et al. Adenovirus-mediated adiponectin expression augments skeletal muscle insulin sensitivity in male Wistar rats. *Diabetes* 2005; 54: 1304–1313.
56. Fernandez-Real JM et al. Adiponectin is associated with vascular function independent of insulin sensitivity. *Diabetes Care* 2004; 27: 739–745.
57. Iwashima Y et al. Hypoadiponectinemia is an independent risk factor for hypertension. *Hypertension* 2004; 43: 1318–1323.
58. Sung SH et al. Adiponectin, but not leptin or high-sensitivity C-reactive protein, is associated with blood pressure independently of general and abdominal adiposity. *Hypertens Res* 2008; 31: 633–640.
59. Devaraj S et al. Adiponectin decreases C-reactive protein synthesis and secretion from endothelial cells: evidence for an adipose tissue–vascular loop. *Arterioscler Thromb Vasc Biol* 2008; 28: 1368–1374.
60. Ohashi KN et al. Adiponectin replenishment ameliorates obesity-related hypertension. *Hypertension* 2006; 47: 1108–1116.
61. Ouchi N et al. Adiponectin stimulates angiogenesis by promoting cross-talk between AMP-activated protein kinase and Akt signaling in endothelial cells. *J Biol Chem* 2004; 279: 1304–1309.
62. Kubota N et al. Disruption of adiponectin causes insulin resistance and neointimal formation. *J Biol Chem* 2002; 277: 25863–25866.
63. Braun M et al. Cellular adhesion molecules on vascular smooth muscle cells. *Cardiovasc Res* 1999; 41: 395–401.
64. Pagano C et al. Plasma adiponectin is decreased in nonalcoholic fatty liver disease. *Eur J Endocrinol* 2005; 152: 113–118.
65. Zou CC et al. Serum adiponectin, resistin levels and non-alcoholic fatty liver disease in obese children. *Endocr J* 2005; 52: 519–524.
66. Aygun C et al. Serum levels of hepatoprotective peptide adiponectin in non-alcoholic fatty liver disease. *Eur J Gastroenterol Hepatol* 2006; 18: 175–180.
67. Ma H et al. Expression of adiponectin and its receptors in livers of morbidly obese patients with non-alcoholic fatty liver disease. *J Gastroenterol Hepatol* 2009; 24: 233–237.
68. Hennige AM et al. Fetuin-A induces cytokine expression and suppresses adiponectin production. *PLoS ONE* 2008; 3: e1765.
69. Haffner SM et al. Effect of rosiglitazone treatment on nontraditional markers of cardiovascular disease in patients with type 2 diabetes mellitus. *Circulation* 2002; 106: 679–684.
70. Katsuki A et al. Troglitazone reduces plasma levels of tumour necrosis factor-α in obese patients with type 2 diabetes. *Diabetes Obes Metab* 2000; 2: 189–191.
71. Khanolkar MP et al. Rosiglitazone produces a greater reduction in circulating platelet activity compared with gliclazide in patients with type 2 diabetes mellitus: an effect probably mediated by direct platelet PPARγ activation. *Atherosclerosis* 2008; 197: 718–724.
72. Combs TP et al. Induction of adipocyte complement-related protein of 30 kilodaltons by PPARγ agonists: potential mechanism of insulin sensitization. *Endocrinology* 2002; 143: 998–1007.
73. Yu JG et al. The effect of thiazolidinediones on plasma adiponectin levels in normal, obese, and type 2 diabetic subjects. *Diabetes* 2002; 51: 2968–2974.
74. Ikeda Y et al. Pioglitazone rapidly increases serum adiponectin levels in men with normal glucose tolerance. *Diabetes Care* 2007; 30: e48.

75. Libby P and Plutzky J. Inflammation in diabetes mellitus: role of peroxisome proliferator-activated receptor-α and peroxisome proliferator-activated receptor-γ agonists. *Am J Cardiol* 2007; 99: 27B–40B.
76. Guo KY et al. Effects of obesity on the relationship of leptin mRNA expression and adipocyte size in anatomically distinct fat depots in mice. *Am J Physiol Regul Integr Comp Physiol* 2004; 287: R112–R119.
77. Trayhurn P and Bing C. Appetite and energy balance signals from adipocytes. *Philos Trans R Soc Lond B Biol Sci* 2006; 361: 1237–1249.
78. Hukshorn CJ et al. Leptin and the proinflammatory state associated with human obesity. *J Clin Endocrinol Metab* 2004; 89: 1773–1778.
79. Clement K and Langin D. Regulation of inflammation-related genes in human adipose tissue. *J Intern Med* 2007; 262: 422–430.
80. van Dielen FM et al. Leptin and soluble leptin receptor levels in obese and weight-losing individuals. *J Clin Endocrinol Metab* 2002; 87: 1708–1716.
81. Lahlou N et al. Soluble leptin receptor in serum of subjects with complete resistance to leptin: relation to fat mass. *Diabetes* 2000; 49: 1347–1352.
82. Ahima RS. Central actions of adipocyte hormones. *Trends Endocrinol Metab* 2005; 16: 307–313.
83. Lafontan M. Fat cells: afferent and efferent messages define new approaches to treat obesity. *Annu Rev Pharmacol Toxicol* 2005; 45: 119–146.
84. Harris RB et al. Metabolic responses to leptin in obese db/db mice are strain-dependent. *Am J Physiol Regul Integr Comp Physiol* 2001; 281: R115–R132.
85. Remesar X et al. Is leptin an insulin counter-regulatory hormone? *FEBS Lett* 1997; 402: 9–11.
86. Otero M et al. Leptin, from fat to inflammation: old questions and new insights. *FEBS Lett* 2005; 579: 295–301.
87. Dunn SL et al. Feedback inhibition of leptin receptor/Jak2 signaling via Tyr1138 of the leptin receptor and suppressor of cytokine signaling 3. *Mol Endocrinol* 2005; 19: 925–938.
88. Zhang F et al. Leptin: structure, function and biology. *Vitam Horm* 2005; 71: 345–372.
89. Emilsson V et al. Leptin treatment increases suppressors of cytokine signaling in central and peripheral tissues. *FEBS Lett* 1999; 455: 170–174.
90. Waki H and Tontonoz P. Endocrine functions of adipose tissue. *Annu Rev Pathol* 2007; 2: 31–56.
91. Slieker LJ et al. Regulation of expression of ob mRNA and protein by glucocorticoids and cAMP. *J Biol Chem* 1996; 271: 5301–5304.
92. Zhang Y et al. Positional cloning of the mouse obese gene and its human homologue. *Nature* 1994; 372: 425–432.
93. Inoue W et al. Leptin induces cyclooxygenase-2 via an interaction with interleukin-1β in the rat brain. *Eur J Neurosci* 2006; 24: 2233–2245.
94. Grunfeld C et al. Endotoxin and cytokines induce expression of leptin, the ob gene product, in hamsters. *J Clin Invest* 1996; 97: 2152–2157.
95. Johnson PM et al. COX-2 inhibition attenuates anorexia during systemic inflammation without impairing cytokine production. *Am J Physiol Endocrinol Metab* 2002; 282: E650–E656.
96. Faggioni R et al. IL-1β mediates leptin induction during inflammation. *Am J Physiol* 1998; 274: R204–R208.
97. Sarraf P et al. Multiple cytokines and acute inflammation raise mouse leptin levels: potential role in inflammatory anorexia. *J Exp Med* 1997; 185: 171–175.
98. Fain JN et al. Stimulation of leptin release by arachidonic acid and prostaglandin E(2) in adipose tissue from obese humans. *Metabolism* 2001; 50: 921–928.

99. Sugiura K et al. Contribution of adipocytokines to low-grade inflammatory state as expressed by circulating C-reactive protein in Japanese men: comparison of leptin and adiponectin. *Int J Cardiol* 2008; 130: 159–164.
100. Singh P et al. Leptin induces C-reactive protein expression in vascular endothelial cells. *Arterioscler Thromb Vasc Biol* 2007; 27: e302–e307.
101. Kang SM et al. Expression of leptin receptor (Ob-R) in human atherosclerotic lesions: potential role in intimal neovascularization. *Yonsei Med J* 2000; 41: 68–75.
102. Hajer GR, van Haeften TW, and Visseren FL. Adipose tissue dysfunction in obesity, diabetes, and vascular diseases. *Eur Heart J* 2008.
103. Hennige AM et al. Leptin downregulates insulin action through phosphorylation of serine-318 in insulin receptor substrate 1. *FASEB J* 2006; 20: 1206–1208.
104. Wang MY et al. OB-Rb gene transfer to leptin-resistant islets reverses diabetogenic phenotype. *Proc Natl Acad Sci USA* 1998; 95: 714–718.
105. Greenberg AS and McDaniel ML. Identifying the links between obesity, insulin resistance and β-cell function. *Eur J Clin Invest* 2002; 32: 24–34.
106. Ktori C, Shepherd PR, and O'Rourke L. TNF-α and leptin activate the α isoform of class II phosphoinositide 3-kinase. *Biochem Biophys Res Commun* 2003; 306: 139–143.
107. Cerman AA et al. Serum leptin levels, skin leptin and leptin receptor expression in psoriasis. *Br J Dermatol* 2008; 159: 820–626.
108. Wang Y et al. Psoriasis is associated with increased levels of serum leptin. *Br J Dermatol* 2008; 158: 1134–1135.
109. Johnston A et al. Obesity in psoriasis: leptin and resistin as mediators of cutaneous inflammation. *Br J Dermatol* 2008; 159: 342–350.
110. Martin SS, Qasim A, and Reilly MP. Leptin resistance: a possible interface of inflammation and metabolism in obesity-related cardiovascular disease. *J Am Coll Cardiol* 2008; 52: 1201–1210.
111. Yu X et al. Making insulin-deficient type 1 diabetic rodents thrive without insulin. *Proc Natl Acad Sci USA* 2008; 105: 14070–14075.
112. Oral EA et al. Leptin replacement therapy for lipodystrophy. *New Engl J Med* 2002; 346: 570–578.
113. Moller DE and Kaufman KD. Metabolic syndrome: a clinical and molecular perspective. *Annu Rev Med* 2005; 56: 45–62.
114. Steppan CM and Lazar MA. Resistin and obesity-associated insulin resistance. *Trends Endocrinol Metab* 2002; 13: 18–23.
115. Holcomb IN et al. FIZZ1, a novel cysteine-rich secreted protein associated with pulmonary inflammation, defines a new gene family. *EMBO J* 2000; 19: 4046–4055.
116. Kim KH et al. A cysteine-rich adipose tissue-specific secretory factor inhibits adipocyte differentiation. *J Biol Chem* 2001; 276: 11252–11256.
117. Steppan CM et al. The hormone resistin links obesity to diabetes. *Nature* 2001; 409: 307–312.
118. Patel SD et al. Disulfide-dependent multimeric assembly of resistin family hormones. *Science* 2004; 304: 1154–1158.
119. Kusminski CM, McTernan PG, and Kumar S. Role of resistin in obesity, insulin resistance and Type II diabetes. *Clin Sci* 2005; 109: 243–256.
120. Banerjee RR and Lazar MA. Resistin: molecular history and prognosis. *J Mol Med* 2003; 81: 218–226.
121. Hartman HB et al. Mechanisms regulating adipocyte expression of resistin. *J Biol Chem* 2002; 277: 19754–19761.
122. Patel L et al. Resistin is expressed in human macrophages and directly regulated by PPARγ activators. *Biochem Biophys Res Commun* 2003; 300: 472–476.
123. Nogueiras R et al. Resistin is expressed in different rat tissues and is regulated in a tissue- and gender-specific manner. *FEBS Lett* 2003; 548: 21–27.

124. Nogueiras R et al. Regulation of resistin by gonadal, thyroid hormone, and nutritional status. *Obes Res* 2003; 11: 408–414.
125. Gui Y, Silha JV, and Murphy LJ. Sexual dimorphism and regulation of resistin, adiponectin, and leptin expression in the mouse. *Obes Res* 2004; 12: 1481–1491.
126. Lee JH et al. Circulating resistin levels are not associated with obesity or insulin resistance in humans and are not regulated by fasting or leptin administration. *J Clin Endocrinol Metab* 2003; 88: 4848–4856.
127. Silha JV et al. Plasma resistin, adiponectin and leptin levels in lean and obese subjects: correlations with insulin resistance. *Eur J Endocrinol* 2003; 149: 331–335.
128. Yannakoulia M et al. Body fat mass and macronutrient intake in relation to circulating soluble leptin receptor, free leptin index, adiponectin, and resistin concentrations in healthy humans. *J Clin Endocrinol Metab* 2003; 88: 1730–1736.
129. Azuma K et al. Correlation between serum resistin level and adiposity in obese individuals. *Obes Res* 2003; 11: 997–1001.
130. Degawa-Yamauchi M et al. Serum resistin (FIZZ3) protein is increased in obese humans. *J Clin Endocrinol Metab* 2003; 88: 5452–5455.
131. Lee JH et al. Circulating resistin in lean, obese, and insulin-resistant mouse models: lack of association with insulinemia and glycemia. *Am J Physiol Endocrinol Metab* 2005; 288: E625–E632.
132. Mahadik SR, Deo SS, and Mehtalia SD. Association of adiposity, inflammation and atherosclerosis: role of adipocytokines and CRP in Asian Indian subjects. *Metab Syndr Relat Disord* 2008; 6: 121–128.
133. Piestrzeniewicz K et al. Resistin increases with obesity and atherosclerotic risk factors in patients with myocardial infarction. *Metabolism* 2008; 57: 488–493.
134. Rajala MW et al. Regulation of resistin expression and circulating levels in obesity, diabetes, and fasting. *Diabetes* 2004; 53: 1671–1679.
135. Schaffler A et al. Identification of variables influencing resistin serum levels in patients with type 1 and type 2 diabetes mellitus. *Horm Metab Res* 2004; 36: 702–707.
136. Vendrell J et al. Resistin, adiponectin, ghrelin, leptin, and proinflammatory cytokines: relationships in obesity. *Obes Res* 2004; 12: 962–971.
137. Chen CC et al. Serum resistin level among healthy subjects: relationship to anthropometric and metabolic parameters. *Metabolism* 2005; 54: 471–475.
138. Farvid MS et al. Association of adiponectin and resistin with adipose tissue compartments, insulin resistance and dyslipidaemia. *Diabetes Obes Metab* 2005; 7: 406–413.
139. Maebuchi M et al. Low resistin levels in adipose tissues and serum in high-fat fed mice and genetically obese mice: development of an ELISA system for quantification of resistin. *Arch Biochem Biophys* 2003; 416: 164–170.
140. Savage DB et al. Resistin / Fizz3 expression in relation to obesity and peroxisome proliferator-activated receptor-γ action in humans. *Diabetes* 2001; 50: 2199–2202.
141. Hirosumi J et al. A central role for JNK in obesity and insulin resistance. *Nature* 2002; 420: 333–336.
142. Fujinami A et al. Enzyme-linked immunosorbent assay for circulating human resistin: resistin concentrations in normal subjects and patients with type 2 diabetes. *Clin Chim Acta* 2004; 339: 57–63.
143. Smith SR et al. A promoter genotype and oxidative stress potentially link resistin to human insulin resistance. *Diabetes* 2003; 52: 1611–1618.
144. Hegele RA et al. Elevated serum C-reactive protein and free fatty acids among nondiabetic carriers of missense mutations in the gene encoding lamin A/C (LMNA) with partial lipodystrophy. *Arterioscler Thromb Vasc Biol* 2003; 23: 111–116.
145. Juan CC et al. Suppressed gene expression of adipocyte resistin in an insulin-resistant rat model probably by elevated free fatty acids. *Biochem Biophys Res Commun* 2001; 289: 1328–1383.

146. Levy JR et al. Lipid metabolism and resistin gene expression in insulin-resistant Fischer 344 rats. *Am J Physiol Endocrinol Metab* 2002; 282: E626–E633.
147. Pfutzner A et al. Evaluation of human resistin assays with serum from patients with type 2 diabetes and different degrees of insulin resistance. *Clin Lab* 2003; 49: 571–576.
148. Stejskal D et al. Resistin concentrations in persons with type 2 diabetes mellitus and in individuals with acute inflammatory disease. *Biomed Pap Med Fac Univ Palacky Olomouc Czech Repub* 2003; 147: 63–69.
149. Youn BS et al. Plasma resistin concentrations measured by enzyme-linked immunosorbent assay using a newly developed monoclonal antibody are elevated in individuals with type 2 diabetes mellitus. *J Clin Endocrinol Metab* 2004; 89: 150–156.
150. Bokarewa M et al. Resistin, an adipokine with potent proinflammatory properties. *J Immunol* 2005; 174: 5789–5795.
151. Fu Y et al. Pro-inflammatory cytokine production and insulin sensitivity regulated by overexpression of resistin in 3T3-L1 adipocytes. *Nutr Metab* 2006; 3: 28.
152. Kaser S et al. Resistin messenger-RNA expression is increased by proinflammatory cytokines in vitro. *Biochem Biophys Res Commun* 2003; 309: 286–290.
153. Lu SC et al. Lipopolysaccharide increases resistin gene expression in vivo and in vitro. *FEBS Lett* 2002; 530: 158–162.
154. Rajala MW et al. Cell type-specific expression and coregulation of murine resistin and resistin-like molecule-α in adipose tissue. *Mol Endocrinol* 2002; 16: 1920–1930.
155. Fasshauer M et al. Tumor necrosis factor α is a negative regulator of resistin gene expression and secretion in 3T3-L1 adipocytes. *Biochem Biophys Res Commun* 2001; 288: 1027–1031.
156. Shojima N et al. Humoral regulation of resistin expression in 3T3-L1 and mouse adipose cells. *Diabetes* 2002; 51: 1737–1744.
157. Al-Daghri N et al. Serum resistin is associated with C-reactive protein and LDL cholesterol in type 2 diabetes and coronary artery disease in a Saudi population. *Cardiovasc Diabetol* 2005; 4: 10.
158. Axelsson J et al. Elevated resistin levels in chronic kidney disease are associated with decreased glomerular filtration rate and inflammation, but not with insulin resistance. *Kidney Int* 2006; 69: 596–604.
159. Bo S et al. Relationships between human serum resistin, inflammatory markers and insulin resistance. *Int J Obes* 2005; 29: 1315–1320.
160. Kawanami D et al. Direct reciprocal effects of resistin and adiponectin on vascular endothelial cells: a new insight into adipocytokine–endothelial cell interactions. *Biochem Biophys Res Commun* 2004; 314: 415–419.
161. Verma S et al. Resistin promotes endothelial cell activation: further evidence of adipokine–endothelial interaction. *Circulation* 2003; 108: 736–740.
162. Burnett MS et al. The potential role of resistin in atherogenesis. *Atherosclerosis* 2005; 182: 241–248.
163. Calabro P et al. Resistin promotes smooth muscle cell proliferation through activation of extracellular signal-regulated kinase 1/2 and phosphatidylinositol 3-kinase pathways. *Circulation* 2004; 110: 3335–3340.
164. Jung HS et al. Resistin is secreted from macrophages in atheromas and promotes atherosclerosis. *Cardiovasc Res* 2006; 69: 76–85.
165. Reilly MP et al. Resistin is an inflammatory marker of atherosclerosis in humans. *Circulation* 2005; 111: 932–939.
166. Senolt L et al. Resistin in rheumatoid arthritis synovial tissue, synovial fluid and serum. *Ann Rheum Dis* 2007; 66: 458–463.
167. Schaffler A et al. Adipocytokines in synovial fluid. *JAMA* 2003; 290: 1709–1710.
168. Migita K et al. The serum levels of resistin in rheumatoid arthritis patients. *Clin Exp Rheumatol* 2006; 24: 698–701.

# 11 Dietary Fatty Acids as Modulators of Adipose Inflammation

*Maximilian Zeyda and Thomas M. Stulnig*

## CONTENTS



## 11.1 INTRODUCTION

Dietary fatty acids, the main constituents of fat, are clearly overabundant in Western diets and due to their high energy contents are certainly major causes for obesity. However, it has become more and more evident that both the quantity and also the quality of dietary fat, i.e., fatty acid composition and bioactive lipid level, are critical for metabolic consequences of overnutrition (Nagao and Yanagita 2008). For an overview of the common dietary fatty acid species, see Table 11.1.

The decisive parameter of fatty acid composition is mainly the balance of saturated and monounsaturated acids on the one side and polyunsaturated fatty acids (PUFAs) on the other. However, fatty acid chain length is also important because of the effects of short and medium chain fatty acids on inflammatory and metabolic processes as reviewed elsewhere (Kles and Chang 2006; Nagao and Yanagita 2008). PUFAs, particularly of the n-3 series (often called omega-3 fatty acids) from marine

**TABLE 11.1**
**Common Dietary Fatty Acids**

| Fatty Acids | Designation | Structure |
|---|---|---|
| Saturated fatty acids: | | |
| Palmitic acid | 16:0 | 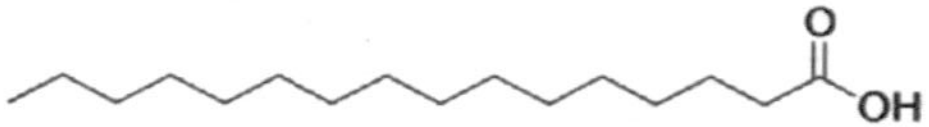 |
| Stearic acid | 18:0 | 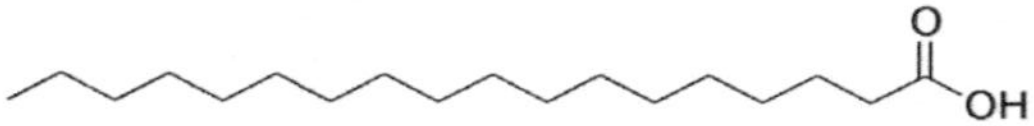 |
| Monounsaturated fatty acids:[a] | | |
| Palmitoleic acid | 16:1n-9 | 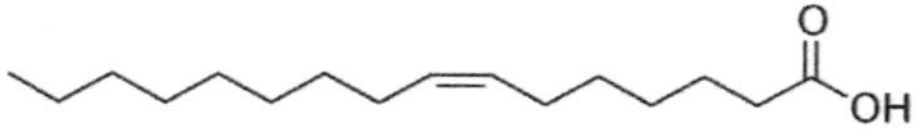 |
| Oleic acid | 18:1n-9 | 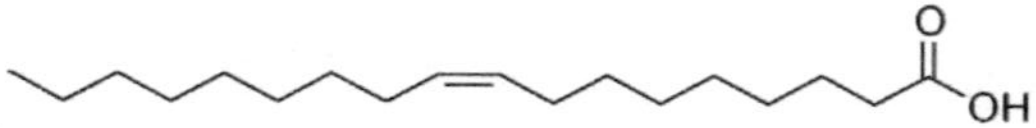 |
| Polyunsaturated fatty acids: | | |
| n-6 PUFAs | | |
| Linoleic acid | 18:2n-6 | 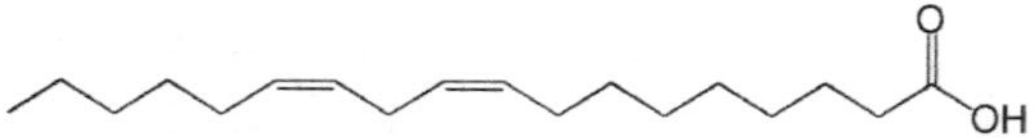 |
| γ-Linolenic acid | 18:3n-6 | 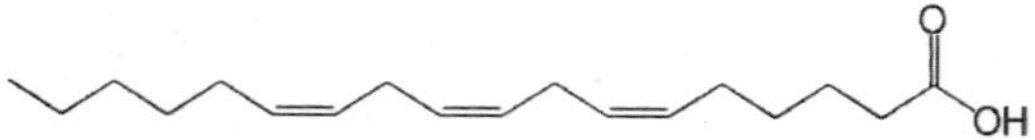 |
| Dihomo-γ-linolenic acid | 20:3n-6 | 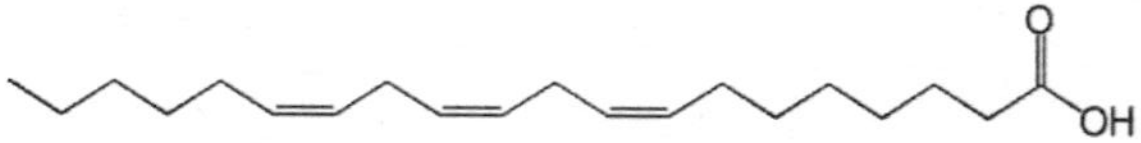 |
| Arachidonic acid | 20:4n-6 | 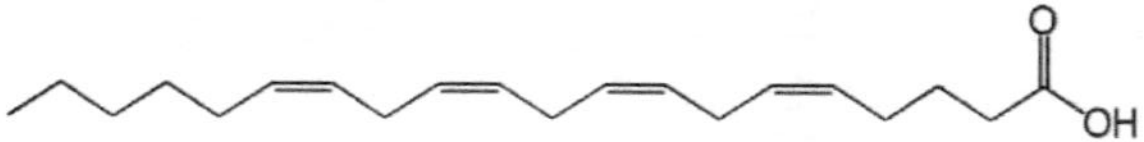 |

| | | |
|---|---|---|
| n-3 PUFAs | | |
| α-Linolenic acid | 18:3n-3 | 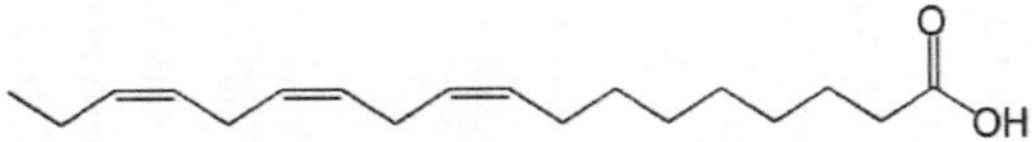 |
| Eicosapentaenoic acid (EPA) | 20:5n-3 | 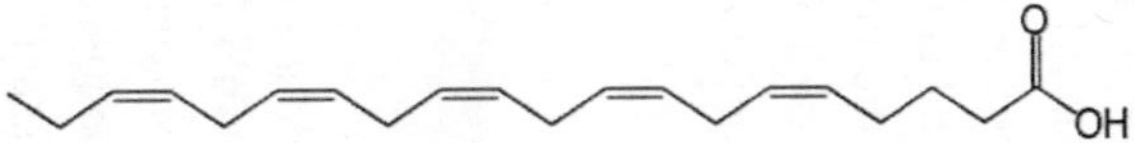 |
| Docosahexaenoic acid (DHA) | 22:6n-3 | 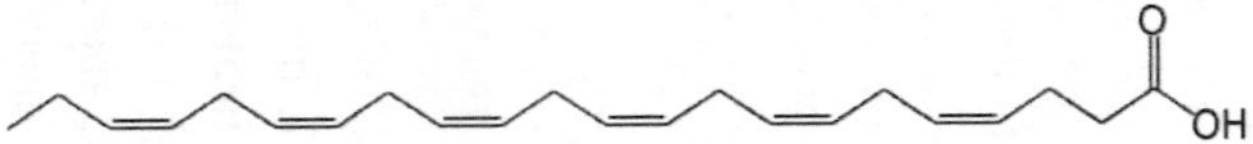 |
| Conjugated linoleic acids: | | |
| *cis*-9, *trans*-11 Linoleic acid | 9c,11t-18:2 | 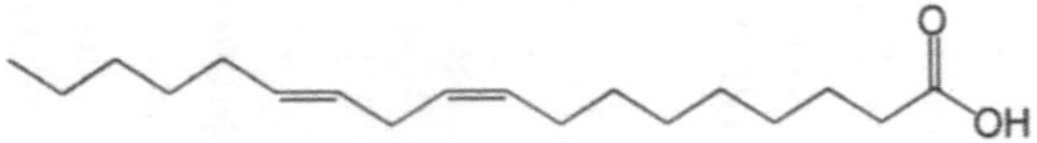 |
| *trans*-10, *cis*-12 Linoleic acid | 10t,12c-18:2 | 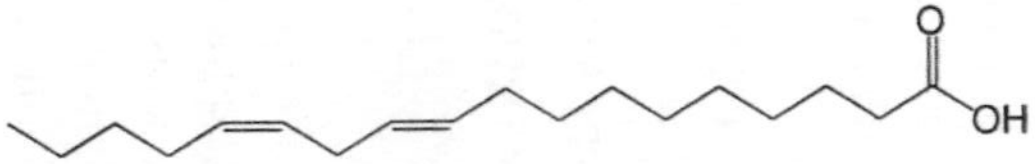 |

[a] All double bonds of listed fatty acids are in cis-configuration unless otherwise specified.

fish oils, are long known to act as anti-inflammatory and immunomodulatory agents (Calder 2006; Stulnig 2003). Moreover, we learned more than ten years ago that n-3 PUFAs from fish oil prevent the development of high fat feeding-induced insulin resistance in animals (Storlien et al. 1987).

Based on the latest findings delineating a mechanistic role of inflammation, particularly adipose tissue (AT) inflammation, in the development of insulin resistance and type 2 diabetes mellitus as discussed throughout this book, a link between anti-inflammatory and insulin sensitizing effects of PUFA appears more than plausible. Investigating the impact and the mechanisms of PUFA action on AT inflammation and insulin resistance is the focus of recent research to find novel strategies in treatment and prevention of obesity-associated insulin resistance and complications linked directly by dietary supplementation of PUFAs or by specific utilization of relevant mechanisms. Research findings of the effects of fatty acids on adipose tissue inflammation will be discussed in this chapter with emphasis on long chain PUFAs.

## 11.2 DIETARY FATTY ACID EFFECTS ON OBESITY AND CARDIOMETABOLIC RISK

### 11.2.1 Obesity

Obesity and associated insulin resistance lead to metabolic disorders such as type 2 diabetes that have become some of the most prominent health issues in industrialized countries. Moreover, type 2 diabetes confers a high risk for cardiovascular morbidity and mortality. Hence, pathophysiology and novel therapeutic approaches targeting obesity and associated metabolic disorders are major issues of medical research and pharmaceutical development.

Obesity may be treated with life style modification such as changes in dietary habits and physical activity. However, life-style interventions require enormous financial and personal efforts to be effective over the long term. Surgical procedures introduced to induce weight loss are invasive and expensive, and pharmacological approaches to reduce weight yield very limited long-term effects. Thus, weight reduction—while desirable—is not achievable in a large population at risk for type 2 diabetes and the metabolic syndrome.

An alternative approach to preventing obesity-related metabolic disorders is interfering with the development of insulin resistance rather than treatment of obesity. Modulation of metabolic and particularly inflammatory pathways by dietary fatty acids may gain importance for the development of broadly applicable preventive and therapeutic strategies for obesity-associated cardiometabolic disorders.

Weight reduction is a highly desirable effect of any dietary intervention in obesity. Indeed, a recent study shows that two months of treatment with fish oil rich in n-3 PUFA decreased body weight in women with type 2 diabetes (Kabir et al. 2007). However, in large epidemiologic studies, consumption of greater amounts of fish did not reduce weight although it did not increase obesity despite higher intake of total calories (Frost and Vestergaard 2005; Iso et al. 2006; Mozaffarian et al. 2004). Another study of dietary mixed n-6/n-3 PUFA compared to a saturated fatty acid-rich diet (mostly dairy products) confirms that total body weight is not affected by PUFAs but suggests that subcutaneous fat may be reduced by PUFA and only in women (Summers et al. 2002).

Common animal models for obesity and diabetes are mice deficient for the leptin receptor (db/db). Feeding these mice and their non-diabetic, lean littermates (db/+) with a diet enriched in re-esterified marine n-3 PUFAs (EPA, DHA) moderately increased body weight compared to an isocaloric high fat-diet rich in saturated and monosaturated fatty acids (lard oil) (Todoric et al. 2006). Another study using fish oil with considerably less n-3 PUFA content than that used in the study by Todoric et al. (2006) revealed higher perigonadal, i.e., visceral, AT mass despite unchanged total body weight (Saraswathi et al. 2007), indicating that fat mass may even be increased in mouse models of obesity.

In contrast to PUFAs found in animal and vegetable oils that include *cis*-double bonds, conjugated PUFAs also include *trans*-unsaturated fatty acids that sterically resemble saturated double bonds and exert biological actions different from those of *cis*-PUFAs. Supplementation with a mixture of conjugated linoleic acids (CLA; *cis*-9, *trans*-11 and *trans*-10, *cis*-12 isomers of CLA) over 6 months significantly reduced body fat of overweight adults (Watras et al. 2006). In addition, mice fed *trans*-10, *cis*-12 CLA exhibited 60% decreases in body fat (Terpstra et al. 2002). However, results of other clinical studies with conjugated fatty acids differed. The lack of effect on body weight of CLA is probably due to the fact that the CLA isomer responsible for reducing body fat is *trans*-10, *cis*-12 CLA, whereas >90% of total CLA intake in humans contains the *cis*-9, *trans*-11 isomer. Moreover, mouse studies showed that *trans*-10, *cis*-12 CLA may cause several undesirable side effects such as aggravation of insulin resistance (Terpstra 2004). In summary, PUFAs including CLA cannot be considered as weight-reducing dietary supplements.

### 11.2.2 Insulin Resistance and Type 2 Diabetes

Dietary unsaturated and particularly n-3 polyunsaturated fatty acids exert preventive effects on the development of insulin resistance and diabetes (Fasching et al. 1991; Storlien et al. 1996; Vessby et al. 1994) in contrast to saturated fatty acids that promote diabetes development (van Dam et al. 2002; Vessby et al. 2001). Supplementation with fish oil markedly decreased insulin responses to oral glucose loads in healthy humans (Delarue et al. 1996) and in overweight women with inflammatory phenotypes (Browning et al. 2007). This effect suggests improved insulin sensitivity from n-3 PUFA ingestion.

Furthermore, n-3 PUFA consumption during energy reduction exerts positive effects on fasting insulin serum concentrations and insulin resistance measured by HOMA-IR in young overweight and obese individuals independently from changes in body weight (Ramel et al. 2008). Also in elderly people, 8 weeks of high n-3 PUFA consumption increased insulin sensitivity (Tsitouras et al. 2008). However, fish oil and n-3 PUFA studies on insulin resistance and glycemic control in subjects with type 2 diabetes led to conflicting results (Friedberg et al. 1998; Woodman et al. 2002; Lombardo and Chicco 2006). Thus, the effects of n-3 PUFA on insulin resistance may be beneficial in obese individuals at risk for type 2 diabetes, indicating a preventive potential for type 2 diabetes development that may be particularly effective in subjects with pronounced obesity-associated inflammatory conditions.

In rats the beneficial effects of PUFAs on insulin sensitivity, particularly under high fat diet-mediated disordered conditions, are well pronounced. Storlien et al. were the first to show that PUFA of the n-3 series markedly prevented the development of insulin resistance induced by diets rich in linoleic acid, an n-6 fatty acid, and saturated fats (Storlien et al. 1987). Accordingly, during a euglycemic–hyperinsulinemic clamp, safflower oil-fed rats were insulin-resistant compared with control and fish oil-fed rats as reflected by markedly reduced glucose infusion rates (Jucker et al. 1999). Also in mice fed a high-fat diet containing safflower oil with a partial (8%) fish oil replacement for 2 weeks euglycemic–hyperinsulinemic clamp experiments demonstrated that oil supplementation preserved hepatic insulin sensitivity (Neschen et al. 2007). In contrast, peripheral insulin sensitivity was not affected, indicating that studies of PUFA effects on insulin sensitivity should discriminate hepatic and peripheral insulin resistance.

A study investigating *cis*-9, *trans*-11 CLA-feeding in ob/ob mice showed reduced insulin resistance in conjunction with reduced inflammatory parameters indicating that *cis*-9, *trans*-11 CLA may improve insulin sensitivity in obesity by counteracting the inflammatory response (Moloney et al. 2007). In contrast, feeding a diet containing *trans*-10, *cis*-12 CLA induced adverse effects, namely hyperlipidemia, insulin resistance, and AT inflammation and macrophage infiltration (Poirier et al. 2006; Roche et al. 2002; Tsuboyama-Kasaoka et al. 2000). These and other data make clear that potentially beneficial effects of CLA are isomer-specific and restricted to *cis*-9, *trans*-11 CLA.

### 11.2.3 Lipid Metabolism and Cardiometabolic Risk

Dietary PUFAs positively modify several factors related to cardiometabolic risk. n-3 PUFAs markedly decrease fasting and postprandial serum triacylglycerol and free fatty acid concentrations (Weintraub et al. 1988). The lipid-lowering effect of n-3 PUFAs is also evident in mice (Todoric et al. 2006). Fish oil profoundly lowers very low density lipoprotein (VLDL) and apolipoprotein B concentrations in healthy and hypertriglyceridemic subjects (Nestel et al. 1984). n-3 PUFA effects on serum LDL are divergent, with slight increases in LDL cholesterol concentrations and reductions in atherogenic small, dense LDL particles (Carpentier et al. 2006).

In addition to their impact on serum lipids, n-3 PUFAs exert anti-atherogenic and anti-thrombotic effects that add to their beneficial effects on cardiovascular risk. They lower blood pressure and heart rate, improve vascular and platelet function, and induce changes in cellular fatty acid partitioning, away from triacylglycerol synthesis pathways and toward fat oxidation (Mori and Woodman 2006). Hence, incident atrial fibrillation and coronary heart disease can be reduced by PUFA uptake (Frost and Vestergaard 2005; Iso et al. 2006; Mozaffarian et al. 2004). Notably, n-3 PUFAs significantly reduce mortality, non-fatal myocardial infarction, and stroke in patients after myocardial infarction (Gruppo Italiano, 1999).

## 11.3 DIETARY FATTY ACID EFFECTS ON ADIPOSE TISSUE INFLAMMATION

Gene expression profiling revealed that the n-3 PUFAs, EPA, and DHA potently counteract saturated fat-induced alterations in obese db/db mice fed isocaloric high-fat diets (Todoric et al. 2006). Highly purified n-3 PUFAs re-esterified to triglycerides were used in these studies, virtually eliminating potential confounding effects from other fish oil components (Huber et al. 2007; Todoric et al. 2006).

Most importantly, a saturated fatty acid-rich diet induces a large number of inflammatory genes encoding, e.g., cytokines and chemokines in AT, whose induction was completely blunted by n-3 PUFAs. These include genes for MCP-1, macrophage markers CD68 and CD11b, and TNF. n-3 PUFAs in a high-fat diet also increased adiponectin gene expression in AT and substantially enhanced circulating adiponectin levels compared to control diet-fed animals (Itoh et al.2007; Neschen et al.2007; Todoric et al. 2006). Similar results were found after fish oil feeding in LDL receptor-deficient mice (Saraswathi et al. 2007).

The anti-inflammatory effect of n-3 PUFAs in db/db mice was evident in the subcutaneous and also in the metabolically important visceral (gonadal) AT, but not in spleen and lung, indicating a specific effect in obesity on inflamed AT. In parallel with inflammatory genes, high-fat diet-induced AT infiltration by macrophages was completely prevented by n-3 PUFAs as was inflammatory JNK activation (Todoric et al. 2006).

One important parameter for insulin resistance and inflammatory alterations in adipose tissue is adipocyte size. Along with their anti-inflammatory actions, n-3 PUFAs reduced adipocyte size in high-fat diet-fed db/db mice (Huber et al. 2007). n-3 PUFAs were significantly enriched in adipose tissue, pointing to direct effects of n-3 PUFAs in adipocytes as discussed below.

Concomitantly, a large number of markers of adipose tissue remodeling that are heavily upregulated by usual high-fat diets rich in saturated fatty acids remained unaltered compared to low-fat diets when the high-fat diet included n-3 PUFAs. Hence n-3 PUFAs exert highly potent effects on adipocyte size, AT inflammation, and remodeling in mice. Also, the addition of *cis*-9, *trans*-11 CLA as a free fatty acid (0.6 g/100 g) to a high-fat diet markedly downregulated several inflammatory markers in adipose tissue (Moloney et al. 2007).

Clinical studies investigating PUFA effects on AT have not been published yet, although the animal experiments discussed above strongly suggest a link between improvement of insulin sensitivity and inflammatory parameters as observed, for example, in healthy elderly individuals (Tsitouras et al. 2008). A hint of a causal link between PUFA action on inflammation and insulin resistance comes from the finding that supplementation of n-3 PUFAs to premenopausal, nondiabetic females markedly decreased insulin responses to oral glucose loads in subjects with high levels of inflammatory indices, but not significantly in those with low inflammatory status (Browning et al. 2007). Treatment with PUFAs may unlock the development of adipose tissue inflammation in obesity (White and Marette 2006). Nevertheless, clinical studies of their impacts on AT inflammation are eagerly awaited to learn whether PUFA will fulfill these expectations in humans.

## 11.4 MECHANISMS OF PUFA ACTION ON ADIPOSE TISSUE INFLAMMATION

### 11.4.1 Reduction of AT Macrophage Numbers

Diminished AT inflammation after n-3 and conjugated PUFA treatment is exemplified by a significant reduction of macrophage numbers in white AT (Todoric et al. 2006; Moloney et al. 2007). AT macrophages are major sources of inflammatory adipokines. The question of which mechanisms are responsible for AT macrophage attraction in obesity is hence of critical importance but is still largely unresolved (Zeyda and Stulnig 2007).

Tissue invasions by macrophages and other inflammatory cells are generally driven by chemokines. Expression of a number of chemokines has been shown to occur in obese human AT (Moraes et al. 2003; Wu et al. 2007; Huber et al. 2008) although secretion has only been shown for MCP-1 (Dahlman et al. 2005). However, chemokines largely act locally so that only an overspill will be found in the circulation. n-3 PUFAs suppress expression of many chemokines in adipose tissue, indicating a possible mechanism of n-3 PUFA interference with AT inflammation. A reduction in AT gene expression by PUFAs has been shown for MCP-1 (Todoric et al. 2006). However, the role of MCP-1 for macrophage recruitment to AT in obesity is still controversial (Chow et al. 2007; Inouye et al. 2007; Kirk et al. 2008). PUFAs may also directly affect macrophage migration. They modulate adhesion and migration of different leukocyte subsets (reviewed in (Pompeia et al. 2000)), but a possible role of such effects in the reduction of AT macrophage infiltration remains to be evaluated.

An important molecular mediator of PUFA effects on macrophages may be osteopontin, a multifunctional protein also involved in monocyte and macrophage migration (Denhardt and Guo 1993) and induction of a variety of cytokines and chemokines in myeloid cells (Xu et al. 2005). The expression of osteopontin is drastically increased in obesity and osteopontin is crucial in the development of insulin resistance (Kiefer et al. 2008; Nomiyama et al. 2007). Of note, n-3 PUFAs markedly downregulate osteopontin expression in AT in parallel with reduced macrophage numbers (Todoric et al. 2006 and unpublished data). However, involvement of osteopontin in the reduction of adipose tissue inflammation by n-3 PUFAs is still speculative.

### 11.4.2 Alteration of Macrophage Signal Transduction

PUFAs exert manifold effects on signal transduction of leukocytes, for example, by altering eicosanoid (prostaglandin, leukotriene; see Section 11.4.3) synthesis, nuclear receptor activation (peroxisome proliferator-activated receptors, PPARs; see Section 11.4.4), and by changing the molecular compositions of special signaling platforms called lipid rafts (Stulnig 2003; Zeyda and Stulnig 2006). Concerning macrophages, an important feature of PUFA action is counteraction of Toll-like receptor-4 (TLR4) signaling activated by saturated fatty acids (Lee et al. 2003b).

Saturated fatty acids induce phosphorylation of AKT and activation of the nuclear factor NF-κB in macrophages (Lee and Hwang 2006). NF-κB activation is also involved in the induction of inflammation during obesity and is inhibited by n-3 PUFAs (Lee et al. 2003a; Lee et al. 2003b). Importantly, NF-κB-independent inhibition of TLR4 signaling in myeloid cells was described by Zeyda et al. (2005). Inhibition of TLR4 signaling by PUFAs would also block the pro-inflammatory effects of high-fat diet-induced endotoxemia in obesity (Cani et al. 2007; Erridge et al. 2007). Elevated circulating endotoxin (LPS) concentrations result in adipose tissue inflammation and elevated fasting glycemia and insulinemia due to hepatic insulin resistance (Cani et al. 2007). Hence, inhibition of the activation of TLR4 pathways is probably an important contribution to the beneficial effects of PUFAs, particularly under conditions of hyperlipidemia and endotoxemia associated with obesity.

### 11.4.3 Eicosanoids

PUFAs are precursors of immunologically active lipid mediators, i.e., eicosanoid messenger molecules such as prostaglandins, leukotrienes, and thromboxanes. These mediators are usually derived from arachidonic acid (n-6) liberated from membrane phospholipids by phospholipase $A_2$. Metabolism of arachidonic acid by cyclooxygenases (COXs) leads to generation of prostaglandins and thromboxanes of the 2 series. PUFAs of the n-3 series interfere with the biosynthesis of arachidonic acid-derived molecules and give rise to chemically different mediator molecules.

When eicosapentaenoic acid (n-3) instead of arachidonic acid is metabolized by COX, the reaction produces prostaglandins and thromboxanes of the 3 series that exert attenuated or partially different biological effects (Calder et al. 2002; Cantrell 2002). Moreover, although the affinity of COX for eicosapentaneoic acid is low, it inhibits COX activity (Wada et al. 2007). In addition to directly interfering with enzymes of eicosanoid synthesis, PUFAs can also affect involved enzymes by altering gene expression as shown for COX-2 in monocytes (Lee et al. 2003a). Although n-6 and n-3 PUFAs affect eicosanoid synthesis differently, the functional outcomes of these changes with respect to immunomodulation, in particular in vivo interactions of the generated messenger molecules, are often not predictable. Differences of in vitro and in vivo eicosanoid production may occur (Knapp et al. 1986; Saito et al. 1997) as do species differences in eicosanoid effects (Morita et al. 1983). Hence, extrapolations of in vitro data to in vivo situations are extremely difficult and depend on the clinical situation, e.g., the inflammatory condition under study, and only in certain circumstances.

Recent research characterized endogenous mediators of resolution, i.e., the actively regulated program of returning from inflammation to a healthy state (Gilroy et al. 2004). These resolving lipid mediators, named resolvins and protectins, are synthesized in several enzymatic steps from n-3 and also n-6 PUFAs (Serhan et al. 2007). The elucidation of their biosynthetic pathways revealed that n-3 PUFAs not only replace analogous n-6-derived inflammatory mediators, but also give rise to special biologically active molecules produced via distinct biosynthetic steps. The contribution of these novel classes of PUFA-derived lipid mediators to beneficial and anti-inflammatory effects remains to be elucidated, but they potently drive the program of resolution in the nanomolar range (Bannenberg et al. 2005; Schwab et al. 2007) and thus their therapeutic potential appears promising.

### 11.4.4 PPARγ Activation in AT Macrophages

Another principal mechanism for modulation of immune responses by PUFAs is through direct alteration of gene expression by binding and activation of nuclear receptors, i.e., ligand-binding transcription factors. PPARγ preferentially binds a variety of PUFAs and their derivatives and has been shown to be critically involved in adipocyte and macrophage differentiation (Clark et al. 2000; Marx et al. 1998; Yang et al. 2000). Activation of PPARγ and other members of the PPAR family is a mechanism by which PUFAs could directly improve insulin resistance, e.g., by inducing adipogenesis to provide more small well-functioning adipocytes.

PPARγ activation provokes a shift to the anti-inflammatory M2 type of macrophage (Castrillo and Tontonoz 2004; Sharma and Staels 2007). Hence, in addition to its metabolic and direct insulin-sensitizing effect mediated through adipocytes, PUFA-mediated PPARγ activation indirectly influences insulin sensitivity by affecting AT macrophages. In accordance with these considerations, recent studies show that myeloid cell-specific disruption of PPARγ results in a significant shift to inflammatory M1 macrophages and decreased insulin sensitivity (Odegaard et al. 2007). Thus the insulin sensitizing role of the classical PPARγ activators, i.e., thiazolidinediones, and also of PUFAs may extend to the anti-inflammatory role of PPARγ in macrophages, particularly in the AT.

### 11.4.5 Induction of Adiponectin

In addition to PPARγ-dependent effects in macrophages, PPAR activation may also indirectly affect AT macrophages and inflammation by induction of adiponectin, a well established regulator of insulin sensitivity with anti-inflammatory properties (Ouchi and Walsh 2007). PUFAs stimulate adiponectin expression in adipocytes via PPARα (Neschen et al. 2006). Adiponectin reduces lipopolysaccharide (LPS)-stimulated TNF-α production in human macrophages (Yokota et al. 2000), while stimulating anti-inflammatory IL-10 production (Kumada et al. 2004). Moreover, adiponectin was shown to inhibit TLR-mediated NF-κB activation in mouse macrophages (Yamaguchi et al. 2005).

N-3 PUFA treatment showed adiponectin-increasing effects in healthy and obese subjects in most but not all clinical studies of this issue (Guebre-Egziabher et al. 2007; Itoh et al. 2007; Ramel et al. 2008; Tsitouras et al. 2008). Also a combination of n-3 PUFA and CLA tended to increase adiponectin levels in a small and heterogeneous study population (Sneddon et al. 2008). Moreover, the proportion of n-3 PUFAs in human plasma was shown to be positively associated with circulating adiponectin concentrations (Fernandez-Real et al. 2005). Animal studies robustly confirm the n-3 PUFA-induced increase of AT adiponectin gene expression and elevation of its serum concentrations (Flachs et al. 2006; Itoh et al. 2007; Neschen

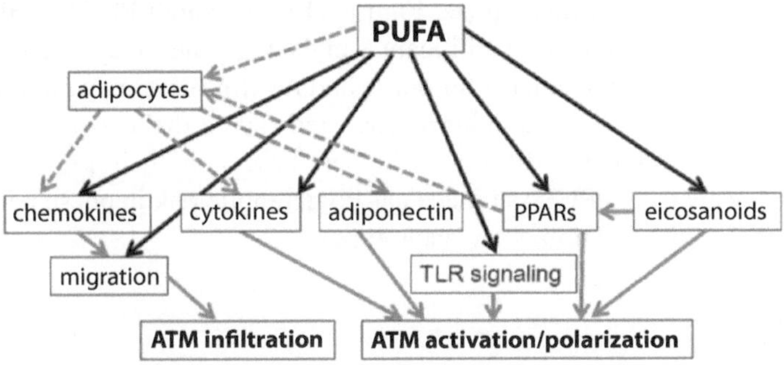


**FIGURE 11.1** Network of PUFA action on adipose tissue inflammation. Black arrows = PUFA action on inflammatory pathways. Dashed arrows = PUFA action via adipocytes. Grey arrows = mechanisms affecting adipose tissue (AT) inflammation. ATM = adipose tissue macrophage.

et al. 2007; Neschen et al. 2006; Todoric et al. 2006). Interestingly, this increase of adiponectin expression may, at least in part, be due to diminished TNF-α expression in AT macrophages, as suggested by adipocyte–macrophage co-culture experiments (Itoh et al. 2007). Thus, PUFA, PPARs, and adiponectin appear to critically regulate the interplay of adipocytes and macrophages to reduce AT inflammation and increase insulin sensitivity.

Improved insulin sensitivity by PUFA-induced adiponectin expression may in turn contribute to the AT inflammation-diminishing effects of PUFAs by restoring adipocyte function. Adipocytes, and particularly pre-adipocytes, contribute to AT inflammation by production of inflammatory cytokines stimulated by LPS (Chung et al. 2006; Hoch et al. 2008) and macrophage-secreted factors (Permana et al. 2006). Hence, dietary fatty acids may interfere with inflammatory responses of adipocytes. 3T3-L1 adipocytes treated with saturated palmitic acid exhibited increases in TNF-α production and decreases in IL-10 production, whereas n-3 PUFA treatment had no effect on TNF-α and increased anti-inflammatory IL-10 production (Bradley et al. 2008). However, the relative contribution of adipocytes to the inflammatory reaction in adipose tissue appears limited in obesity when AT has been infiltrated by a large number of macrophages.

## 11.5 SUMMARY AND CONCLUSIONS

A marked inhibition of AT inflammation with dietary fatty acids, in particular long-chain PUFAs including n-3 PUFA and CLA, can be achieved in animal experiments, but clinical studies to confirm these data are still missing. Moreover, molecular mechanisms of PUFA action on AT inflammation and their relevance in obesity must be evaluated in greater detail.

Inhibition of TLR activation in macrophages and adipocytes, PPAR activation, alteration in eicosanoid production including novel classes of PUFA-derived lipid mediators, and regulation of adiponectin expression are currently the most promising candidate mechanisms of PUFA action on AT inflammation. For an overview of

PUFA action on AT inflammation see Figure 11.1. Although PUFAs will not cure obesity by leading to clinically significant weight loss, they may become particularly attractive dietary supplements for obese and insulin-resistant patients, reducing several risk factors for cardiovascular disease and type 2 diabetes in this high-risk population. Future studies will show whether, based on their pronounced anti-inflammatory effects, n-3 PUFAs reduce cardiometabolic risk in obese patients, particularly in those with inflammatory phenotypes.

## REFERENCES

Bannenberg GL et al. 2005. Molecular circuits of resolution: formation and actions of resolvins and protectins. *J Immunol* 174: 4345–4355.

Bradley RL, Fisher FF, and Maratos-Flier E. 2008. Dietary fatty acids differentially regulate production of TNF-α and IL-10 by murine 3T3-L1 adipocytes. *Obesity* 16: 938–944.

Browning LM et al. 2007. The impact of long chain n-3 polyunsaturated fatty acid supplementation on inflammation, insulin sensitivity and CVD risk in a group of overweight women with an inflammatory phenotype. *Diabetes Obes Metab* 9: 70–80.

Calder PC. 2006. Polyunsaturated fatty acids and inflammation. *Prostagl Leukot Essent Fatty Acids* 75: 197–202.

Calder PC et al. 2002. Fatty acids and lymphocyte functions. *Br J Nutr* 87: S31–S48.

Cani PD et al. 2007. Metabolic endotoxemia initiates obesity and insulin resistance. *Diabetes* 56: 1761–1722.

Cantrell DA. 2002. T cell antigen receptor signal transduction. *Immunology* 105: 369–374.

Carpentier YA, Portois L, and Malaisse WJ. 2006. n-3 fatty acids and the metabolic syndrome. *Am J Clin Nutr* 83: 1499S–1504S.

Castrillo A and Tontonoz P. 2004. Nuclear receptors in macrophage biology: at the crossroads of lipid metabolism and inflammation. *Annu Rev Cell Dev Biol* 20: 455–480.

Chow FY et al. 2007. Monocyte chemoattractant protein-1-induced tissue inflammation is critical for the development of renal injury but not type 2 diabetes in obese db/db mice. *Diabetologia* 50: 471–480.

Chung S et al. 2006. Preadipocytes mediate LPS-induced inflammation and insulin resistance in primary cultures of newly differentiated human adipocytes. *Endocrinology* 147: 5340–5351.

Clark RB et al. 2000. The nuclear receptor PPARγ and immunoregulation: PPARγ mediates inhibition of helper T cell responses. *J Immunol* 164: 1364–1371.

Dahlman I et al. 2005. A unique role of monocyte chemoattractant protein 1 among chemokines in adipose tissue of obese subjects. *J Clin Endocrinol Metab* 90: 5834–5840.

Delarue J et al. 1996. Effects of fish oil on metabolic responses to oral fructose and glucose loads in healthy humans. *Am J Physiol* 270: E353–362.

Denhardt DT and Guo X. 1993. Osteopontin: a protein with diverse functions. *FASEB J* 7: 1475–1482.

Erridge C et al. 2007. A high-fat meal induces low-grade endotoxemia: evidence of a novel mechanism of postprandial inflammation. *Am J Clin Nutr* 86: 1286–1292.

Fasching P et al. 1991. Metabolic effects of fish oil supplementation in patients with impaired glucose tolerance. *Diabetes* 40: 583–589.

Fernandez-Real JM, Vendrell J, and Ricart W. 2005. Circulating adiponectin and plasma fatty acid profile. *Clin Chem* 51: 603–609.

Flachs P et al. 2006. Polyunsaturated fatty acids of marine origin induce adiponectin in mice fed a high-fat diet. *Diabetologia* 49: 394–397.

Friedberg CE et al. 1998. Fish oil and glycemic control in diabetes: a meta-analysis. *Diabetes Care* 21: 494–500.

Frost L and Vestergaard P. 2005. n-3 Fatty acids consumed from fish and risk of atrial fibrillation or flutter: Danish Diet, Cancer, and Health Study. *Am J Clin Nutr* 81: 50–54.

Gilroy DW et al. 2004. Inflammatory resolution: new opportunities for drug discovery. *Nat Rev Drug Discov* 3: 401–416.

Gruppo Italiano per lo Studio della Sopravivenza nell'Infarto Miocardico. 1999. Dietary supplementation with n-3 polyunsaturated fatty acids and vitamin E after myocardial infarction: results of GISSI Prevenzione trial. *Lancet* 354: 447–455.

Guebre-Egziabher et al. 2008. Nutritional intervention to reduce the n-6/n-3 fatty acid ratio increases adiponectin concentration and fatty acid oxidation in healthy subjects. *Eur J Clin Nutr* 62: 1287–1293.

Hoch M et al. 2008. LPS induces interleukin-6 and interleukin-8 but not tumor necrosis factor-α in human adipocytes. *Cytokine* 41: 29–37.

Huber J et al. 2008. CC chemokine and CC chemokine receptor profiles in visceral and subcutaneous adipose tissue are altered in human obesity. *J Clin Endocrinol Metab* 93: 3215–3221.

Huber J et al. 2007. Prevention of high-fat diet-induced adipose tissue remodeling in obese diabetic mice by n-3 polyunsaturated fatty acids. *Int J Obes* 31: 1004–1013.

Inouye KE et al. 2007. Absence of CC chemokine ligand 2 does not limit obesity-associated infiltration of macrophages into adipose tissue. *Diabetes* 56: 2242–2250.

Iso H et al. 2006. Intake of fish and n-3 fatty acids and risk of coronary heart disease among Japanese: Japan Public Health Center–Based (JPHC) Study Cohort I. *Circulation* 113: 195–202.

Itoh M et al. 2007. Increased adiponectin secretion by highly purified eicosapentaenoic acid in rodent models of obesity and human obese subjects. *Arterioscler Thromb Vasc Biol* 27: 1918–1925.

Jucker BM et al. 1999. Differential effects of safflower oil versus fish oil feeding on insulin-stimulated glycogen synthesis, glycolysis, and pyruvate dehydrogenase flux in skeletal muscle: a $^{13}C$ nuclear magnetic resonance study. *Diabetes* 48: 134–140.

Kabir M et al. 2007. Treatment for two months with n 3-polyunsaturated fatty acids reduces adiposity and some atherogenic factors but does not improve insulin sensitivity in women with type 2 diabetes: a randomized controlled study. *Am J Clin Nutr* 86: 1670–1679.

Kiefer FW et al. 2008. Osteopontin expression in human and murine obesity: extensive local up-regulation in adipose tissue but minimal systemic alterations. *Endocrinology* 149: 1350–1357.

Kirk EA et al. 2008. Macrophage chemoattractant protein-1 deficiency fails to restrain macrophage infiltration into adipose tissue. *Diabetes* 57: 1254–1261.

Kles KA and Chang EB. 2006. Short-chain fatty acids impact on intestinal adaptation, inflammation, carcinoma, and failure. *Gastroenterology* 130: S100–105.

Knapp HR et al. 1986. In vivo indexes of platelet and vascular function during fish-oil administration in patients with atherosclerosis. *New Engl J Med* 314: 937–942.

Kumada M et al. 2004. Adiponectin specifically increased tissue inhibitor of metalloproteinase-1 through interleukin-10 expression in human macrophages. *Circulation* 109: 2046–2049.

Lee JY and Hwang DH. 2006. Modulation of inflammatory gene expression by lipids: mediation through Toll-like receptors. *Mol Cells* 21: 174–185.

Lee JY et al. 2003a. Differential modulation of Toll-like receptors by fatty acids: preferential inhibition by n-3 polyunsaturated fatty acids. *J Lipid Res* 44: 479–486.

Lee JY et al. 2003b. Reciprocal modulation of Toll-like receptor-4 signaling pathways involving MyD88 and phosphatidylinositol 3-kinase/AKT by saturated and polyunsaturated fatty acids. *J Biol Chem* 278: 37041–37051.

Lombardo YB and Chicco AG. 2006. Effects of dietary polyunsaturated n-3 fatty acids on dyslipidemia and insulin resistance in rodents and humans: a review. *J Nutr Biochem* 17: 1–13.

Marx N et al. 1998. Macrophages in human atheroma contain PPARγ: differentiation-dependent peroxisomal proliferator-activated receptor γ (PPARγ) expression and reduction of MMP-9 activity through PPARγ activation in mononuclear phagocytes in vitro. *Am J Pathol* 153: 17–23.

Moloney F et al. 2007. Antidiabetic effects of cis-9, trans-11 conjugated linoleic acid may be mediated via anti-inflammatory effects in white adipose tissue. *Diabetes* 56: 574–582.

Moraes RC et al. 2003. Study of the alteration of gene expression in adipose tissue of diet-induced obese mice by microarray and reverse transcription polymerase chain reaction analyses. *Endocrinology* 144: 4773–4782.

Mori TA and Woodman RJ. 2006. The independent effects of eicosapentaenoic acid and docosahexaenoic acid on cardiovascular risk factors in humans. *Curr Opin Clin Nutr Metab Care* 9: 95–104.

Morita I et al. 1983. Effects of eicosapentaenoic acid on arachidonic acid metabolism in cultured vascular cells and platelets: species difference. *Thromb Res* 31: 211–217.

Mozaffarian D et al. 2004. Fish intake and risk of incident atrial fibrillation. *Circulation* 110: 368–373.

Nagao K and Yanagita T, 2008. Bioactive lipids in metabolic syndrome. *Progr Lipid Res* 47: 127–146.

Neschen S et al. 2007. n-3 Fatty acids preserve insulin sensitivity in vivo in a peroxisome proliferator-activated receptor-α-dependent manner. *Diabetes* 56: 1034–1041.

Neschen S et al. 2006. Fish oil regulates adiponectin secretion by a peroxisome proliferator-activated receptor-γ-dependent mechanism in mice. *Diabetes* 55: 924–928.

Nestel PJ et al. 1984. Suppression by diets rich in fish oil of very low density lipoprotein production in man. *J Clin Invest* 74: 82–89.

Nomiyama T et al. 2007. Osteopontin mediates obesity-induced adipose tissue macrophage infiltration and insulin resistance in mice. *J Clin Invest* 117: 2877–2888.

Odegaard JI et al. 2007. Macrophage-specific PPARγ controls alternative activation and improves insulin resistance. *Nature* 447: 1116–1120.

Ouchi N and Walsh K. 2007. Adiponectin as an anti-inflammatory factor. *Clin Chim Acta* 380: 24–30.

Permana PA, Menge C, and Reaven PD. 2006. Macrophage-secreted factors induce adipocyte inflammation and insulin resistance. *Biochem Biophys Res Commun* 341: 507–514.

Poirier H et al. 2006. Nutritional supplementation with trans-10, cis-12 conjugated linoleic acid induces inflammation of white adipose tissue. *Diabetes* 55: 1634–1641.

Pompeia C et al. 2000. Effect of fatty acids on leukocyte function. *Braz J Med Biol Res* 33: 1255–1268.

Ramel A et al. 2008. Beneficial effects of long-chain n-3 fatty acids included in an energy-restricted diet on insulin resistance in overweight and obese European young adults. *Diabetologia* 51: 1261–1268.

Roche HM et al. 2002. Isomer-dependent metabolic effects of conjugated linoleic acid: insights from molecular markers sterol regulatory element-binding protein-1c and LXRα. *Diabetes* 51: 2037–2044.

Saito J et al. 1997. Mechanisms of enhanced production of PGI2 in cultured rat vascular smooth muscle cells enriched with eicosapentaenoic acid. *Atherosclerosis* 131: 219–228.

Saraswathi V et al. 2007. Fish oil increases cholesterol storage in white adipose tissue with concomitant decreases in inflammation, hepatic steatosis, and atherosclerosis in mice. *J Nutr* 137: 1776–1782.

Schwab JM et al. 2007. Resolvin E1 and protectin D1 activate inflammation resolution programmes. *Nature* 447: 869–874.
Serhan CN et al. 2007. Resolution of inflammation: state of the art, definitions and terms. *FASEB J* 21: 325–332.
Sharma AM and Staels B. 2007. Review: peroxisome proliferator-activated receptor γ and adipose tissue: understanding obesity-related changes in regulation of lipid and glucose metabolism. *J Clin Endocrinol Metab* 92: 386–395.
Sneddon AA et al. 2008. Effect of a conjugated linoleic acid and omega-3 fatty acid mixture on body composition and adiponectin. *Obesity* 16: 1019–1024.
Storlien LH et al. 1996. Dietary fats and insulin action. *Diabetologia* 39: 621–631.
Storlien LH et al. 1987. Fish oil prevents insulin resistance induced by high-fat feeding in rats. *Science* 237: 885–888.
Stulnig TM. 2003. Immunomodulation by polyunsaturated fatty acids: mechanisms and effects. *Int Arch Allergy Immunol* 132: 310–321.
Summers LK et al. 2002. Substituting dietary saturated fat with polyunsaturated fat changes abdominal fat distribution and improves insulin sensitivity. *Diabetologia* 45: 369–377.
Terpstra AH. 2004. Effect of conjugated linoleic acid on body composition and plasma lipids in humans: an overview of the literature. *Am J Clin Nutr* 79: 352–361.
Terpstra AH et al. 2002. The decrease in body fat in mice fed conjugated linoleic acid is due to increases in energy expenditure and energy loss in the excreta. *J Nutr* 132: 940–945.
Todoric J et al. 2006. Adipose tissue inflammation induced by high-fat diet in obese diabetic mice is prevented by n-3 polyunsaturated fatty acids. *Diabetologia* 49: 2109–2119.
Tsitouras PD et al. 2008. High omega-3 fat intake improves insulin sensitivity and reduces CRP and IL-6, but does not affect other endocrine axes in healthy older adults. *Horm Stoffwechselforsch* 40: 199–205.
Tsuboyama-Kasaoka N et al. 2000. Conjugated linoleic acid supplementation reduces adipose tissue by apoptosis and develops lipodystrophy in mice. *Diabetes* 49: 1534–1542.
van Dam RM et al. 2002. Dietary fat and meat intake in relation to risk of type 2 diabetes in men. *Diabetes Care* 25: 417–424.
Vessby B et al. 1994. The risk to develop NIDDM is related to the fatty acid composition of the serum cholesterol esters. *Diabetes* 43: 1353–1357.
Vessby B et al. 2001. Substituting dietary saturated for monounsaturated fat impairs insulin sensitivity in healthy men and women: KANWU Study. *Diabetologia* 44: 312–319.
Wada M et al. 2007. Enzymes and receptors of prostaglandin pathways with arachidonic acid-derived versus eicosapentaenoic acid-derived substrates and products. *J Biol Chem* 282: 22254–22266.
Watras AC et al. 2006. The role of conjugated linoleic acid in reducing body fat and preventing holiday weight gain. *Int J Obes* 31: 481–487.
Weintraub MS et al. 1988. Dietary polyunsaturated fats of the W-6 and W-3 series reduce post-prandial lipoprotein levels: chronic and acute effects of fat saturation on postprandial lipoprotein metabolism. *J Clin Invest* 82: 1884–1893.
White PJ and Marette A, 2006. Is omega-3 key to unlocking inflammation in obesity? *Diabetologia* 49: 1999–2001.
Woodman RJ et al. 2002. Effects of purified eicosapentaenoic and docosahexaenoic acids on glycemic control, blood pressure, and serum lipids in type 2 diabetic patients with treated hypertension. *Am J Clin Nutr* 76: 1007–1015.
Wu H et al. 2007. T cell accumulation and regulated on activation, normal T cell expressed and secreted upregulation in adipose tissue in obesity. *Circulation* 115: 1029–1038.
Xu G et al. 2005. Role of osteopontin in amplification and perpetuation of rheumatoid synovitis. *J Clin Invest* 115: 1060–1067.

Yamaguchi N et al. 2005. Adiponectin inhibits Toll-like receptor family-induced signaling. *FEBS Lett* 579: 6821–6826.

Yang XY et al. 2000. Activation of human T lymphocytes is inhibited by peroxisome proliferator-activated receptor γ (PPARγ) agonists: PPARγ co-association with transcription factor NFAT. *J Biol Chem* 275: 4541–4544.

Yokota T et al. 2000. Adiponectin, a new member of the family of soluble defense collagens, negatively regulates the growth of myelomonocytic progenitors and the functions of macrophages. *Blood* 96: 1723–1732.

Zeyda M et al. 2005. Polyunsaturated fatty acids block dendritic cell activation and function independently of NF-κB activation. *J Biol Chem* 280: 14293–14301.

Zeyda M and Stulnig TM. 2006. Lipid Rafts & Co.: an integrated model of membrane organization in T cell activation. *Progr Lipid Res* 45: 187–202.

Zeyda M and Stulnig TM. 2007. Adipose tissue macrophages. *Immunol Lett* 112: 61–67.

# 12 Anti-Inflammatory Properties of Plant Sterols and Phytoestrogens

## *Experimental and Clinical Evidence*

*Rgia A. Othman and Mohammed H. Moghadasian*

**CONTENTS**



## 12.1 INTRODUCTION

Dietary phytosterols and phytoestrogens may reduce the incidence of cardiovascular diseases through their anti-inflammatory effects in adipose tissue. Phytosterols are plant sterols similar in structure to cholesterol. Studies show that phytosterol-based diets reduce the levels of the inflammatory mediators including C-reactive protein (CRP), interleukin-6 (IL-6), tumor necrosis factor (TNF-α), phospholipase A-1 (PLA-1), and fibrinogen. These mediators are elevated in adipose tissue inflammation (Bouic et al., 1999; Devaraj et al., 2006; Jones et al., 2007; Nashed et al., 2005). In addition, phytosterols exhibit immunomodulatory effects, increasing T-cell proliferation and modifying cytokine profiles in favor of T-helper type 1 (Th1-type) response and decreasing cytokines associated with the Th2-type cells (Bouic et al., 1996; Bouic et al., 1999; Breytenbach et al., 2001).

Phytoestrogen compounds are found in many foodstuffs and are so named because they possess weak estrogenic or anti-estrogenic activity (Hutchins et al., 2001). They include certain isoflavones, lignans, and saponins. Dietary phytoestrogens may reduce risk of cardiovascular disease (Pan et al., 2001; van der Schouw et al., 2002; van der Schouw et al., 2005), by improving lipid profiles (MerzDemlow et al., 2000; Francis et al., 2002), reducing platelet aggregation (Gottstein et al., 2003), and enhancing endothelial function (Squadrito et al., 2002). Phytoestrogens also demonstrate some anti-inflammatory (Regal et al., 2000; Jenkins et al., 2003; Hallund et al., 2008), immunosuppressive (Park et al., 2007), and anti-parasitic activities (de Andrade-Neto et al., 2007). This chapter reviews the current knowledge of the effects of dietary phytosterols and phytoestrogens on inflammation and their potential roles in reducing inflammatory-based diseases.

## 12.2 PHYTOSTEROLS

Phytosterols in the diet lower total cholesterol (TC) and low-density lipoprotein (LDL) cholesterol (Moghadasian, 2006). The cholesterol-lowering effects of phytosterols may be due to interruptions in bile salt micelle formation (Jones et al., 2007, Pritchard et al., 2003). Plant sterols structurally resemble cholesterol and are derived from three main sources: by-products of the pulp and paper industry (tall oil-derived phytosterols), vegetables, and transgenic plants (Moghadasian, 2000).

The five main plant sterols in nature are β-sitosterol, stigmasterol, campesterol, sitostanol, and campestanol (Figure 12.1). Sitosterol (65%) and campesterol (30%) are the principal dietary phytosterols; stigmasterol, sitostanol, and campestanol account for the remaining 5% (Salen et al., 1985). The consumption of plant sterols interferes with cholesterol absorption and consequently decreases serum cholesterol levels (Miettinen and Gylling, 2006). Plant sterols have also been shown to reduce CRP, IL-6, TNF-α, PLA-1 and fibrinogen concentrations (Bouic et al., 1999; Devaraj et al., 2006; Jones et al., 2007; Nashed et al., 2005). They also increase T-cell proliferation and modify cytokine profiles toward Th1-type responses (Bouic et al., 1996; Bouic et al., 1999; Breytenbach et al., 2001). The following sections review studies of the effects of plant sterols on the inflammatory systems of both animals and humans.

### 12.2.1 Experimental Studies of Phytosterols

Table 12.1 presents a summary of experimental studies addressing the potential anti-inflammatory effects of dietary phytosterols. Our previous work demonstrated that incorporating a phytosterol mixture in the diets (2% wt/wt) of apolipoprotein E-knockout (apoE-KO) mice significantly reduced plasma TC levels and prevented atherosclerotic lesions (Moghadasian et al., 1999; Moghadasian et al., 2001; Nashed et al., 2005; Yeganeh et al., 2005; Moghadasian, 2006). These lipid changes and atherosclerotic protection were associated with reduced levels of pro-inflammatory cytokines, including IL-6 and TNF-α, and increased production of anti-inflammatory IL-10 after lipopolysaccharide (LPS) stimulation in phytosterol-fed apoE-KO mice compared to controls (Nashed et al., 2005). Immunocompetence was not

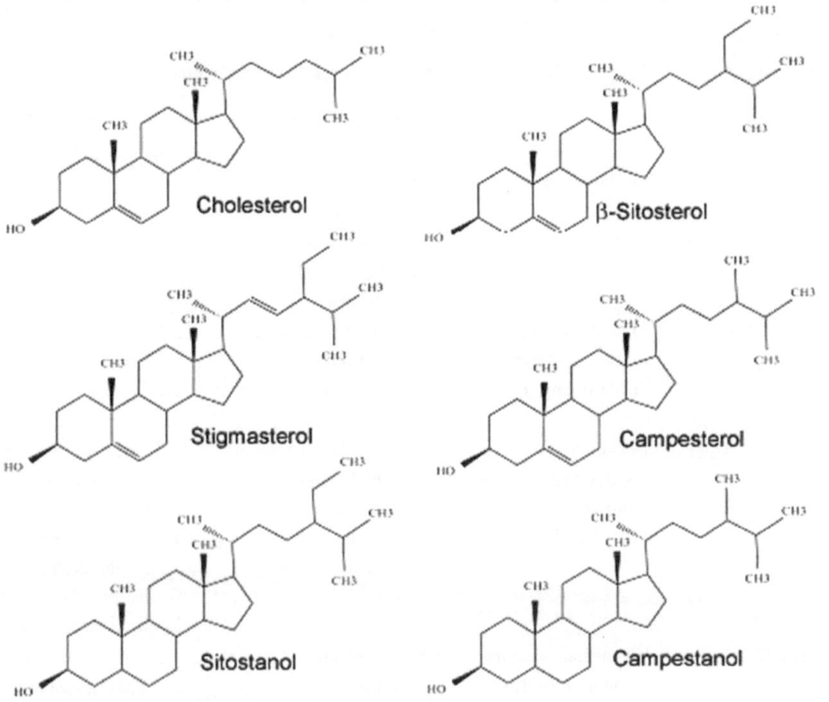


**FIGURE 12.1** Chemical structures of cholesterol and common plant sterols and stanols.

completely impaired by the lowered capacity to transmit pro-inflammatory cytokine and chemokine responses to inflammatory stimuli.

We speculated that attenuations in plasma cholesterol concentrations may have lowered the extent of LDL oxidation, resulting in diminished recruitment of immune cells to the arterial intima and consequently, both reduced expression of adhesion molecules and diminished secretion and production of cytokines and reactive oxygen species (ROS) (Nashed et al., 2005).

Treating male ICR mouse models of chronic inflammation with intraperitoneal injections of cactus ethanol extract containing β-sitosterol (200 mg/kg) for five consecutive days promoted an anti-inflammatory response as monitored by reducing granuloma weight and its content (Park et al., 2001). Navarro et al. (2001) showed that a mixture of plant sterols in the diet (30 and 60 mg/kg) decreased carrageenan paw edema in mice.

Topical application of the sterol fraction profoundly inhibited adjuvant-induced ear edema in mice in a dose-dependent manner. β-sitosterol suppressed edema formation by 52% at a 0.5 mg dose and blocked leukocyte granular enzyme release (β-glucuronidase) and superoxide generation but it did not suppress histamine release from mast cells. Correspondingly, oral doses (0.25 to 2.0 g/kg) of stigmasterol-containing extracts from *L. inflata,* a perennial herb, significantly decreased the edema thickness in the paws of mice (Al-Yousuf et al., 2002). It is conceivable that other

**TABLE 12.1**
**Animal Studies of Anti-Inflammatory Effects of Phytosterols**

| Study | Model | Dose and Duration | Observations |
|---|---|---|---|
| Awad (2000) | Mouse topical inflammation | 1.0 and 1.5 mg/ear (topical); green alga, *Ulva lactuca*, extract containing 3-O-β–D glucopranosyl clerosterol | ↓ 62 and 72% in ear edema (P <0.05) |
| Park et al. (2001) | Male ICR mice with induced chronic inflammation (n = 5) | 200 mg/kg i.p. cactus extract containing β-sitosterol (5 days) | ↓ 22% (P <0.05) adjuvant-induced pouch granuloma weight |
| Navarro et al. (2001) | Female Swiss mice with carrageenan-induced paw edema (n = 8) | 0.25, 0.5 and 1 mg/ear (topical); mix of campesterol, stigmasterol and β-sitosterol (4 hours) | ↓ 41, 43 and 59% (P <0.01) inflammatory response at 0.25, 0.5 and 1 mg/ear, respectively; ↓ 52% in edema formation (P <0.01) by β-sitosterol (0.5 mg/ear); |
| Awad et al. (2001) | Wistar rats, vascular smooth muscle cells from thoracic aorta | 4 to 32 μM/ml campesterol and β-sitosterol (3 days) | ↓ 16 and 30% (P <0.05) in VSMC growth by campesterol and β-sitosterol, respectively |
| Villaseñor et al. (2002) | 7- to 12-week old Swiss Webster albino mice with induced paw edema (n = 5) | 100 mg/kg i.p. β-sitosterol and β-sitosterol-β-D-glucoside (3 hours) | ↑ swelling of induced edema by β-sitosterol (0.028) and ↓ edema by glucoside (0.20) |
| Al-Yousuf et al. (2002) | Male albino mice with induced paw edema (n = 10) | 500, 1000, and 2000 mg/kg (oral) *Leucas inflata* extract containing stigmasterol (2 to 4 hours) | ↓ paw thickness after carrageenan administration (P <0.05) |
| Choi et al. (2002) | Mongolian gerbils with induced brain damage (n = 6) | 500 μg/kg/day i.p. β-sitosterol (19 days) | ↑ expression of VEGF, VEGF receptor Flk-1, angiogenic genes (P <0.05) |
| Nashed et al. (2005) | 4-week old apo E-deficient mice (n = 7) | 2% (w/w) ~0.1g/kg/ day; mix of β-sitosterol, campesterol, stigmasterol and dihydrobrassica sterol (98 days) | ↓ plasma IL-6 and TNF-α (P <0.05); ↑ 8 to 10 times in IL-10 levels |

**TABLE 12.1 (continued)**
**Animal Studies of Anti-Inflammatory Effects of Phytosterols**

| Study | Model | Dose and Duration | Observations |
|---|---|---|---|
| Calpe-Berdiel et al. (2007) | 8-week old Apo-E-deficient mice (n = 6 to 8) | 2% (w/w) ~0.1g/kg/day; mix of campesterol, stigmasterol and β-sitosterol (28 days) | ↑ 1.9- and 3.3-fold in IL-2 and IFN-γ respectively; no change in IL-4 and IL-10 levels ($P < 0.05$) |
| Yuk et al. (2007) | 6- to 8-week old female BALB/c asthmatic mice (n = 5) | 1 mg/kg i.p. β-sitosterol and β-sitosterol glucoside (13 days) | ↓ in mRNA and protein expression of IL-4 and IL-5 ($P < 0.05$); ↑ IFN-γ levels and ↓ ROS release |
| Lee et al. (2007) | 6-week old BALB/c female mice with candidiasis (n = 5) | 200µg/ml i.p. (2×/day, 3 days) | ↑ expression of IL-2 and IFN-γ vs IL-4 and IL-10 production ($P < 0.01$) |

extract components such as chromone and coumarins may have contributed to these anti-inflammatory effects.

Awad (2000) first reported that topical administration (1 to 1.5 mg/ear) to mice of the sterol 3-O-b-D-glucopyranosyl-stigmasta-5,25-dien from the marine green alga *Ulva lactuca* significantly inhibited the development of edema. The results of this sterol treatment were similar to results from topical indomethacin (1000 mg/ear) treatment. In similar studies, treatment of Swiss Webster albino mice with 100 mg/kg β-sitosterol and β-sitosteryl-β-D-glucoside extracted from the leaves of the Philippine mint (*Mentha cordifolia* Opiz) produced 70 and 73% reductions in the number of squirms induced by acetic acid (Villaseñor et al., 2002). These analgesic effects of β-sitosterol and its glucoside derivative were further supported by 300 and 157% increments in pain tolerance, respectively, relative to mefenamic acid, a known analgesic (Villaseñor et al., 2002). Similarly, Al-Yousuf et al. (2002) reported a reduction in formalin-induced pain in male albino mice after oral doses (0.25, 0.5, 1.0, or 2.0 g/kg) of a stigmasterol-containing extract. The mechanisms underlying these analgesic effects are uncertain.

It has been suggested that the immunomodulating effects of plant sterols may be mediated by inhibition of the classical or alternative complement pathways. A plant sterol fraction was shown to inhibit hemolytic activity of the classical complement pathway in mice (Navarro et al., 2001). Phytosterols may also affect angiogenesis. The intraperitoneal (i.p.) injection of β-sitosterol isolated from the flowering herb (*Aloe vat*) (500 µg/kg daily for 19 days) enhanced the expression of vascular endothelial growth factor (VEGF), VEGF receptor Flk-1, and blood vessel matrix laminin in the brains of an ischemia–reperfusion model (Choi et al., 2002).

In asthmatic mice, β-sitosterol and β-sitosterol glucoside administration significantly inhibited the leukocytosis and eosinophilia in the bronchoalveolar lavage (BAL) fluid, attenuated ROS production, and mitigated the increased mRNA and protein expression of IL-4 and IL-5 in the lung tissue and BAL fluid compared to the asthmatic control mice (Yuk et al., 2007). Histologic evidence showed less eosinophil infiltration and mucus hypersecretion in the airway remodeling process after plant sterols treatment (Yuk et al., 2007). In a parallel manner, phytosterols extracted from transgenic *P. ginseng* improved inflammatory profiles in human mast cells. Phytosterol-treated mast cells showed suppressed synthesis of TNF-α, IL-6, and IL-8 as well as reduced expression of cyclooxygenase-2 (COX-2) (Kim et al., 2007). On the other hand, the β-sitosterol glycoside daucosterol isolated from a medicinal plant *(Astragalus membranaceus)* had no effect on growth of *Candida albicans*; although it increased mean survival times compared to untreated controls (29 ± 11 versus 13 ± 4 days) when administered i.p. in mice with disseminated candidiasis (Lee et al., 2007).

In a cell culture model, β-sitosterol and campesterol reduced vascular smooth muscle cell (VSMC) growth by 30 and 16%, with DNA synthesis reduced 25% by β-sitosterol but remaining unchanged by campesterol (Awad et al., 2001). This reduction in cell growth may be related to inhibitory effects of phytosterols on cholesterol metabolism with impaired cell membrane synthesis and cell proliferation. In macrophages, β-sitosterol or campesterol treatment was associated with significant inhibition of PGE2 release (68 and 55%) and prostacyclin (PGI2) production (67 and 52%) (Awad et al., 2004). This decline in prostaglandin release was not accompanied by any alteration in the expression of cytosolic cPLA-2 or COX-2 enzymes—the key rate-limiting enzymes in prostaglandin synthesis. This study concluded that incorporation of phytosterols into macrophages may protect against atherosclerosis by diminishing prostaglandin secretion and ultimately atherosclerotic plaque formation.

Plant sterols may improve blood vessel tone and function by enhancing PGI2 secretion. Basal PGI2 release was increased 43% by campesterol and 81% by β-sitosterol, whereas calcium ionophore (A23187)-stimulated PGI2 release was increased 25% by campesterol and 54% by β-sitosterol. This suggests a significant impact of β-sitosterol supplementation on PGI2 release from VSMC (Awad et al., 2001).

Purified fungal and mushroom sterols (ergosterol and ergosterol peroxide) have been shown to inhibit LPS-induced TNF-α secretion, IL-1α/β expression, and DNA binding activity of nuclear factor-κB (NF-κB) in macrophage-like cells and to suppress interferon-inducible genes in colorectal adenocarcinoma cells after as little as 5 days of treatment (Kobori et al., 2007). Likewise, plant sterol derived from the gum resin *(guggulu)* of the tree (*Commiphora mukul*) inhibited NF-κB activation and downregulated the expression of inflammatory gene products such as COX-2 and matrix metalloproteinase-9 (MMP-9) in human lung cancer cells (Shishodia and Aggarwal, 2004). Furthermore, this sterol, guggulsterone, completely impeded cytokine-mediated cytotoxicity and NO and PGE2 production. These effects were associated with decreased levels of the inducible form of NO synthase (iNOS) and COX-2 mRNA and protein expressions.

These diverse anti-inflammatory effects appeared mediated by inhibition of NF-κB activation (Lv et al., 2008). However, the precise mechanisms by which these phytosterols induce anti-inflammatory effect are not fully understood. It has been suggested that incorporation of plant sterols into cell membranes may modulate cell membrane composition, leading to alterations in fluidity (Awad et al., 2004) and sensitivity (Feigin et al., 1995). It is significant that phytosterols also affect innate immune response by serving as ligands for the nuclear receptor LXR (Kaneko et al., 2003). The LXR signaling pathway has been shown to play a significant role in the control of gene expression involved in Th1 responses to bacterial pathogens such as *Listeria monocytogenes* (Joseph et al., 2004).

### 12.2.2 Clinical Trials of Phytosterols

Table 12.2 summarizes clinical trials addressing the anti-inflammatory effects of phytosterols. As a result of positive outcomes of clinical trials, phytosterols are now added to margarines, butters, spreads, and breakfast cereals. In a recent intervention study of 200 healthy 20- to 50-year-old mildly hypercholesterolemic subjects, daily intake of 2 g plant stanols from stanol ester spread with camelina, rapeseed, or sunflower oil spread for 3 months resulted in a 9% reduction in LDL cholesterol compared to sunflower oil spread without plant stanols (Raitakari et al., 2008). Despite this, there was no effect on vascular function in the treatment group.

However, subgroup analysis of the data showed that in subjects with low baseline values for carotid artery compliance and brachial artery flow-mediated dilatation (FMD), dietary plant stanols significantly improved arterial elasticity and endothelial function. However, a double-blind study with diabetic patients reported by Hallikainen et al. (2007) showed a significant diminution in serum cholesterol levels after 12 weeks of a diet rich in plant stanol esters (2 g/day stanols) but no improvements in endothelial function as evaluated by brachial artery FMD.

Likewise, in a double-blind intervention trial, 45 patients on stable statin treatment received margarine with or without added plant sterols for one month. The participants were then randomly assigned into three groups of 15, a control group receiving margarine only and two treatment groups receiving margarine enriched with either a plant sterol or stanol (2.5 g per day). Despite significant attenuations in LDL cholesterol serum levels, no significant changes were observed in levels of soluble adhesion molecules, CRP, or monocyte chemotactic protein-1 (MCP-1) (De Jong et al., 2008). It appears that decreases in plasma LDL cholesterol levels, at least under the time course studied, were insufficient to alter endothelial dysfunction or low grade inflammatory markers in these populations. Another explanation for this observation may be the fact that the impacts of dietary phytosterols on endothelial dysfunction may not be cholesterol-dependent. The authors suggested that statin treatment may have concealed the beneficial effects of plant sterols on endothelium function.

In another study, endothelial function was not improved in hypercholesterolemic subjects without diabetes given a Mediterranean-inspired diet rich in sterol esters (Ambring et al., 2004) or in prepubertal children with familial hypercholesterolemia regardless of lowering LDL cholesterol concentrations (de Jongh et al., 2003). It should be noted that endothelial dysfunction was assessed by measuring brachial arterial dilator response to increased blood flow and cellular adhesion molecules. Additional markers may reflect improvements of dysfunctional endothelium.

In a randomized, double-blind, 20-week study with a cross-over design, 76 moderately hypercholesterolemic men and women participated. The subjects in the intervention group consumed plant stanol ester and plant sterol ester spreads (total plant

**TABLE 12.2**
**Investigative Clinical Trials of Anti-Inflammatory Effects of Dietary Phytosterols**

| Study | Model | Dose and Duration | Observations |
|---|---|---|---|
| Jenkins et al. (2005) | Healthy hyperlipidemic subjects aged 36 to 71 years (n = 34) | 1 g sterols/day/1000 kcal diet in plant sterol-enriched margarine (4 weeks) | ↓ 23.8% CRP ( <0.001) |
| Hallikainen et al. (2006) | Subjects with total cholesterol <8 mmol/l aged 21 to 72 years (n = 80) | Total sterols and stanols, 2.07 g in STAEST spread and 2.00 g in STEEST spread (20 weeks) | No significant changes in serum CRP, IL-6 or TNF-α |
| Devaraj et al. (2006) | Healthy subjects aged 19 to 74 years (n = 72) | 2 g sterol/day, containing 40% β-sitosterol, 25% campesterol, 20% stigmasterol (2×/day) | ↓ 12% (P <0.02) in CRP levels; ↓ 23%(P <0.01) in CRP levels in earlier study |
| Hallikainen et al. (2007) | Type 1 diabetic subjects aged 29 to 53 years (n = 22) | 2.0 g sterols/day (1.66 g sitostanol and 0.38 g campestanol) in STAEST spread (12 weeks) | No significant change of brachial artery diameter, FMD or endothelial function |
| Jones et al. (2007) | Overweight, hypercholesterolemic subjects (n = 21) | 1.7 g sterol/day using fish oil, olive oil, soybean oil as carriers; (28 days, 4 weeks washout) | PS-FO ↓ PAI-1 concentration compared to FA-SO (P = 0.0282) |
| Madsen et al. (2007) | Mildly hypercholesterolemic subjects aged 20 to 70 years (n = 50) | 2.3 g plant sterols/day, 3×/day, 69% β-sitosterol, 15% campesterol and 8% stigmasterol (8 weeks) | ↓ 4.6% apo-B (P <0.05); no significant change in apo A-1 or CRP |
| Acuff et al. (2007) | Healthy volunteers (n = 16) | 1.3 g sterols/day, 2×/day (4 weeks) | No significant changes in serum CRP |
| Hansel et al. (2007) | Moderately hypercholesterolemic subjects (n = 194) | 1.6 g plant sterols in low-fat fermented milk, 75% β-sitosterol and 8.4% campesterol (6 weeks) | No significant changes in plasma CRP |
| De Jong et al. (2008) | Satin treatment patients aged 18 to 65 years (n = 41) | 2.5 g sterols/day, 49% β-sitosterol, 31% campesterol, 16% stigmasterol (4 weeks) | No significant change in inflammatory markers |

sterols and stanols 1 to 2 g/day) for 10 weeks each. The control group received spread without added stanols/sterols for 20 weeks. The results revealed no changes in serum levels of CRP, IL-6, and TNF-α and no effects on brachial artery FMD. However, a significant reduction (–2.2%) in brachial artery diameter was observed during the plant sterol ester period as compared to STAEST (Hallikainen et al. 2006), suggesting a minor hemodynamic influence.

Although brachial artery diameters are larger in subjects with coronary artery disease than in normal subjects (Holubkov et al., 2002), the clinical significance remains to be elucidated. In a head-to-head comparison with statins, the inclusion of plant sterols (1.0 g/1000 kcal) in the portfolio diet along with soy protein, viscous fiber, and almonds improved atherogenic lipid profiles and attenuated CRP concentrations in hyperlipidemic men and women (Jenkins et al., 2003; Jenkins et al., 2005). Interestingly, a significant reduction (–42%) in serum levels of CRP was noted among hypercholesterolemic men when plant stanols (3 g/day) were added to statin therapy (1.7 ± 1.2 mg/L versus 3.0 ± 2.4 mg/L, mean values) (Cater et al., 2005). However, no significant changes in CRP concentrations were observed with plant stanol esters alone.

To expand the potential benefits of phytosterols, the effectiveness of plant sterol fortification was tested in a nonfat medium. Devaraj et al. (2004) reported a significant cholesterol-lowering effect of plant sterols incorporated in nonfat orange juice in a population of mildly hypercholesterolemic healthy subjects. Drinking plant sterol-fortified orange juice (2 g/day) for 8 weeks was associated with significant diminutions in levels of TC (7.2%), LDL (12.4%), non-HDL cholesterol (7.8%), and apolipoprotein B (apo-B; 9.5%) compared to placebo orange juice and baseline. This indicates that plant sterols do not only reduce the cholesterol content of LDL cholesterol particles, but also reduce the total number of circulating atherogenic particles.

Moreover, in a similar study by the same group in 2006, incorporation of 1 g phytosterol into a reduced-calorie (50 calories/240 mL) nonfat orange juice beverage significantly reduced CRP concentrations (–12%) in 72 healthy subjects who drank the juice containing plant sterols twice a day with meals for 8 weeks (Devaraj et al., 2006).

Anti-inflammatory mechanisms of phytosterols may result from a decline in the pro-inflammatory burden in the liver. However, daily consumption of 1.6 g plant sterol-supplemented low-fat fermented milk for 6 weeks appeared to reduce CRP levels in hypercholesterolemic subjects but not significantly different in comparison to the control fermented milk and baseline (0.93 versus 1.02 and 1.07 mg/L, respectively) (Hansel et al., 2007).

On the other hand, plant sterol esters in capsule form (1.3 g/day) did not alter CRP levels in free-living hypercholesterolemic subjects during a 4-week intervention (Acuff et al., 2007). It should be noted that all study participants had low or average CRP levels at baseline (0.8 ± 0.5 mg/L). These findings paralleled recent findings in subjects with mild (Madsen et al., 2007), mild-to-moderate (AbuMweis et al., 2006), and moderate hypercholesterolemia (Jones et al., 2007).

Although most studies focused on the cholesterol-lowering activity of phytosterols, some evidence proposes that these plant sterols may exert additional biological actions that include immunomodulatory properties. In supporting this concept, Bouic et al. (1996) revealed that T-cell proliferative responses increased both in vivo and in vitro by several hundred percent after the use of β-sitosterol and its glucoside. A mixture of plant sterols (20 mg β-sitosterol and 0.2 mg β-sitosterol glucoside per day) consumed by healthy volunteers for 4 weeks showed a higher in vivo proliferation of T-cells (20–920%) compared to baseline value. In vitro stimulation of T-cells with phytohemagglutinin (PHA) and treatment with 1 μg/ml of plant sterol mixture significantly enhanced the expression of CD25 and HLA-Dr activation antigens on T-cells and expression of IL-2 and IFN-γ.

In addition, natural killer cell activity was also increased by individual sterols, but even higher activity was attained with a combination (Bouic et al., 1996). In agreement with this study, HIV-infected patients who received a mixture of plant sterols showed beneficial Th1 responses (IFN-γ) while HIV-infected patients on no sterol therapy exhibited predominant Th2 responses (IL-4 secretion) (Breytenbach et al., 2001). This may suggest a dietary means to modulate immune profiles in HIV patients and maintain significant Th1 responses that help suppress viral replication over the long run. Similarly, a pilot study of marathon runners in South Africa showed that the ingestion of plant sterols and sterolins (BSS:BSSG mixture) was associated with less neutrophilia, lymphopenia, and leukocytosis in comparison to counterparts who received placebo capsules. The supplemental BSS:BSSG also decreased the plasma levels of IL-6 in the runners and significantly reduced their levels of cortisol (Bouic et al., 1999).

## 12.3 PHYTOESTROGENS

Phytoestrogens are natural compounds found in many dietary plants. They resemble endogenous estrogens in chemical structure and have shown weak estrogenic activity in numerous bioassays (Kurzer and Xu, 1997). The major groups of phytoestrogens are isoflavones, lignans, and saponins.

Isoflavones include the common dietary genistein and daidzein compounds (Kurzer and Xu, 1997). Lignans are relatively minor plant compounds; matairesinol and secoisolariciresinol (SDG) are the principal lignans. Saponins are protective against microbes and are found enriched in plants such as soapwort, soapberry, oats, and spinach. Soy is a principal dietary source of isoflavones and saponins, and flaxseed is an abundant source of lignin glycosides (Thompson et al., 2006). These dietary sources have been used in most studies investigating the potential health effects of phytoestrogens.

High dietary consumption of phytoestrogens has been associated with a decrease in fasting insulin and C-peptide (an index of insulin secretion) in men (van der Schouw et al., 2005) and reductions in aortic stiffness and lowered cancer incidence in women (van der Schouw et al., 2002; Adlercreutz et al., 1991). Although the mechanisms by which dietary phytoestrogens induce such effects are still unclear, improving lipid profile (MerzDemlow et al., 2000), inhibiting LDL oxidation (Tikkanen et al., 1998), suppressing proliferation of aortic smooth muscle cells (Pan et al., 2001), and decreasing inflammatory markers are potentially involved (Regal et al., 2000). Recent animal and human studies focusing on anti-inflammatory actions of dietary phytoestrogens including isoflavones, lignans, and saponins are discussed below.

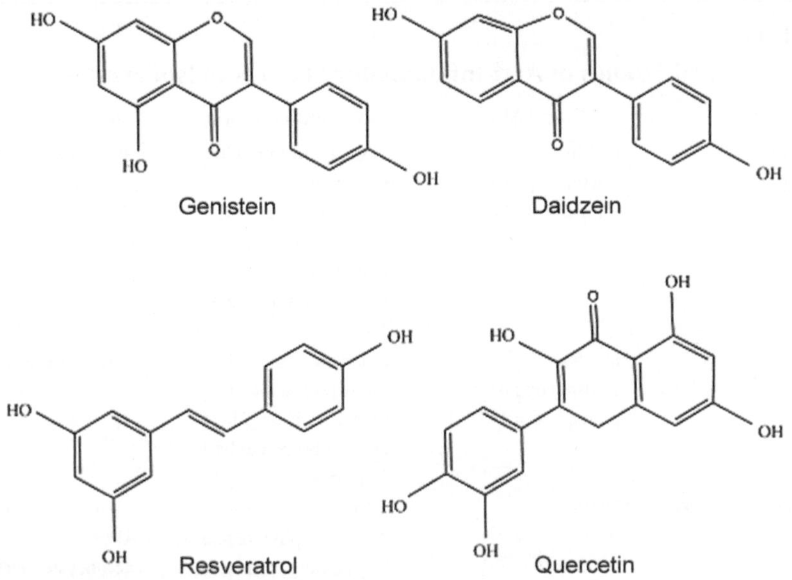


**FIGURE 12.2** Chemical structures of most common isoflavones.

### 12.3.1 Isoflavones

Isoflavones are found primarily in foods containing soy, but also found as resveratrol in red grapes and peanuts and quercetin in capers, citrus, and berries. Figure 12.2 depicts chemical structures of selected isoflavones. Table 12.3 summarizes clinical studies of isoflavones. After ingestion, the isoflavone glucosides found in soy are hydrolyzed by both intestinal mucosa and bacterial β-glucosidases, releasing isoflavone aglycons that are absorbed directly (Setchell et al., 2002; McMahon et al., 1997). Isoflavones may then be reconjugated with sulfate and glucuronide and excreted into the bile or urine (Bingham et al., 1998; Cassidy et al., 2006) or further metabolized in the intestine to other metabolites, such as equol (Hall et al., 2005) and O-desmethylandgolensin (Joannou et al., 1995).

Consumption of isoflavones has been associated with lowered risk for cardiovascular disease (CVD). Sixty healthy postmenopausal women without clinical atherosclerosis or diabetes participated in a randomized cross-over design in which they consumed daily a half-cup of soy nuts (25 g soy protein and 101 mg aglycone isoflavones) for 8 weeks (Nasca et al. 2008). While no significant changes were observed in normotensive women, soy nut treatment decreased sVCAM-1 levels in hypertensive women (624 ± 154 versus 554 ± 114 ng/ml, respectively) and showed a trend of decline in CRP levels in both normotensive (23%) and hypertensive women (5%). Hall et al. (2005) noticed that isoflavone treatment was less likely to alter CRP concentrations relative to controls. A similar finding was reported by Hermansen et al. (2001). An opposing study showed that ET-1 concentrations decreased in postmenopausal women when supplemented with 54 mg genistein/day (Squadrito et al., 2002

## TABLE 12.3
## Recent Clinical Studies of Anti-Inflammatory Effects of Isoflavones

| Study | Clinical Model | Dose and Duration | Major Findings |
|---|---|---|---|
| Steinberg et al. (2003) | Healthy postmenopausal women (n = 28) | 25 g soy protein with isoflavones, ethanol-washed soy protein with trace isoflavones, and total milk protein (6 weeks each) | ↓ 37% PFV(P = 0.03) for isoflavone group, no significant change in endothelial markers |
| Hall et al. (2005) | Healthy postmenopausal women (n = 117) | Isoflavone-enriched (genistein:daidzein ratio 2:1; 50 mg/day) or placebo cereal bars (8 weeks) | ↓ CRP (P <0.001) |
| Hanson et al. (2006) | Healthy postmenopausal women (n = 55) | 4 SPI treatments (40 g/day): low phytate and low isoflavone, native phytate and low isoflavone, low phytate and native isoflavone, or native phytate/native isoflavone (6 weeks) | No significant change in CRP; ↓ homocysteine (P = 0.017) |
| Clerici et al. (2007) | Subjects with hypercholesterolemia (n = 62) | 33 mg isoflavones/mg pasta (10 weeks) | ↓ 2.2% CRP (P = 0.03); ↑ 2.3% FMV (P = 0.003) |
| Cupisti et al. (2007) | Renal transplant recipients with stable renal unction (n = 20) | 25 g soy protein (5 weeks) | ↓ 6.3% FMV (P <0.001); no significant change in CRP |
| Azadbakht et al. (2007) | Postmenopausal women with metabolic syndrome (n = 42) | 30 g soy nut, 30 g soy protein, controls (8 weeks) | ↓ 11 and 5% on E-selectin for nut (P <0.01) and protein (P = 0.19); ↓ 9% IL-18 for nut (P <0.01), ↓ 9 and 2% CRP for nut (P <0.01) and protein (P 0.01); ↓ 11% TNF-α (P <0.05) for protein |
| Evans et al. (2007) | Postmenopausal women (n = 22) | 25 g soy protein and 20 g soy lecithin; 25 g soy protein and placebo; 20 g soy lecithin and placebo, double placebo (4 weeks) | No significant change in endothelial function |

**TABLE 12.3 (continued)**
**Recent Clinical Studies of Anti-Inflammatory Effects of Isoflavones**

| Study | Clinical Model | Dose and Duration | Major Findings |
|---|---|---|---|
| Nasca et al. (2008) | Postmenopausal women, 12 with hypertension (n = 60) | 25 g soy protein and 101 mg aglycone isoflavones (8 weeks) | ↓ 11.2% sVCAM-1 in hypertensive patients ($p = 0.003$); ↓ 27.3% CRP in normotensive ($p = 0.07$) and ↓ 5% in hypertensive patients ($p = 0.90$) |
| Walker et al. (2008) | Female cynomolgus macaques (n = 79) | Diet containing 1.88 mg aglycone FI/g protein and control diet | ↓ MCP-1 and ICAM-1; IL-6 mRNA transcripts ($P < 0.0001$) |

and 2003). In assent, Teede et al. (2003) reported a remarkable decrease in VCAM-1 levels and arterial stiffness after consumption of 80 mg formononetein, a precursor of daidzien, daily by 80 healthy subjects. A similar study found that 54 mg/day of pure genistein for a year led to an improvement of 5.5% in FMD in healthy postmenopausal women (Squadrito et al., 2003).

A related study found significant decreases in NO, serum E-selectin, IL-18, TNF-α, and CRP in 42 postmenopausal women in a randomized trial following DASH diet guidelines. Ingestion of 30 g soy nut protein, with a total phytoestrogen content of 340 mg, raised plasma phytoestrogen by 64%, increased NO by 9.8%, decreased serum E-selectin by 11.4%, decreased serum IL-18 by 9.2%, decreased CRP by 8.9%, and decreased TNF-α by 11% (Azadbakht et al., 2007).

Equol is an intestinal bacterial metabolite of daidzein. It is thought to be an important bioactive metabolite of isoflavones due to its greater binding affinity for estrogen receptors and antioxidant capacity (Bingham et al., 2003). Equol generation among individuals varies, making it difficult to assess effects of dietary isoflavones. Contradictory evidence exists. One study suggests no significant differences in isoflavone reduction of inflammatory biomarkers between equol producers and non-equol producers (Hall et al., 2005). Conversely, an equol-producing cohort consuming 33 mg of isoflavones/day exhibited significant changes in CRP and brachial artery FMD compared to non-equol producers (Clerici et al., 2007).

Isoflavones may indirectly improve endothelial vasodilatation by reducing total homocysteine levels (Fu et al., 2002). In a randomized, double-blind study, 55 healthy postmenopausal women were randomly assigned into four groups: (1) low phytate and low isoflavone (LP/LI), (2) native phytate and low isoflavone (NP/LI), (3) low phytate and native isoflavone (LP/NI), and (4) native phytate and native isoflavone (NP/NI). After 6 weeks, total homocysteine declined from baseline by 19% in the NP/LI group, 8% in the LP/NI group, 12% in the NP/NI group, and 3% in the LP/LI group (Steinberg et al., 2003).

In accordance with those results, consumption of isoflavone-containing SPI (108 mg) by 28 postmenopausal women for 5 weeks with a 5 week wash-out period was accountable for a 37% decrease ($P = 0.03$) in PFV (Cupisti et al., 2007). Health-associated benefits of isoflavones were reported by Verma et al. (2003) in a study of renal transplant patients who tend to have a greater risk of endothelial dysfunction. The study showed that receiving 25 g of soy protein/day significantly improved FMD by modulating the L-arginine/ADMA ratio but did not change CRP levels. Conversely, 33 mg of isoflavones taken by 62 hypercholesterolemic subjects for 8 weeks was attributable to 42% attenuation in CRP levels and a 2.3% increase in endothelial function (Clerici et al., 2007). However, these improvements in endothelial function were not reported for otherwise healthy individuals (Evans et al., 2007; Blum et al., 2003) suggesting beneficial effects of isoflavones may be observable only when endothelial function is abnormal.

The exact mechanisms by which isoflavones modulate the inflammatory process are unclear. It is possible that isoflavones exhibit direct antioxidant activities (Bingham et al., 2003; Lin et al., 2008) or that their antioxidant activities are mediated via interactions with common co-antioxidants (Hwang et al., 2000).

Also, the anti-inflammatory effects of isoflavones may involve endothelial-dependent vasodilatation (Walker et al., 2001; Squadrito et al., 2003) and endothelial-independent actions. The latter has been suggested to include calcium channel antagonism (Figtree et al., 2000), inhibition of COX-2 transcription along with inhibition of NF-κB (Subbaramaiah et al., 1998; Manna et al., 2000), suppression of inflammatory cytokines that mediate recruitment of inflammatory leukocytes (Donnelly et al., 2004), inhibition monocyte rolling, and adhesion to cytokine-activated endothelial cells by activation of peroxisome proliferator-activated receptor-γ (PPARγ), a nuclear receptor that mediates a variety of anti-inflammatory actions (Chacko et al., 2005).

### 12.3.2 Lignans

Lignans are phenolic compounds found in many plants and foods (Thompson et al., 1991; Meagher and Beecher, 2000). Table 12.4 summarizes anti-inflammatory effects of lignans. Flaxseed is an excellent source of the secoisolariciresinol diglucoside (SDG) lignin precursor (Thompson et al., 1991). Besides SDG, flaxseed contains lower but significant amounts of isolariciresinol, pinoresinol, and matairesinol lignans (Meagher et al., 1999). Such lignans undergo fermentation by microflora in the proximal colon to remove glycoconjugates, resulting in the generation of bioactive absorbable mammalian lignans including enterodiol and enterolactone (Thompson et al., 1991). Figure 12.3 illustrates the in vivo metabolism of dietary lignans to mammalian lignans.

Mammalian lignans may reduce the risk of CVD (van der Schouw et al., 2002), perhaps because of their structural similarity to estrogens (Hutchins et al., 2001). In an observational study of 403 women aged 49 to 70 years who underwent natural menopause and were followed up for 4 years, habitual intake of plant lignans (>0.9 mg/day) correlated with reduced aortic pulse wave velocity. These results were even more pronounced in older women (van der Schouw et al., 2002). Conversely, consumption of a low-fat muffin supplemented with SDG isolated from flaxseed (500 mg/day), a lignin

**TABLE 12.4**
**Recent Clinical Studies of Anti-Inflammatory Effects of Lignans**

| Study | Model | Dose and Duration | Major Findings |
|---|---|---|---|
| Hallund et al. (2006a) | Healthy postmenopausal women 61 ± 7 years (n = 22) | 500 mg/day SDG in low-fat muffin (6 weeks) | No significant changes of FMD and plasma NOx, ET-1 and ADMA |
| Hallund et al. (2006b) | Healthy postmenopausal women 61 ± 7 years (n = 22) | 500 mg/day SDG in low-fat muffin (6 weeks) | No significant changes of TEAC and FRAP; ↑ serum and urinary concentrations of enterolactone ($P < 0.001$) |
| Ogborn et al. (2006) | Han:SPRD-cy rats with polycystic kidney disease (n = 13) | AIN-93G diet plus 7% flax oil and 20 mg/kg diet SDG (12 weeks) | ↓ Macrophage infiltration ($P = 0.017$) and renal release of PGE2 in both sexes ($P < 0.01$) |
| Prasad (2007) | New Zealand White rabbits on high cholesterol diet (n = 12) | 0.25% cholesterol diet for 2 months followed by regular diet and FLC 40 mg/kg (4 months) | ↓ 35% in atherosclerotic lesions ($P < 0.05$) |
| Bergman et al. (2007) | 6- to 8-week old female athymic mice, BALB/c nu/nu with estradiol-induced tumors (n = 7 to 9) | AIN-93G diet plus 10% flaxseed, AIN-93G diet and s.c. injection of enterodiol or enterolactone, 15 mg/kg body weight (3 weeks) | ↓ VEGF ($P < 0.01$); ↓ in tumor vessel area; 1.1 ± 0.2% in flaxseed; 1 ± 0.4% in enterodiol, 0.6 ± 0.1% in enterolactone vs 3 ± 0.7% in basal diet tumors ($P < 0.01$) |
| Nakano et al. (2008) | 6-week old male un-nephrectomized SD rats (n = 6) | Sesamin-containing diet (1% w/w) with twice weekly DOCA (15 mg/kg s.c) with 1% NaCl added to drinking water (5 weeks) | ↓ DOCA salt-induced hypertension; ↓ aortic O2 production and ↓ increase in NADPH oxidase activity and elevated aortic mRNA expression of its subunits; |
| Hallund et al. (2008) | Healthy postmenopausal women 61 ± 7 years (n = 22) | 500 mg/day SDG in low-fat muffin (6 weeks) | ↓ 15% in CRP ($P < 0.02$); no changes in IL-6. TNF-α, VCAM-1, ICAM-1, or MCP-1 vs placebo |
| Lee et al. (2008) | 6- to 8-week old female C57Bl/6 mice with pulmonary ischemia–reperfusion injury (n = 6) | AIN-93G diet plus 10% flaxseed (3 weeks); incubation of pulmonary microvascular endothelial cells with 5μM (5 hours) | ↓ MDA, lipid oxidation and ROS ($P < 0.05$); ↓ ROS production vs control |

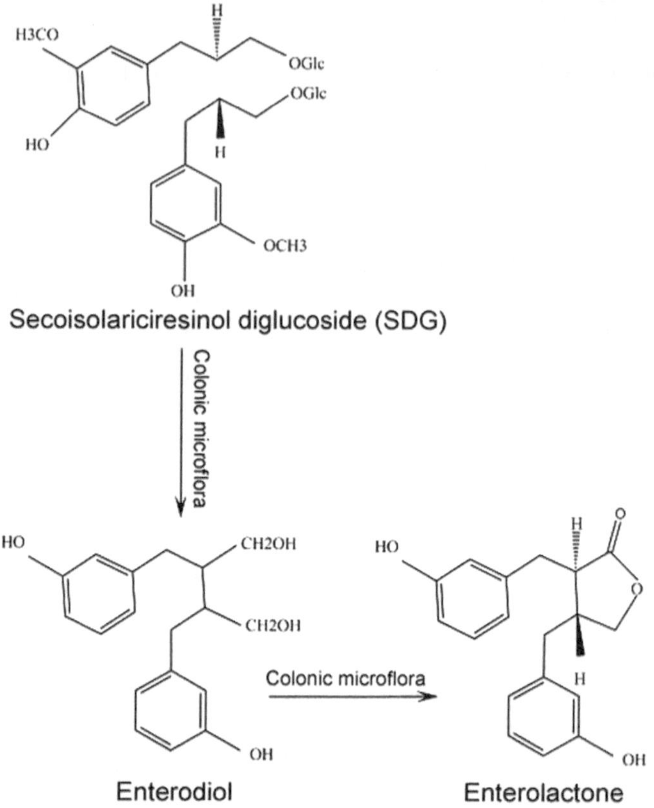


**FIGURE 12.3** Metabolism of dietary lignans to mammalian lignans.

precursor, showed no impact on FMD, endothelium-dependent vasodilatation FMD or plasma levels of ET-1, asymmetric dimethylarginine (ADMA), and NO in healthy postmenopausal women over 6 weeks of treatment (Hallund et al., 2006a). In a more recent study with the same design by Hallund et al. (2008), a significant reduction of 15% ($P < 0.03$) was observed for plasma CRP levels in the lignan-treated group compared to placebo; however, no effects were observed regarding other inflammatory markers such as IL-6, TNF-α, VCAM-1, ICAM-1, and MCP-1.

Lignans are metabolized in the gut to produce the estrogenic enterodiol and enterolactone that exert antioxidant activity and are absorbed into the serum. This serum increase in these compounds may explain, at least in part, the observed attenuation in CRP levels after lignan intervention (Hallund et al., 2008). Data of Lee et al. (2008) supported this contention. A 10% flaxseed-supplemented diet reduced oxidative lung damage even at 3 weeks as assessed by malondialdehyde (MDA) levels and lipid oxidation measured by immunohistochemical staining of lungs for iPF2α-III F2 isoprostane, an in vivo marker of lipid peroxidation.

Interestingly, increased serum enterolactone levels were correlated with lowered plasma F2-isoprostane concentrations in 100 mildly hypercholesterolemic men (Vanharanta et al., 2002). Supplementing diets of polycystic rats with 20 mg/kg of

SDG for 12 weeks was related to less macrophage infiltration, PGE2 release, lipid peroxide formation, and thus reduced renal inflammation (Ogborn et al., 2006). Lung MDA levels, BAL neutrophils, and overall alveolar white blood cell influx were significantly reduced in 10% flaxseed-treated mice relative to results from isocaloric controls following hyperoxia and acid aspiration-induced lung injury but not after LPS instillation (Kinniry et al., 2006). This amelioration was accompanied by rapid serum accumulation of both enterodiol and enterolactone as early as 1 week after initiation of the flaxseed diet.

This study provides evidence that dietary flaxseed may provide a protective strategy against selected types of pro-oxidant-induced tissue damage in vivo, although supplementing a low fat muffin with lignan isolated from flaxseed (500 mg/day) did not alter lipoprotein oxidation lag time, plasma trolox-equivalent antioxidant capacity (TEAC), or ferric reducing ability of plasma (FRAP) despite significant increases in serum and urinary concentrations of enterolactone in healthy postmenopausal women (Hallund et al., 2006b).

Feeding one group of ovariectomized mice a diet enriched with 10% flaxseed and two groups a basal diet along with subcutaneous (s.c.) injection of enterodiol or enterolactone (15 mg/kg body weight) for 3 weeks significantly ($P < 0.01$) reduced extracellular VEGF release in vivo and also in vitro (Bergman-Jungeström et al., 2007). The diminutions in VEGF secretion in vivo paralleled decreased tumor vessel areas in all the treated groups relative to the basal diet treated-group (Bergman-Jungeström et al., 2007).

In parallel with these findings, Saarinen et al. (2008) investigated the influence of lignans isolated from knots of *Pinus cembra* in models of in vivo dimethylbenz(a) anthracene-induced mammary cancer in rats and of human MCF-7 breast cancer xenografts in athymic mice. The results showed that administration of lariciresinol (3 or 15 mg/kg of body weight perorally) to rats for 9 weeks or consumption of a lariciresinol diet (20 or 100 mg/kg) for 5 weeks by athymic mice suppressed tumor growth and tumor angiogenesis. While enterolactone significantly inhibited estradiol-stimulated VEGF secretion in MCF-7 cells, lariciresinol administration enhanced tumor cell apoptosis and increased ERβ expression in MCF-7 xenografts (Saarinen et al., 2008).

Of 12 lignans isolated from *Saururus chinensis*, 10 mg/kg i.p. of lignans 9 (sauchinone) and 10 (sauchinone B) equally augmented mice survival rates (80 versus 20%), and significantly reduced plasma levels of TNF-α (16 and 18 ng/ml versus 24 ng/ml) relative to untreated controls. Furthermore, lignans isolated from *Pterocarpus santalinus* revealed inhibitory activities on TNF-α production in RAW264.7 cells challenged with LPS and on proliferation of concanavalin-A (Con-A)-induced mouse splenocytes without showing cytotoxicity (Cho et al., 2001). It has been suggested that dietary lignans may exert anti-inflammatory effects by inhibition of δ-(5)-desaturation of n-6 fatty acids, resulting in significant reductions in plasma PGE2 concentrations and TNF-α (Utsunomiya et al., 2000).

Although the mechanisms by which dietary lignans induce anti-inflammatory effects are not well understood, antioxidant properties may be involved. Lignans suppress lipid peroxidation (Kitts et al., 1999), exhibit direct hydroxyl radical scavenging action (Prasad, 2000; Nakano et al., 2006), reduce ROS secretion (Lee et al., 2008), and suppress metabolic vitamin E catalysis (Sontag and Parker, 2002). However, other mechanisms may be responsible and remain to be elucidated. Lignans may have indirect antioxidative effects.

In this regard, Kivelä et al. (2008) have shown that enterolactone induced heme oxygenase-1 (HO-1) expression. This stress inducible enzyme stimulates heme degradation to bilirubin, biliverdin, and carbon monoxide; such metabolites promote antioxidant and anti-inflammatory activities of HO-1 in blood vessels (Stocker and Perrella, 2006). Products of HO-1 interfere with the expression of pro-inflammatory cytokines and chemokines (Stocker and Perrella, 2006). Some lignans also possess other relevant pharmacological activities such as analgesic (Lin et al., 2007), anti-parasitic (de Andrade-Neto et al., 2007), antiviral (Swarup et al., 2008), and immunosuppressive properties (Park et al., 2007).

### 12.3.3 Saponins

Saponins are glycosides characterized by a triterpene or steroid nucleus with one or more oligosaccharide chains attached to an aglycone core (Shi et al., 2004). Table 12.5 summarizes anti-inflammatory effects of saponins. Figure 12.4 shows chemical structures of select saponins. Plant saponins have been used in food manufacturing applications mainly as surface active and foaming agents (San Martin and Briones, 1999) and have been identified as anti-nutritional factors (Thompson, 1993) due to their toxicity and hemolytic activity (Oda et al., 2000). However, soy isolated saponins did not show toxic effects in humans (Shi et al., 2004) and in vitro and in vivo studies support anti-inflammatory activity of soy saponins.

For example, co-incubation of peritoneal macrophages with 100 ng/ml of LPS and 30 to 100 mg/ml of soy saponins for 16 hours significantly inhibited the LPS-stimulated release of PGE2, NO, TNF-α, and MCP-1 (Kang et al., 2005). This activity was accompanied by the downregulation of COX-2 and iNOS expression at both the mRNA and protein levels as well as inhibition of NF-κB activation, iNOS, COX-2, TNF-α, IL-1β, and IL-6 expression by blocking degradation of IkB-α. These data suggest a potential anti-inflammatory activity of soybean saponin. However, other animal studies revealed that dietary saponins promote enteritis, perhaps by increasing intestinal permeability (Knudsen et al., 2007).

Changkil saponins (CKSs) isolated from the roots of *Platycodon grandiflorum* significantly reduced protein and mRNA expression of VCAM-1 and ICAM-1 in endothelial cells and inhibited the TNF-α-induced production of intracellular ROS and activation of NF-κB by preventing IκB degradation (Kim et al., 2006a). The inhibitory effects of CKSs on the TNF-α-induced activation of NF-κB may be due to its antioxidant properties. Anti-inflammatory effects of CKSs were observed in inflammatory responses induced by carrageenan in an acute air pouch inflammation rat model (Kim et al., 2006b). A crude saponin extract (CSE) of *Hedera helix* (English ivy) at 100 and 200 mg/kg doses exhibited 77% acute anti-inflammatory effects as shown by a decrease in volume of carrageenan-induced paw edema in rats. Saponin purified extracts (SPEs) showed acute anti-inflammatory effects of 60% for 100 mg/kg doses and 68% for 200 mg/kg (Süleyman et al., 2003). Conversely, SPE

**TABLE 12.5**
**Recent Studies of Anti-Inflammatory Effects of Saponins**

| Study | Model | Dose/Duration | Major Findings |
|---|---|---|---|
| Kang et al. (2005) | Peritoneal macrophages from BALB/c mice treated with LPS | 30 to 100 mg/ml saponins in serum-free medium (16 hours) | ↓ PGE2, NO, TNF-α and MCP-1 release ($P < 0.01$) |
| Kim et al. (2006) | Cytokine-induced monocyte/human endothelial cell interaction | Cells pretreated with different saponin concentrations: 0.2, 0.5, 2,5 μg/ml, for 1 hour, then stimulated with TNF-α 10 ng/ml (6 hours) | ↓ protein and mRNA expression of VCAM-1 and ICAM-1; ↓ in ROS release |
| Knudsen et al. ( 2007) | Seawater-adapted Atlantic salmon (n = 12) | Diet plus 10% saponin containing subfractions of soybean molasses (62 days) | ↑ intestinal inflammation and enteritis; ↓ IkB-α phosphorylation |
| Suh et al. (2007) | LPS-induced raw 264.7 murine macrophages | Triterpenoid saponin, from *Aralia elata*, 0, 0.4, 0.9 and 1.8 μM (24 hours) | ↓ NO, PGE2 synthesis; ↓ COX-2 mRNA and protein; |
| Dang et al. (2007) | Male SD rats with induced hepatic fibrosis (n = 15) | Saikosaponin-d (SSD), 1.0, 1.5 and 2.0 mg/kg i.p. (6 weeks) | 56 and 55% ↓ hepatic TNF-α and IL-6; ↓ NF-κB p65 expression; ↑ hepatic activity of I-κBα, ($P < 0.01$) |
| Toshkova et al. ( 2007) | 2- to 4-month old Golden Syrian hamsters (n = 10) | Saponin mixture, 50 mg/kg s.c. (4 days) | ↑ number, migration and phagocytic indexes of peritoneal macrophages |
| Wu et al. (2007) | Murine N9 microglial cells | Ginsenosides 0.01 to 100 μM/ml; LPS (1 μg/ml) added to N9 microglial cells (48 hours) | ↓ LPS-induced TNF-α and NO production; inhibition of NF-κB |
| Yang et al. (2007) | 5 week-old male ICR mice (n = 5 to 10) | Immunized s.c. with OVA, LPS or ConA with ginsenoside-Rd,10, 25 or 50 μg (2 weeks) | ↑ OVA-specific IgG, IgG1, IgG2b ($P < 0.05$); ↑ production of Th1 and Th2 cytokines in OVA-immunized mice; ↑ IL-2, IFN-γ, IL-4 and IL-10 mRNA in splenocytes induced by Con A ($P < 0.05$) |

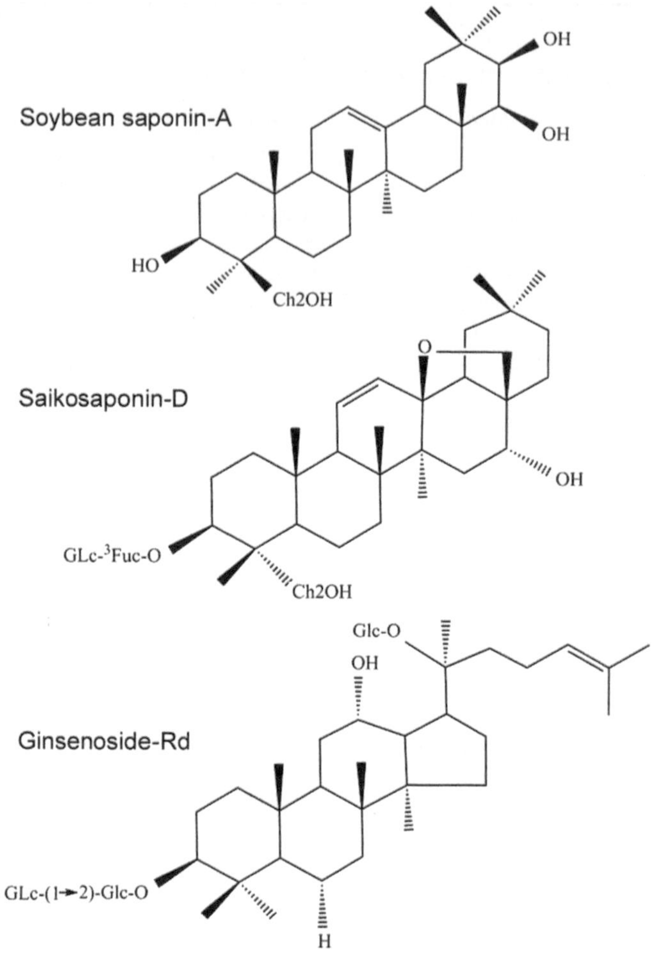


**FIGURE 12.4** Chemical structures of selected saponins.

exhibited more potent chronic anti-inflammatory activity than CSE as measured by reductions in cotton pellet granuloma tissue formation (Süleyman et al., 2003).

Since the most potent effect of saponins was observed in the fourth hour, this study suggests that acute anti-inflammatory effects of *H. helix* extracts may be mediated by blocking histamine and/or serotonin release rather than prostaglandin and/or bradykinin. Won et al. (2006) supported these results by showing inhibitions in NO, PGE2, and TNF-α production, reductions in LPS-induced levels of iNOS and COX-2 at protein levels, and iNOS, COX-2, TNF-α, IL-1β and IL-6 mRNA expression along with blocking of NF-κB activation after incubating RAW 264.7 macrophages with 2.5 to 10 μM saponins isolated from *Pleurospermum kamtschatidum*. Recently, saponins from *Aralia elata* were found to suppress the activity of NO, PGE2 synthesis, iNOS, and COX-2 mRNA and its protein expression in a dose-dependent manner at 0.4 to 1.8 μM concentrations in LPS-induced raw 264.7 murine macrophages (Suh et al., 2007). These observations were associated with decreased NF-κB translocation and the inhibition of LPS-induced phosphorylation of IkB-α, suggesting that the anti-inflammatory activities of these saponins may be mediated by downregulation of NF-κB.

Three days after intravenous injection of sasanqua saponin (SQS; 2.5, 5, and 10 mg/kg), extracted from *Camellia oleifera* Abel, a Chinese traditional herb, Huang et al. (2005) noted decreased serum levels of sICAM-1 and suppressed expression and transcription of transmembrane ICAM-1 (mICAM-1) in aortae of rats with burn-induced inflammation. Of six saponins extracted from *Polygala japonica*, another Chinese herb, saponins 1, 4, and 5 showed anti-inflammatory effects at a dose of 0.1 μmol/kg on acute and delayed phases of carageenan-induced paw edema in mice. The effects were not noted with saponins 2, 3 and 6 (Wang et al., 2008). This study suggests a structure–activity relationship among saponins in determining their anti-inflammatory effects.

Furthermore, ginseng saponins, ginsenosides-Rd, -Rb2, -Rg1 and -Re, over a range of concentrations (0.01 to 100 μM), exerted a notable inhibitory effect on LPS-induced TNF-α production in N9 microglial cells. The inhibitory effect of ginseng saponins on TNF-α and NO production was consistent with the observed down-regulation of IκB, suggesting that anti-inflammatory effects of ginsenosides may be mediated by inhibition of NF-κB.

Saponins of *P. notoginseng* showed immunomodulatory effects on murine macrophages in vitro (Rhule et al., 2006). Feeding rats a high fat diet for 9 weeks along with i.p. injections of *P. notoginseng* saponins (100 mg/kg/day) was associated with significant reductions in the gene expression of adhesion molecules (VCAM-1 and ICAM-1), inflammatory factors (integrins, IL-18, IL-β1, MMP-2, and MMP-9) and the expression of NF-κB/p65 along with an increase in the expression of I-κBα (Zhang et al., 2008). In addition, saponin treatment remarkably lowered serum TC, TG, and blood viscosity. Such changes may explain alleviation of typical pathological changes associated with atherosclerosis in animals treated with saponins. Analysis of this broad array of studies suggests that the anti-inflammatory influence of ginseng saponins may involve regulation of the NF-κB pathway and the expression of I-κBα.

Evidence indicates that inflammation may cause abnormal lipid metabolism, which in turn may lead to inflammation (Hotamisligil, 2006). *P. notoginseng* saponins may reduce inflammation by lessening TC and TG levels along with lowering Apo-E expression that causes attenuation in lipid transfer from blood vessels to macrophages below the endothelium (reviewed by Zhang et al., 2008). Although the precise mechanisms involved in hypolipidemic effects of saponins remain unclear, the limited experimental evidence suggests that saponins may reduce cholesterol by forming insoluble complexes with cholesterol in the intestine and impede its absorption.

Saponins may lower cholesterol indirectly by increasing the excretion of bile acids (Sidhu and Oakenfull, 1986). Huang et al. (2007) reported that saponins isolated from *Gynostemma pentaphyllum* (GP) may act as potent activators of PPARα and in this way protect against atherosclerosis, inflammation, and cancer. These data indicate that saponins may lower expression of inflammatory mediators. The mechanisms behind these beneficial properties of saponins are unknown, but may involve inhibition NF-κB activation and inhibition of IkB-α degradation.

## 12.4 CONCLUSIONS

Increasing evidence suggests that phytosterols and phytoestrogens have anti-inflammatory properties. It is likely that these compounds may reduce or mitigate the secretion of inflammatory mediators from visceral adipose tissue into the portal system and thus reduce risk of CVD and other inflammatory disorders. The mechanisms by which these phytochemicals affect cytokine production may center on altered NF-κB activity but this requires further investigation

## ACKNOWLEDGMENTS

The support of Heather Johannson is greatly appreciated.

## REFERENCES

AbuMweis SS et al. 2006. Intake of a single morning dose of standard and novel plant sterol preparations for 4 weeks does not dramatically affect plasma lipid concentrations in humans. *J Nutr* 136: 1012–1016.

Acuff RV et al. 2007. The lipid lowering effect of plant sterol ester capsules in hypercholesterolemic subjects. *Lipids Health Dis* 9: 6–11.

Adlercreutz H et al. 1991. Urinary excretion of lignans and isoflavonoid phytoestrogens in Japanese men and women consuming traditional Japanese diet. *Am J Clin Nutr* 54: 1093–1100.

Al-Yousuf MH et al. 2002. Central nervous system activity of *Leucas inflata* Benth. in mice. *Phytomedicine* 9: 501–3507.

Ambring A, Friberg P, and Axelsen M. 2004. Effects of a Mediterranean-inspired diet on blood lipids, vascular function and oxidative stress in healthy subjects. *Clin Sci* 106: 519–525.

Awad NE. 2000. Biologically active steroid from the green alga *Ulva lactuca. Phytother Res* 14: 641–643.

Awad AB, Smith AJ, and Fink CS. 2001. Plant sterols regulate rat vascular smooth muscle cell growth and prostacyclin release in culture. *Prostagl Leukot Essent Fatty Acids* 64: 323–330.

Awad AB, Toczek J, and Fink CS. 2004. Phytosterols decrease prostaglandin release in cultured P388D(1)/MAB macrophages. *Prostagl Leukot Essent Fatty Acids* 70: 511–520.

Azadbakht L et al. 2007. Soy consumption, markers of inflammation, and endothelial function. *Diabetes Care* 30: 967–973.

Bergman-Jungeström M, Thompson LU, and Dabrosin C. 2007. Flaxseed and its lignans inhibit estradiol-induced growth, angiogenesis, and secretion of vascular endothelial growth factor in human breast cancer xenografts in vivo. *Clin Cancer Res* 13: 1061–1067.

Bingham M et al. 2003. Gut metabolism and cardioprotective effects of dietary isoflavones. *Curr Top Nutraceutical Res* 1: 31–48.

Bingham SA et al. 1998. Phytoestrogens: where are we now? *Br. J. Nutr* 79: 393–406.

Blum A et al. 2003. Effects of oral soy protein on markers of inflammation in postmenopausal women with mild hypercholesterolemia. *Am Heart J* 145: e7.

Bouic PJ et al. 1996. β-Sitosterol and β-sitosterol glucoside stimulate human peripheral blood lymphocyte proliferation: implications for their use as an immunomodulatory vitamin combination. *Int J Immunopharmacol* 18: 693–700.

Bouic PJ et al. 1999. The effects of β-sitosterol (BSS) and β -sitosterol glucoside (BSSG) mixture on selected immune parameters of marathon runners: inhibition of post marathon immune suppression and inflammation. *Int J Sports Med* 20: 258–262.

Breytenbach U et al. 2001. Flow cytometric analysis of the Th1–Th2 balance in healthy individuals and patients infected with the human immunodeficiency virus (HIV) receiving a plant sterol/sterolin mixture. *Cell Biol Int* 25: 43–49.

Calpe-Berdiel L et al. 2007. Dietary phytosterols modulate T-helper immune response but do not induce apparent anti-inflammatory effects in a mouse model of acute, aseptic inflammation. *Life Sci* 80: 1951–1956.

Cassidy A et al. 2006. Factors affecting the bioavailability of soy isoflavones in humans after ingestion of physiologically relevant levels from different soy foods. *J Nutr.* 136: 24–51.

Cater NB et al. 2005. Responsiveness of plasma lipids and lipoproteins to plant stanol esters. *Am J Cardiol* 96: 23D–28D.

Chacko BK et al. 2005. Revealing anti-inflammatory mechanisms of soy isoflavones by flow: modulation of leukocyte–endothelial cell interactions. *Am J Physiol Heart Circ Physiol* 289: H908–H915.

Cho JY et al. 2001. Savinin, a lignan from *Pterocarpus santalinus* inhibits tumor necrosis factor-α production and T cell proliferation. *Biol Pharm Bull* 24: 167–171.

Choi S et al. 2002. Angiogenic activity of β-sitosterol in the ischaemia/reperfusion-damaged brain of Mongolian gerbil. *Planta Med* 68: 330–335.

Clerici C et al. 2007. Pasta naturally enriched with isoflavone aglycons from soy germ reduces serum lipids and improves markers of cardiovascular risk. *J Nutr* 137: 2270–2278.

Cupisti A et al. 2007. Soy protein diet improves endothelial dysfunction in renal transplant patients. *Nephrol Dial Transplant* 22: 229–234.

Dang SS et al. 2007. Inhibitory effects of saikosaponin-d on CCl4-induced hepatic fibrogenesis in rats. *World J Gastroenterol* 13: 557–563.

de Andrade-Neto VF et al. 2007. Antiplasmodial activity of aryltetralone lignans from *Holostylis reniformis*. *Antimicrob Agents Chemother* 51: 2346–2350.

De Jong A et al. 2008. Effects of plant sterol and stanol ester consumption on lipid metabolism, antioxidant status and markers of oxidative stress, endothelial function and low-grade inflammation in patients on current statin treatment. *Eur J Clin Nutr* 62: 263–273.

de Jongh S et al. 2003. Plant sterols lower LDL cholesterol without improving endothelial function in prepubertal children with familial hypercholesterolaemia. *J Inherit Metab Dis* 26: 343–351.

Devaraj S, Jialal I, and Vega-López S. 2004. Plant sterol-fortified orange juice effectively lowers cholesterol levels in mildly hypercholesterolemic healthy individuals. *Arterioscler Thromb Vasc Biol* 24: E25–E28.

Devaraj S, Autret BC, and Jialal I. 2006. Reduced-calorie orange juice beverage with plant sterols lowers C-reactive protein concentrations and improves the lipid profile in human volunteers. *Am J Clin Nutr* 84: 756–7561.

Donnelly LE et al. 2004. Anti-inflammatory effects of resveratrol in lung epithelial cells: molecular mechanisms. *Am J Physiol Lung Cell Mol Physiol* 287: 774–783.

Evans M et al. 2007. Effect of soy isoflavone and soy lecithin on endothelial function in healthy post menopausal women. *Menopause* 14: 141–149.

Feigin AM, Teeter JH, and Brand JG. 1995. The influence of sterols on the sensitivity of lipid bilayers to melittin. *Biochem Biophys Res Commun* 211: 312–317.

Figtree GA et al. 2000. Plant-derived estrogens relax coronary arteries in vitro by a calcium antagonistic mechanism. *J Am Coll Cardiol* 35: 1977–1985.

Francis G et al. 2002. The biological action of saponins in animal systems: a review. *Br J Nutr* 88: 587–605.

Fu WY et al. 2002. Homocysteine attenuates hemodynamic responses to nitric oxide in vivo. *Atherosclerosis* 161: 169–176.

Gottstein N et al. 2003. Effect of genistein and daidzein on platelet aggregation and monocyte and endothelial function. *Br J Nutr* 89: 607–616.

Hall WL et al. 2005. Soy-isoflavone-enriched foods and inflammatory biomarkers of cardiovascular disease risk in postmenopausal women: interactions with genotype and equol production. *Am J Clin Nutr* 82: 1260–1268.

Hallikainen M et al. 2006. Endothelial function in hypercholesterolemic subjects: efects of plant stanol and sterol esters. *Atherosclerosis* 188: 425–432.

Hallikainen M et al. 2007. Effects of plant stanol esters on serum cholesterol concentrations, relative markers of cholesterol metabolism and endothelial function in type 1 diabetes. *Atherosclerosis* (epub).

Hallund J et al. 2006a. Daily consumption for six weeks of a lignan complex isolated from flaxseed does not affect endothelial function in healthy postmenopausal women. *J Nutr* 136: 2314–2318.

Hallund J et al. 2006b. A lignan complex isolated from flaxseed does not affect plasma lipid concentrations or antioxidant capacity in healthy postmenopausal women. *J Nutr* 136: 112–116.

Hallund J et al. 2008. The effect of a lignan complex isolated from flaxseed on inflammation markers in healthy postmenopausal women. *Nutr Metab Cardiovasc Dis* (epub).

Hansel B et al. 2007. Effect of low-fat, fermented milk enriched with plant sterols on serum lipid profile and oxidative stress in moderate hypercholesterolemia. *Am J Clin Nutr* 86: 790–796.

Hanson LN et al. 2006. Effects of soy isoflavones and phytate on homocysteine, C-reactive protein, and iron status in postmenopausal women. *Am J Clin Nutr* 84: 774–780.

Hermansen K et al. 2001. Beneficial effects of a soy-based dietary supplement on lipid levels and cardiovascular risk markers in type 2 diabetic subjects. *Diabetes Care* 24: 228–233.

Holubkov R et al. 2002. Large brachial artery diameter is associated with angiographic coronary artery disease in women. *Am Heart J* 143: 802–807.

Hotamisligil GS. 2006. Inflammation and metabolic disorders. *Nature* 444: 860–867.

Huang Q et al. 2005. Inhibitory effects of sasanquasaponin on over-expression of ICAM-1 and on enhancement of capillary permeability induced by burns in rats. *Burns* 31: 637–642.

Huang TH et al. 2007. Gypenoside XLIX, a naturally occurring PPAR-α activator, inhibits cytokine-induced vascular cell adhesion molecule-1 expression and activity in human endothelial cells. *Eur J Pharmacol* 565: 158–165.

Hutchins AM et al. 2001. Flaxseed consumption influences endogenous hormone concentrations in postmenopausal women. *Nutr Cancer* 39: 58–65.

Hwang J et al. 2000. Synergistic inhibition of LDL oxidation by phytoestrogens and ascorbic acid. *Free Radic Biol Med* 29: 79–89.

Jenkins DJ et al. 2003. Effects of a dietary portfolio of cholesterol-lowering foods versus lovastatin on serum lipids and C-reactive protein. *JAMA* 290: 502–510.

Jenkins DJ et al. 2005. Direct comparison of dietary portfolio versus statin on C-reactive protein. *Eur J Clin Nutr* 59: 851–860.

Joannou GE et al.1995. A urinary profile study of dietary phytoestrogens: the identification and mode of metabolism of new isoflavonoids. *J Steroid Biochem Mol Biol* 54: 167–184.

Jones PJ et al. 2007. Fish-oil esters of plant sterols differ from vegetable-oil sterol esters in triglycerides lowering, carotenoid bioavailability and impact on plasminogen activator inhibitor-1 (PAI-1) concentrations in hypercholesterolemic subjects. *Lipids Health Dis* 6: 28–32.

Joseph SB et al. 2004. LXR-dependent gene expression is important for macrophage survival and the innate immune response. *Cell* 119: 299–309.

Kaneko E et al. 2003. Induction of intestinal ATP-binding cassette transporters by a phytosterol-derived liver X receptor agonist. *J Biol Chem* 278: 36091–36098.

Kang JH et al. 2005. Soybean saponins suppress the release of proinflammatory mediators by LPS-stimulated peritoneal macrophages. *Cancer Lett* 230: 219–227.

Kim JY et al. 2006a. Inhibition of tumor necrosis factor-α-induced expression of adhesion molecules in human endothelial cells by the saponins derived from roots of *Platycodon grandiflorum*. *Toxicol Appl Pharmacol* 210: 150–156.

Kim JY et al. 2006b. Inhibitory effect of the saponins derived from roots of *Platycodon grandiflorum* on carrageenan-induced inflammation. *Biosci Biotechnol Biochem* 70: 858–864.

Kim SJ et al. 2007. Transgenic panax ginseng inhibits the production of TNF-α, IL-6, and IL-8 as well as COX-2 expression in human mast cells. *Am J Clin Med* 35: 329–339.

Kinniry P et al. 2006. Dietary flaxseed supplementation ameliorates inflammation and oxidative tissue damage in experimental models of acute lung injury in mice. *J Nutr* 136: 1545–1551.

Kitts DD et al. 1999. Antioxidant activity of the flaxseed lignan secoisolariciresinol diglycoside and its mammalian lignan metabolites enterodiol and enterolactone. *Mol Cell Biochem* 202: 91–100.

Kivelä AM et al. 2008. Enterolactone induces heme oxygenase-1 expression through nuclear factor-E2-related factor 2 activation in endothelial cells. *J Nutr* 138: 1263–1268.

Knudsen D et al. 2007. Saponin-containing subfractions of soybean molasses induce enteritis in the distal intestine of Atlantic salmon. *J Agric Food Chem* 55: 2261–2267.

Kobori M et al. 2007. Ergosterol peroxide from an edible mushroom suppresses inflammatory responses in RAW264.7 macrophages and growth of HT29 colon adenocarcinoma cells. *Br J Pharmacol* 150: 209–219.

Kurzer MS and Xu X. 1997. Dietary phytoestrogens. *Annu Rev Nutr* 17: 353–381.

Lee JC et al. 2008. Dietary flaxseed enhances antioxidant defenses and is protective in a mouse model of lung ischemia–reperfusion injury. *Am J Physiol Lung Cell Mol Physiol* 294: L255–L265.

Lee JH et al. 2007. Immunoregulatory activity by daucosterol, a β-sitosterol glycoside, induces protective Th1 immune response against disseminated candidiasis in mice. *Vaccine* 25: 3834–3840.

Lin CW et al. 2008. Quercetin inhibition of tumor invasion via suppressing PKCδ/ERK/AP-1-dependent matrix metalloproteinase-9 activation in breast carcinoma cells. *Carcinogenesis* (epub).

Lin YR et al. 2007. Effects of honokiol and magnolol on acute and inflammatory pain models in mice. *Life Sci* 81: 1071–1080.

Lv N et al. et al. 2008. Guggulsterone, a plant sterol, inhibits NF-γB activation and protects pancreatic β cells from cytokine toxicity. *Mol Cell Endocrinol* 289: 49–59.

Madsen MB, Jensen AM, and Schmidt EB. 2007. The effect of a combination of plant sterol-enriched foods in mildly hypercholesterolemic subjects. *Clin Nutr* 26: 792–798.

Manna SK, Mukhopadhyay A, and Aggarwal BB. 2000. Resveratrol suppresses TNF-α-induced activation of nuclear transcription factors NF-κB, activator protein-1, and apoptosis: potential role of reactive oxygen intermediates and lipid peroxidation. *J Immunol* 164: 6509–6519.

McMahon LG et al. 1997. Cytosolic pyridoxineb-D-glucoside hydrolase from porcine jejunal mucosa: purification, properties and comparison with broad specificity β-glucosidase. *J Biol Chem* 272: 32025–32033.

Meagher LP et al. 1999. Isolation and characterization of the lignans, isolariciresinol and pinoresinol, in flaxseed meal. *J Agric Food Chem* 47: 3173–3180.

Meagher LP and Beecher GR. 2000. Assessment of data on the lignan content of foods. *J Food Comp Anal* 13: 935–947.

MerzDemlow BE et al. 2000. Soy isoflavones improve plasma lipids in normocholesterolemic, premenopausal women. *Am J Clin Nutr* 71: 1462–1469.

Miettinen TA and Gylling H. 2006. Plant stanol and sterol esters in prevention of cardiovascular diseases: a review. *Int J Clin Pharmacol Ther* 44: 247–250.

Moghadasian MH. 2000. Pharmacological properties of plant sterols in vivo and in vitro observations. *Life Sci* 67: 606–615

Moghadasian MH. 2006. Dietary phytosterols reduce cyclosporine-induced hypercholesterolemia in apolipoprotein E-knockout mice. *Transplantation* 81: 207–213.

Moghadasian MH et al. 1999. Proatherogenic and antiatherogenic effects of probucol and phytosterols in apolipoprotein E-deficient mice: possible mechanisms of action. *Circulation* 99: 1733–1739.

Moghadasian MH et al. 2001. Hepatic cholesterol and bile acid synthesis, low-density lipoprotein receptor function, and plasma and fecal sterol levels in mice: effects of apolipoprotein E deficiency and probucol or phytosterol treatment. *Metabolism* 50: 708–714.

Nakano D et al. 2006. Sesamin metabolites induce an endothelial nitric oxide-dependent vasorelaxation through their antioxidative property-independent mechanisms. *J Pharmacol Exp Ther* 318: 328–335.

Nasca MM, Zhou JR, and Welty FK. 2008. Effect of soy nuts on adhesion molecules and markers of inflammation in hypertensive and normotensive postmenopausal women. *Am J Cardiol* 102: 84–86.

Nashed B et al. 2005. Antiatherogenic effects of dietary plant sterols are associated with inhibition of proinflammatory cytokine production in apo E-KO mice. *J Nutr* 135: 2438–2444.

Navarro A, De las Heras B, and Villar A. 2001. Anti-inflammatory and immunomodulating properties of a sterol fraction from *Sideritis foetens* Clem. *Biol Pharm Bull* 24: 470–473.

Oda K et al. 2000. Adjuvant and haemolytic activities of 47 saponins derived from medicinal and food plants. *Biol Chem* 381: 67–74.

Ogborn MR et al. 2006. Effects of flaxseed derivatives in experimental polycystic kidney disease vary with animal gender. *Lipids* 41: 1141–1149.

Pan W et al. 2001.Genistein, daidzein and glycitein inhibit growth and DNA synthesis of aortic smooth muscle cells from stroke-prone spontaneously hypertensive rats. *J Nutr* 131: 1154–1158.

Park EH et al. 2001. An anti-inflammatory principle from cactus. *Fitoterapia* 72: 288–290.

Park SY et al. 2007. Immunosuppressive lignans isolated from *Saururus chinensis*. *Planta Med* 73: 674–678.

Prasad K. 2000. Antioxidant activity of secoisolariciresinol diglucoside-derived metabolites, secoisolariciresinol, enterodiol, and enterolactone. *Int J Angiol* 9: 220–225.

Prasad K. 2007. A study on regression of hypercholesterolemic atherosclerosis in rabbits by flax lignan complex. *J Cardiovasc Pharmacol Ther* 12: 304–313.

Pritchard PH et al. 2003. Comparison of cholesterol-lowering efficacy and anti-atherogenic properties of hydrogenated verus non-hydrogenated (Pytrol™) tall oil-derived phytosterols in apo-E defcient mice. *Cardiovasc Drugs Ther* 17: 443–449.

Raitakari OT et al. 2008. Plant stanol ester consumption and arterial elasticity and endothelial function. *Br J Nutr* 18: 1–6.

Regal JF et al. 2000. Dietary phytoestrogens have anti-inflammatory activity in a guinea pig model of asthma. *Proc Soc Exp Biol Med* 223: 372–378.

Rhule A et al. 2006. Panax notoginseng attenuates LPS-induced pro-inflammatory mediators in RAW264.7 cells. *J Ethnopharmacol* 106: 121–128.

Saarinen NM et al. 2008. Dietary lariciresinol attenuates mammary tumor growth and reduces blood vessel density in human MCF-7 breast cancer xenografts and carcinogen-induced mammary tumors in rats. *Int J Cancer* 123: 1196–1204.

Salen G et al. 1985. Increased plasma cholestanol and 5 α-saturated plant sterol derivatives in subjects with sitosterolemia and xanthomatosis. *J Lipid Res* 26: 203–209.

San Martin R and Briones R. 1999. Industrial uses and sustainable supply of *Quillaja saponaria* (rosaceae) saponins. *Econ Bot* 53: 302–311.

Setchell KD et al. 2002. Evidence for lack of absorption of soy isoflavone glycosides in humans: supporting the crucial role of intestinal metabolism for bioavailability. *Am J Clin Nutr* 76: 447–453.

Shi J et al. 2004. Saponins from edible legumes: chemistry, processing, and health benefits. *J Med Food* 7: 67–78.

Shishodia S and Aggarwal BB. 2004. Guggulsterone inhibits NF-κB and IκBα kinase activation, suppresses expression of anti-apoptotic gene products, and enhances apoptosis. *J Biol Chem* 279: 47148–47158.

Sidhu GS and Oakenfull DG. 1986. A mechanism for the hypocholesterolaemic activity of saponins. *Br J Nutr* 55: 643–649.

Sontag TJ and Parker RS. 2002. Cytochrome P450ω-hydroxylase pathway of tocopherol catabolism: novel mechanism of regulation of vitamin E status. *J Biol Chem* 277: 290–296.

Squadrito F et al. 2002. The effect of the phytoestrogen genistein on plasma nitric oxide concentrations, endothelin-1 levels and endothelium dependent vasodilation in postmenopausal women. *Atherosclerosis* 163: 339–347.

Squadrito F et al. 2003. Effect of genistein on endothelial function in postmenopausal women: a randomized, double-blind, controlled study. *Am J Med* 114: 470–476.

Steinberg FM et al. 2003. Soy protein with isoflavones has favourable effects on endothelial function that are independent of lipid and antioxidant effects in healthy postmenopausal women. *Am J Clin Nutr* 78: 123–130.

Stocker R and Perrella MA. 2006. Heme oxygenase-1: novel drug target for atherosclerotic diseases? *Circulation* 114: 2178–2189.

Subbaramaiah K et al. 1998. Resveratrol inhibits cyclooxygenase-2 transcription and activity in phorbol ester-treated human mammary epithelial cells. *J Biol Chem* 273: 21875–21882.

Suh SJ et al. 2007. Triterpenoid saponin, oleanolic acid 3-O-β-d-glucopyranosyl(1>3)-α-l-rhamnopyranosyl(1>2)-α-larabinopyranoside (OA) from *Aralia elata* inhibits LPS-induced nitric oxide production by downregulated NF-κB in raw 264.7 cells. *Arch Biochem Biophys* 467: 227–233.

Süleyman H et al. 2003. Acute and chronic antiinflammatory profile of the ivy plant, *Hedera helix*, in rats. *Phytomedicine* 10: 370–374.

Swarup V et al. 2008. Novel strategy for treatment of Japanese encephalitis using arctigenin, a plant lignan. *J Antimicrob Chemother* 61: 679–688.

Teede HJ et al. 2003. Isoflavones reduce arterial stiffness: a placebo-controlled study in men and postmenopausal women. *Arterioscler Thromb Vasc Biol* 23: 1066–1071.

Thompson LU et al. 1991. Mammalian lignan production from various foods. *Nutr Cancer* 16: 43–52.

Thompson LU. 1993. Potential health benefits and problems associated with antinutrients in foods. *Food Res Int* 26: 131–149.

Thompson LU et al. 2006. Phytoestrogen content of foods consumed in Canada, including isoflavones, lignans, and coumestan. *Nutr Cancer* 54: 184–201.

Tikkanen MJ et al. 1998. Effect of soybean phytoestrogen intake on low density lipoprotein oxidation resistance. *Proc Natl Acad Sci USA* 95: 3106–3110.

Toshkova RA et al. 2007. Influence of purified saponin mixture from *Astragalus corniculatus* Bieb. on phagocytic cells in Graffi tumor bearing hamsters. *J Ethnopharmacol* 109: 394–399.

Utsunomiya T et al. 2000. Effects of sesamin-supplemented dietary fat emulsions on the ex vivo production of lipopolysaccharide-induced prostanoids and tumor necrosis factor-α in rats. *Am J Clin Nutr* 72: 804–808.

van der Schouw YT, Pijpe A, and Lebrun CE. 2002. Higher usual dietary intake of phytoestrogens is associated with lower aortic stiffness in postmenopausal women. *Arterioscler Thromb Vasc Biol* 22: 1316–1322.

van der Schouw YT et al. 2005. The usual intake of lignans but not that of isoflavones may be related to cardiovascular risk factors in U.S. men. *J Nutr* 135: 260–266.

Vanharanta M et al. 2002. Association between low serum enterolactone and increased plasma F2-isoprostanes, a measure of lipid peroxidation. *Atherosclerosis* 160: 465–469.

Verma S, Buchanan MR, and Anderson TJ. 2003. Endothelial function testing as a biomarker of vascular disease. *Circulation* 108: 2054–2059.

Villaseñor IM et al. 2002. Bioactivity studies on β-sitosterol and its glucoside. *Phytother Res* 16: 417–421.

Walker HA et al. 2001. The phytoestrogen genistein produces acute nitric oxide-dependent dilation of human forearm vasculature with similar potency to 17β estradiol. *Circulation* 103: 258–262.

Walker SE, Register TC, and Appt SE. 2008. Plasma lipid-dependent and -independent effects of dietary soy protein and social status on atherogenses in premenopausal monkeys. *Menopause* 15: 950–957.

Wang H et al. 2008. Anti-inflammatory activities of triterpenoid saponins from *Polygala japonica*. *Phytomedicine* 15: 321–326.

Won JH et al. et al. 2006. Anti-inflammatory effect of buddlejasaponin IV through the inhibition of iNOS and COX-2 expression in RAW 264.7 macrophages via NF-κB inactivation. *Br J Pharmacol* 148: 216–225.

Wu CF et al. 2007. Differential effects of ginsenosides on NO and TNF-α production by LPS-activated N9 microglia. *Int Immunopharmacol* 7: 313–320.

Yang Z et al. 2007. Ginsenoside Rd elicits Th1 and Th2 immune responses to ovalbumin in mice. *Vaccine* 25: 161–169.

Yeganeh B et al. 2005. Combination of dietary phytosterols plus niacin or fenofibrate: effects on lipid profile and atherosclerosis in apo E-KO mice. *J Nutr Biochem* 16: 222–228.

Yuk JE et al. 2007. Effects of lactose-β-sitosterol and β-sitosterol on ovalbumin-induced lung inflammation in actively sensitized mice. *Int Immunopharmacol* 7: 1517–1527.

Zhang YG et al. 2008. *Panax notoginseng* saponins attenuate atherosclerosis in rats by regulating the blood lipid profile and an anti-inflammation action. *Clin Exp Pharmacol Physiol* (epub).

# 13 Antioxidants, Polyphenols, and Adipose Inflammation

*Fereidoon Shahidi and Ying Zhong*

## CONTENTS



## 13.1 INTRODUCTION

Excessive release of reactive oxygen species (ROS) leads to oxidative stress which has been implicated in the pathogenesis of numerous chronic diseases and health conditions. Although ROS can be beneficial in immune response by killing pathogens and removing injured tissues, they act as a major tissue destructive force when overproduced as in various inflammations. Inflammation is an important component of many chronic diseases and age-related disorders. Metabolic syndrome, for example, has been associated with chronic inflammation, especially in adipose tissues. Therefore, effective control of inflammation is important in the prevention and treatment of such chronic diseases. Antioxidant strategies have been proposed for suppressing inflammation by scavenging the pro-inflammatory ROS, and hence reducing the risk of many inflammation-mediated diseases.

## 13.2 OXIDATIVE STRESS

### 13.2.1 Lipid Oxidation

Lipids are naturally occurring water-insoluble molecules. They include diverse compounds such as triacylglycerols, diacylglyerols, monoacylglycerols, phospholipids, waxes, sterols, and fat-soluble vitamins. Lipids in biological systems function mainly as energy storage and structural components of cell membranes and also serve as important signalling agents. Lipids, especially polyunsaturated fatty acids (PUFAs) in membranes, are susceptible to oxidation. Oxidative change in vivo is thought to exert destructive cellular effects and has been associated with pathophysiology of a number of diseases and health conditions.

Oxidation of lipids occurs via autoxidation, photooxidation, thermal oxidation, and enzymatic means, most of which involve free radicals and/or other reactive species as intermediates (Vercellotti et al., 1992; Shahidi, 2000). Autoxidation—defined as the spontaneous reaction of atmospheric oxygen with lipids—is the most common process leading to oxidative deterioration and damage in food and biological systems (Shahidi and Zhong, 2005a). Lipids in living organisms undergo oxidation reactions during normal aerobic metabolism (Beckman and Ames, 1998). Unsaturated fatty acids in triacylglycerols (as well as diacyglycerols or monoacylglycerols), phospholipids, free fatty acids, and cholesterol (especially LDL cholesterol) are generally the major reactants affected by such reactions. It is widely accepted that lipid oxidation occurs via a free radical chain reaction mechanism. The chain reaction proceeds through three stages: initiation, propagation, and termination (Shahidi and Zhong, 2005a). Figure 13.1 illustrates a simplified scheme of the oxidative free radical chain reaction.

Lipid molecules, in the presence of initiators such as heat, light, ionizing radiation, and metal ions and metalloproteins, lose a hydrogen atom and produce free radicals. In the propagation stage, the lipid radicals react with oxygen to form peroxyl radicals. In turn, the peroxyl radicals act as the chain carriers of the rapidly progressing reaction by further oxidizing the lipid. The hydrogen of the methylene group between two double bonds in a PUFA is very active toward free radical attack (bond energies for bisallylic, allylic, and alkyl hydrogens are 75, 88 and 101 kcal/mol, respectively). This explains why highly unsaturated fatty acids are more susceptible to oxidation compared to saturated or less saturated counterparts (Erickson, 2002).

During propagation, lipid hydroperoxides are produced as primary oxidation products. These hydroperoxides are unstable compounds and break down to a range of secondary oxidation products including alcohols, aldehydes, ketones, hydrocarbons, volatile organic acids, and epoxy compounds. Hydroperoxides and many of the secondary oxidation products are highly reactive and may initiate chain reactions of oxidation, and for that reason they are also considered toxic to tissues and cells (Yanishlieva and Marinova, 2001). In the termination stage of oxidation, free radicals neutralize each other and stable non-radical products are formed.

Free radicals play a key role in the chain reaction of lipid oxidation. Free radicals are unstable, highly reactive, and energized molecules with one or more unpaired electrons. They quickly capture electrons from other compounds to gain stability. Examples of oxygen-derived free radicals are superoxide anion, hydroxyl,

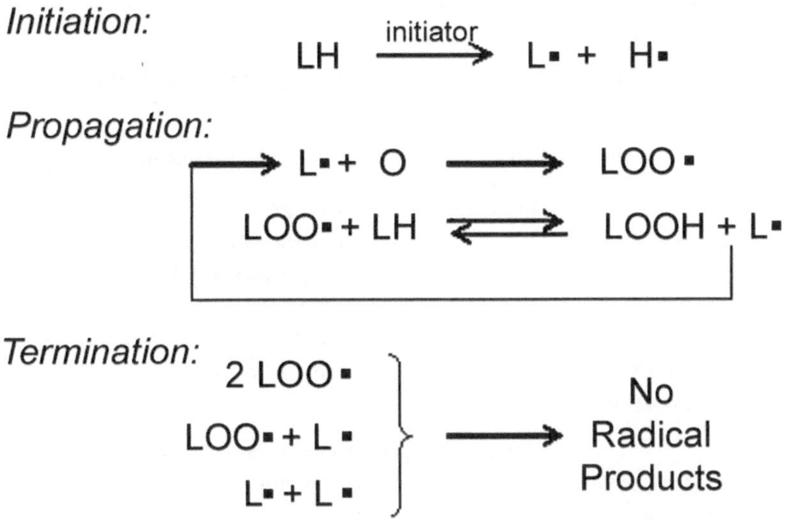


**FIGURE 13.1** Chain reactions involved in free radical chemistry. The processes are initiation, in which reactions create free radicals; propagation, in which free radicals are transferred among chemical species; and termination, in which free radicals are eliminated as the result of combining with each other. Primary antioxidants are able to break the chain reaction of oxidation through free radical scavenging.

hydroperoxyl, peroxyl, and alkoxyl radicals, all of which are referred to as reactive oxygen species (ROS). The primary ROS is superoxide anion ($O_2^{\bullet-}$), which is formed from the single-electron reduction of molecular oxygen through a variety of sources under both physiological and pathophysiological conditions. It has been estimated that 2 to 5% of oxygen consumed by a cell is reduced to oxygen radicals (Floyd and Hensley, 2002).

In addition to free radicals, many non-radical ROS such as hydrogen peroxide ($H_2O_2$) also participate in the oxidation process as strong oxidants. For example, superoxide anion can be neutralized by superoxide dismutase (SOD) to yield hydrogen peroxide. Hydrogen peroxide is a classic ROS and, although less reactive than superoxide anion, has a longer lifetime and is more highly diffusible and can freely cross biological membranes (Tylicki et al., 2003). Moreover, it can participate in Fenton reactions with reduced metal ions such as $Fe^{2+}$ and $Cu^{+}$ to form a hydroxyl radical (•OH) that is considered the most harmful ROS (Lubec, 1996). Other ROS include hydroxyalkenals, singlet oxygen, and ozone.

Nitrogen dioxide and peroxynitrite anion are formed from the interaction of nitric acid with superoxide anion, and together with nitric oxide (NO) are known as reactive nitrogen species (RNS). Singlet oxygen is photochemically generated from molecular oxygen and is usually involved in photooxidation processes. These non-radical ROS and RNS are of particular importance because of their longevity and the lack of efficient endogenous systems to protect against them (Tylicki et al., 2003). Table 13.1 summarizes the reactive species involved in food and biological oxidation. Although free radical-mediated oxidation reactions occur predominantly

**TABLE 13.1**
**Major Reactive Oxygen and Nitrogen Species in Food and Biological Systems**

| Free Radicals | Non-Radicals |
|---|---|
| **Reactive Oxygen Species (ROS)** | |
| Superoxide anion ($O_2^{\bullet -}$) | Hydrogen peroxide ($H_2O_2$) |
| Hydroxyl (•OH) | Hydrochlorous acid (HOCl) |
| Alkyl (L•) | Singlet oxygen ($^1O_2$) |
| Alkoxyl (LO•) | Ozone ($O_3$) |
| Peroxyl (LOO•) | Hydroxyalkenals |
| Hydroperoxyl (HOO•) | |
| **Reactive Nitrogen Species (RNS)** | |
| Nitric oxide (•NO) | Peroxynitrite ($ONOO^-$) |
| Nitrogen dioxide ($NO_2$•) | Alkyl peroxynitrite (LOONO) |
| | Dinitrogen trioxide ($N_2O_3$) |
| | Nitrous acid ($HNO_2$) |

in lipids, particularly PUFAs, the substrate may consist of other lipids or non-lipid substances. Cholesterol, proteins, enzymes, lipoproteins, and DNA can also serve as targets for free radical attack.

### 13.2.2 Oxidative Damage

Free radicals and other reactive species are produced in the body under normal conditions during metabolism and play an important role in a number of regulatory activities and physiological processes. Nevertheless, enhanced production of ROS originating from endogenous or exogenous sources results in oxidative stress, which accounts for many pathological alterations. ROS production in vivo is normally controlled by cellular antioxidant defense systems. However, when the antioxidant defenses are not efficient, ROS escape elimination and may cause oxidative cellular damage (Berger, 2005).

Excessive ROS/RNS generated during oxidation may react, in principle, with all cellular and extracellular components causing diffuse damage to cells and tissues. Oxidative stress arises from an imbalance of ROS production and antioxidant defense (Halliwell and Gutteridge, 1985), and has been associated with pathogenesis of a number of diseases including cardiovascular disease, cancers, hypertension, diabetes, inflammation, autoimmune disorders, as well as aging.

Polyunsaturated fatty acids in membrane lipid bilayers are the major targets for attack by ROS, particularly by •OH, to initiate the process of lipid peroxidation (Kruidenier and Verspaget, 2002). The chain reaction of lipid oxidation promotes the accumulation of hydroperoxides in the cell membrane, which affects fluidity and the activities of many transmembrane enzymes, transporters, receptors, and other membrane proteins (Jourd'Heuil et al., 1993). Oxidatively modified cell membranes demonstrate altered permeability and may consequently give rise to changes in cell volume homeostasis and cellular metabolism (Chen et al., 1995). Moreover, oxidation products such as hydroperoxides and aldehydes are directly toxic to cells and organelles (Aw, 1998). These oxidative products have also been implicated in the regulation of inflammation processes through their neutrophil chemotactic properties and promotion of cytokine production (Curzio et al., 1987; Jayatilleke and Shaw, 1998).

Proteins are significant targets for ROS attack. Protein radicals formed by a free radical attack (e.g., by •OH) cause polypeptide chain scission, cross-linking, oxidation, and modification of amino acids, all of which lead to increased susceptibility to proteolysis, denaturation, and alteration or loss of biological function (Niki, 1997). Protein oxidation frequently introduces new functional groups such as hydroxyls and carbonyls that may markedly affect biological activities (Dean et al., 1997). Oxidatively modified proteins are more susceptible to degradation and aggregation; an important example is the damage to the lens of a cataract patient (Taylor, 1993). In addition, tyrosine residues of proteins have been observed to be irreversibly modified by the peroxynitrite RNS through nitration that may adversely affect cell regulation (Ischiropoulos and Al-Mehdi, 1995; Ischiropoulos, 1998).

DNA, including nuclear and mitochondrial DNA, is also susceptible to modification by ROS and RNS. Base hydroxylation and strand cleavage are among the most common ROS-induced DNA modifications. They have been implicated in ATP depletion, gene mutations, malignant transformation, and cell death (Kruidenier and Verspaget, 2002). Hydroxyl radical produced in situ can effectively attack nearby DNA residues, while peroxynitrite and nitric oxide can directly damage chromatin (Kruidenier and Verspaget, 2002). Oxidatively modified DNA is thought to play a role in human carcinogenesis (Poulsen et al., 1998). In addition to the damage to these macromolecules, oxidative stress has also been implicated to alter calcium homeostasis, cellular signalling cascades, and gene expression, among other events (Dalton et al., 1999; Davies, 2000; Waring, 2005).

### 13.2.3 Oxidative Stress and Inflammation

Oxidative stress is considered a fundamental tissue destructive mechanism. ROS is capable of causing reversible and irreversible damage to oxidizable biomolecules. Excessive ROS production by neutrophils, macrophages, monocytes, and other immune cells constitutes the major tissue destructive force in vivo and accounts for the pathogenesis of many inflammatory diseases. Under non-inflammatory conditions, superoxide radical is produced mainly in mitochondria by endogenous intracellular enzyme systems (Richter and Schweizer, 1997).

Superoxide generation by phagocytes, and to a lesser extent by eosinophils, lymphocytes, and fibroblasts, is essential for effective host defense against bacterial infection. However, excessive liberation of $O_2^{\bullet-}$ is proposed as a major cause of cellular damage and apoptosis leading to extensive tissue destruction, as found in many chronic inflammatory diseases (Weiss, 1989; Rahman and MacNee, 1996; Morcillo et al., 1999). Superoxide is an important mediator in the infiltration and accumulation of neutrophils at sites of inflammation and is also involved in mobilizing arachidonic acid (Kruidenier and Verspaget, 2002). Superoxide can be neutralized, for example by SOD, but also may be rapidly converted following a series of oxidant reactions to much more powerful ROS such as $H_2O_2$, •OH, and hypochlorous acid (HOCl).

Hydrogen peroxide is believed to act as a neutrophil chemoattractant and plays a key role in leukocyte rolling, activation of T lymphocytes, and induction of angiogenesis (Kruidenier and Verspaget, 2002). Moreover, $H_2O_2$ has been shown to cause non-specific irreversible damage to epithelial cells (Mulier et al., 1998). $H_2O_2$ can be metabolized to form HOCl by myeloperoxidase, an abundant enzyme of phagocytes and secretory protein of activated neutrophils (Cohen et al., 1982; King et al., 1997). The resultant HOCl is a reactive chlorine species (RCS) and strong oxidizing/chlorinating agent. It is estimated to be 100 to 1000 times more toxic than $O_2^{\bullet-}$ or $H_2O_2$ (Conner and Grisham, 1996).

Another important group of oxidative and inflammatory mediators is RNS, including NO and peroxynitrite. NO is a lipid-soluble free radical with a significant half-life. It is produced by the nitric oxide synthase (NOS) enzyme and is capable of diffusing several cell diameters from its synthesis site (Kruidenier and Verspaget, 2002). NO alone at nanomolar concentrations is not particularly harmful and in some situations may even exert beneficial effects such as protecting epithelial cells against $H_2O_2$, protecting macrophages from cytokine-induced cytotoxicity, limiting leukocyte binding to endothelial cells, and acting as a vasodilator and neurotransmitter (Kruidenier and Verspaget, 2002). However, it is the precursor of a more damaging RNS peroxynitrite anion ($ONOO^-$), a stable and reactive oxidizing and nitrating agent that can damage a broad array of biomolecules in cells. These reactive species (ROS, RNS, and RCS) are generated by activated inflammatory cells and may contribute to oxidative stress and inflammatory processes in cells and tissues.

The mechanism of inflammatory injury is attributed in part to the release of ROS from activated phagocytes. Excessively generated ROS leads to tissue destruction by damaging macromolecules and promoting membrane lipid oxidation (Winrow et al., 1993; Gutteridge, 1995). ROS can also, in turn, promote inflammation by stimulating production of cytokines, such as tumor necrosis factor (TNF)-α, interleukin-1 (IL-1), IL-6 and IL-8, interferon-γ, vascular cell adhesion molecule (VCAM)-1, and monocyte chemoattractant protein (MCP)-1. These all stimulate recruitment of additional phagocytes to an inflammation site. Therefore, oxidative stress plays an essential role in provoking and sustaining inflammatory processes.

Evidence has shown that ROS are involved in activation of kinases and transcription factors, whose regulation is dependent on redox changes. Nuclear factor (NF)-κB and activator protein (AP)-1, for example, are redox-sensitive and become activated under oxidative or nitrosative stress. Activation of NF-κB and AP-1 leads to upregulation of numerous inflammatory genes, including those coding for inflammatory mediators TNF-α, IL-1, IL-6 and IL-8, VCAM-1, MCP-1, E-selectin, inducible nitric oxide synthase (iNOS), cyclooxygenase (COX-2), and granulocyte-macrophage colony-stimulating factor, as well as major histocompatibility complex class I antigens (Ward, 1996; Akira and Kishimoto, 1997; Barnes and Karin, 1997; Kamata and Hirata, 1999).

Some cytokine products such as TNF-α and IL-1 act as activators of NF-κB (Janssen-Heininger et al., 1999), and are also known to induce cellular ROS production (Adamson and Billings, 1992). Hence, ROS and pro-inflammatory cytokines work in a synergistic manner through a ROS–cytokine–transcription factor regulatory loop, thereby augmenting the inflammatory response and tissue damage (Fiocchi, 1998). This synergistic loop is summarized in Figure 13.2.

Enhanced ROS production has been observed in inflammatory diseases. Increased levels of oxidative biomarkers such as malonaldehyde, 4-hydroxynonenal (Selley, 1997), protein carbonyls and 8-hydroxy-2'-deoxyguanosine (Lih-Brody et al., 1996) have been reported in colonic biopsies of inflammatory bowel disease patients, indicating accelerated oxidation of lipid, protein, and DNA. In addition to cytokine production control, ROS also appear to regulate activity of the tissue destructive proteases called matrix metalloproteinases (MMPs). ROS can activate multiple MMPs while inactivating their inhibitors (Frears et al., 1996; Brenneisen et al., 1997), therefore allowing intensified tissue damage at an inflammation site. Oxidative stress also promotes inflammation by downregulating protective antioxidant genes such as those coding for glutathione (Rahman and MacNee, 2000) and by depleting antioxidant compounds in vivo. Therefore, supplementation of exogenous antioxidants from food or medicinal sources may be significant for better health by

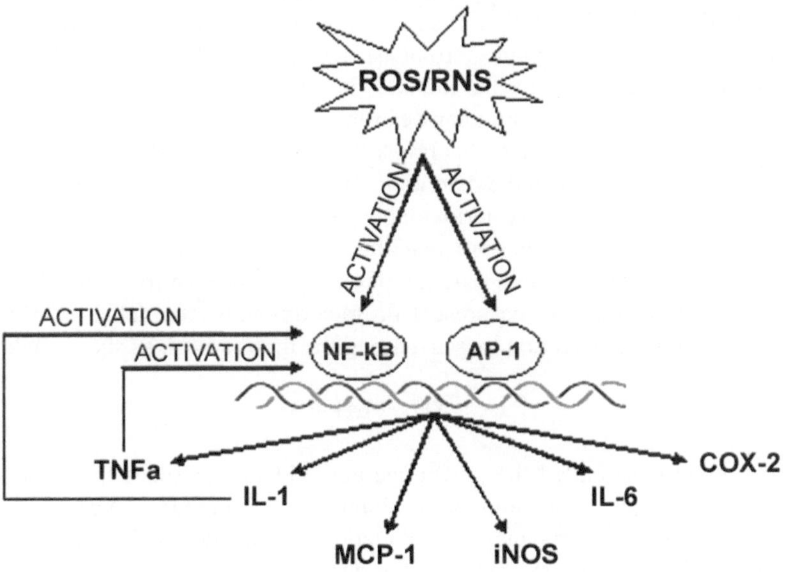


**FIGURE 13.2** Reactive oxygen species (ROS) and reactive nitrogen species (RNS) promote synthesis of inflammatory cytokines. ROS and RNS activate nuclear transcription factors NF-κB and AP-1 that in turn induce the expression of genes encoding potent inflammatory cytokines and enzymes including tumor necrosis factor (TNF), interleukin-1 (IL-1), monocyte chemotactic protein-1 (MCP-1), inducible nitric oxide synthase (iNOS), interleukin-6 (IL-6), and cyclooxygenase-2 (COX-2).

maintaining the balance among oxidants, antioxidants, pro-inflammatory, and anti-inflammatory agents.

## 13.3 ANTIOXIDANTS AND POLYPYHENOLS

### 13.3.1 Introduction of Antioxidants

Antioxidants, by definition, are substances that, when present at low concentrations compared to those of oxidizable substrates significantly delay or prevent oxidation of these substrates. Antioxidants that fit this definition include free radical scavengers, singlet oxygen quenchers, inactivators of peroxides and other ROS, metal ion chelators, quenchers of secondary oxidation products, and inhibitors of pro-oxidative enzymes. These substances play protective roles through different mechanisms and varied activities against ROS-mediated oxidation processes both in vitro and in vivo.

Food manufacturers use antioxidants to stabilize food lipids and thus prevent quality deterioration and increase shelf lives of products. In the health arena, antioxidants are used for their ability to protect against oxidative damage caused by ROS, RNS, and RCS. Although living organisms are constantly exposed to oxidative stress, they are protected by antioxidant systems. Endogenous antioxidant systems consist of both enzymatic and non-enzymatic components. Antioxidant enzymes such as SOD and glutathione peroxidase inhibit oxidation by catalyzing ROS degeneration and thus enhancing elimination of the harmful reactive species.

SOD catalyzes the dismutation of superoxide anion to hydrogen peroxide, which is further degraded by glutathione peroxidase and catalase (Tylicki et al., 2003). Non-enzymatic antioxidant components include vitamins E and C that originate mainly from dietary sources. Antioxidants can be broadly classified according to their modes of action as primary antioxidants that break the chain reaction of oxidation, and secondary antioxidants that slow the oxidation rate by such mechanisms as chelation of metals, regeneration of primary antioxidants, decomposition of hydroperoxides, and suppression of oxygen and other pro-oxidants (Shahidi and Zhong, 2005b). Primary antioxidants can break the chain reaction of oxidation through free radical scavenging. The AH antioxidant, when present in trace amounts, can delay or inhibit the initiation step of lipid oxidation by scavenging the lipid radical, as shown below:

$$L\bullet + AH \rightarrow LH + A\bullet$$

The antioxidant (also called the preventive antioxidant) stabilizes the lipid radical by donating a hydrogen atom, and the resultant antioxidant radical A•, because of its low reactivity, does not propagate the oxidation chain. These chain-breaking antioxidants can also trap peroxyl radicals, the chain-propagating species, by donating an electron to the radical and receiving an electron from the radical to form stable by-products (Young and Woodside, 2001). A simplified scheme is shown below:

$$LOO\bullet + AH \rightarrow LOOH + A\bullet$$

$$A\bullet + LOO\bullet \rightarrow LOOA$$

Secondary antioxidants (e.g., reducing agents) slow the oxidation rate. They reduce lipid peroxides and related oxidants via redox reactions. For instance, thioethers convert hydroperoxides into stable components through a non-radical pathway (Pokorny, 2007). Other antioxidants inhibit oxidation by suppression of oxidation promoters. These include singlet oxygen quenchers (Min and Boff, 2002), metal chelators (Leopoldini et al., 2006), and pro-oxidative enzyme inhibitors (Sud'ina et al., 1993).

When two or more antioxidants act in a cooperative manner, synergism occurs. For example, ascorbic acid (vitamin C) can regenerate tocopherols (vitamin E) after they become oxidized. An antioxidant may protect against oxidation by one or a combination of mechanisms. The operative or dominant mechanism in a particular situation that depends on the conditions will affect the kinetics and hence the activity of the antioxidant (Antolovich et al., 2002). Likewise, structural characteristics, concentration, type of oxidation substrate, physical state of the system, and the presence of pro-oxidants and synergists will influence the effectiveness of antioxidants (Yanishlieva-Malslarova, 2001).

### 13.3.2 Antioxidant Sources

Antioxidants are widely distributed in plant materials, animal tissues, and microorganisms. Edible higher plants provide rich sources of natural antioxidants such as tocopherols and polyphenols. Spices and herbs are known to contain arrays of antioxidant compounds consisting mainly of polyphenols. Fruits, vegetables, cereals, grains, oils, seeds, and teas are also important sources of plant-derived antioxidants. Antioxidants can be derived from marine origin such as algae, fish, shellfish, and marine bacteria (Shahidi and Amarowicz, 1996; Amarowicz et al., 1999; Athukorala et al., 2003).

By-products of the food and agricultural industries have also been explored for their antioxidant potential. For example, hulls, shells, and skins of nuts and cereals, citrus peels and seeds, canola meal and fish viscera extracts possess antioxidant activity (Liyana-Pathirana et al., 2006; Shahidi et al., 2007; Cumby et al., 2008). Naturally occurring antioxidants are important constituents of organisms. They can be isolated as pure compounds and constitute resources for uses in food preservation, nutrition supplements, and therapeutic drugs. Among these are tocopherols, ascorbic acid, carotenoids, polyphenols, antioxidant peptides, and enzymes (Table 13.2).

Phenolic compounds, mainly polyphenols, are the most abundant antioxidants from natural sources. Polyphenols range from simple phenols to highly polymerized compounds. Many complex polyphenols are components of the pigmentation and sensory systems of plants. The antioxidant activity of polyphenols arises from their unique structures that render superior hydrogen-donating ability—more than many other compounds. They act as free radical scavengers, metal chelators, reducing agents, and excellent synergists to other antioxidants, and are potent inhibitors of oxidation in biological systems. Polyphenols are present abundantly in fruits, vegetables, cereals, nuts, herbs, and spices. These plant-derived polyphenols may be classified into non-flavonoids (phenolic acids, stilbenes, and lignans) and flavonoids (flavones, isoflavones, flavonols, flavonones, flavanols, and anthocyanins).

**TABLE 13.2**
**Major Antioxidants from Natural Sources**

| Antioxidants | Examples | Sources |
|---|---|---|
| Tocopherols | α-, β-, γ-, and δ-tocopherols | Seeds, grains, nuts, vegetable oils |
| Ascorbic acid | Ascorbic acid, ascorbate | Fruits, vegetables. |
| Carotenoids | β-carotene, lycopene, canthaxanthin | Carrots, tomato, marine algae, fish, shellfish |
| Phenolics | Ferulic acid, quercetin, catechin, resveratrol, cyanidin | Fruits, vegetables, cereals |
| Peptides | Ferritin, transferrin, lactoferrin | Milk, eggs |
| Enzymes | Superoxide dismutase, catalase, glutathione peroxidase | Plant and animal organisms |

Shahidi and Naczk (2003) published a comprehensive overview of phenolic compounds and their sources, analyses, properties and health benefits. Fruits such as apples, citrus, berries, and grapes are rich in phenolic acids (mainly in esterified form, e.g., gallic, ferulic, sinapic, coumaric, caffeic, and chlorogenic acids) and flavonols (quercetin, kaempferol, myricetin, and isorhamenetin). Herbs are rich sources of flavanols, more specifically, a number of catechin derivatives. In green tea, for example, flavonoids comprise 30% of dry weight of leaves (Graham, 1992). Recent studies of cereals and nuts revealed that phenolic compounds are more concentrated in shells, hulls and skins removed during processing than in the edible grains and kernels (Wijeratne et al., 2006; Shahidi et al., 2006; Shahidi et al., 2007). The antioxidant activity of polyphenols is dependent on chemical structure. The degree of glycosylation significantly influences the antioxidant efficiencies of phenolic compounds. It has been demonstrated that aglycones of quercetin and myricetin were much more effective antioxidants than their glycosides (Hopia and Heinonen, 1999).

### 13.3.3 Health Effects of Antioxidants

Antioxidants are important to human health. They are vital for the integrity of biomolecules such as lipids, proteins, and DNA. Antioxidants can inhibit detrimental oxidation in the body, and are therefore effective in preventing and/or treating numerous free radical-mediated chronic diseases. Experimental and clinical data have shown that antioxidants exert protective effects on a number of pathological conditions including cardiovascular disease, diabetes, infections, inflammation, cancer, and neurodegenerative disorders, among others.

Epidemiological studies have established an inverse relationship between the intake of fruits and vegetables and risk of cancer. Antioxidant polyphenols from dietary sources are believed to exert chemopreventive effects against cancers via various mechanisms. They act as antimutagenic and antimetastatic agents and may promote apoptosis in tumor cells (Surh, 2003). In addition, polyphenols, particularly phenolic acids, are effective in the inhibition of carcinogen uptake, suppression of formation and activation of carcinogens, deactivation or detoxification of carcinogens, prevention of carcinogen binding to DNA, and elevation of efficient DNA repair (Surh, 2003; Nichenametla et al., 2006). Flavonols such as quercetin and kaempferol exhibit anticancer activity through control over the cell cycle, cell growth, and cell proliferation (Nichenametla et al., 2006). Anthocyanins can protect DNA from single strand breaks and resveratrol displays antitumor effect by reducing tumor neovascularization and angiogenesis (Nichenametla et al., 2006).

Antioxidant strategy has also been proposed for preventing and alleviating metabolic syndrome. Oxidative stress has been suggested as one of multiple factors causing metabolic syndrome which is characterized by obesity, diabetes, hypertension, and atherosclerotic cardiovascular disease. Polyphenols, particularly catechins from green tea, have been reported to lower serum and liver cholesterol levels, therefore reducing the risk of atheroscelerosis (Ikeda, 2008). Catechins can also suppress postprandial hypertriacylglycerolemia, which is an important risk factor for coronary heart disease (Unno et al., 2005).

The underlying mechanism by which polyphenols combat the detrimental effects of obesity may relate to the reduction of visceral fat deposition through inhibition of fatty acid synthase (Wang and Tian, 2001) and the enhancement of the energetic expenditure through modulation of gastric and pancreatic lipases (Juhel et al., 2000). In addition, tea polyphenols have been found to increase insulin activity and may reduce insulin resistance in diabetes (Anderson and Polansky, 2002). Antioxidants are also effective antihypertensive agents. Many ACE (angiotensin converting enzyme) inhibitors, angiotensin II type-1 receptor antagonists, and calcium channel blockers have in common the fact that they display antioxidant activity (Tylicki et al., 2003). Antioxidants appear also to exert neuroprotective activity in neurodegenerative diseases such as Alzheimer's and Parkinson's diseases (Foley and White, 2002; Ishihara and Brayne, 2005). ROS in the brain affect the production of neurotrophins, neurotransmitters, and steroids, all of which may promote neurodegenerative effects. Antioxidants such as polyphenols efficiently eliminate ROS and thus may protect neurons from oxidative damage.

## 13.4 ADIPOSE INFLAMMATION

### 13.4.1 Inflammation and Chronic Diseases

Inflammation is a normal response of the immune system to repair injury and combat infection. It is characterized by the movement of phagocytes and plasma from blood vessels to the inflamed tissue as a result of increased capillary permeability and altered expression of adhesion molecules in endothelial and immune cells (Aquilano et al., 2008). However, growing evidence indicates that unregulated inflammation is an underlying component of many chronic diseases such as cancer and neurodegenerative disorder as well as diabetes and cardiovascular disease.

Chronic local inflammation is believed to contribute to multistage carcinogenesis, particularly the promotion stage in which tissues suffer from persistent oxidative and inflammatory damage (Surh, 2002). The ROS produced during inflammation can initiate a series of reactions that eventually cause malignant transformation of the inflamed tissue (Cerutti, 1985). ROS are mediators for cell proliferation by interfering with the intracellular signalling cascade (Surh, 2002). In addition, it has been suggested that the COX-2 enzyme that catalyzes the production of prostaglandins plays an important role in tumor development (Prescott and Fitzpatrick, 2000).

In neurodegenerative diseases, ROS and RNS are generated at increased levels in response to microglia-mediated inflammation. Microglia are phagocytic cells participating in the physiological immune control of the central nervous system, and their activation leads to hyperproduction of ROS and RNS and hence decreased neuronal viability (Aquilano et al., 2008). Persistent ROS and RNS activity is considered one of the factors eliciting neuronal cell death through oxidative and nitrosative damage to neuronal molecules (Mattson and Magnus, 2006). In the vasculature, overproduction of ROS and RNS by inflammatory cells causes endothelial dysfunction and contributes to the development of atherosclerosis (Ross, 1999). Accumulation of leukocytes is detected within discrete regions of vasculature in the early stage of atherosclerotic plaque development (Sundel et al., 2003). Chronic inflammation has also been implicated in obesity-induced complications, such as metabolic syndrome and type 2 diabetes.

### 13.4.2 Adipose Tissue and Metabolic Syndrome

Chronic inflammation has been implicated in diseases including obesity. Obesity is generally considered a chronic disease characterized by excess accumulation of fat stores in adipocytes and has been linked to certain pathologies such as diabetes and atherosclerosis, presumably through oxidative stress and chronic inflammation. Obesity is characterized by persistent low-grade inflammation and this may contribute to the development of metabolic syndrome (Kang et al., 2007). Metabolic syndrome increases the risk of developing diabetes and cardiovascular disease, and the symptoms generally include central or abdominal obesity, hyperglycemia or impaired glucose tolerance, dyslipidemia, and hypertension. Obesity is the most common nutritional disorder and a growing medical problem worldwide. Chronic inflammation in adipose tissue has been proposed to contribute to the development of obesity-related pathologies.

Adipose tissue, although once thought to function simply as a simple and inert storage depot for excess fat in the form of triacylglycerols, has been found to exert broader systemic metabolic control as a major endocrine organ (Mohamed-Ali et al., 1998). Under normal conditions, secreted biomolecules effectively control energy preservation though lipogenesis during the postprandial period and energy mobilization through lipolysis in response to increased energy expenditure (Gonzales and Orlando, 2008). Adipose tissue, particularly central or abdominal adipose tissue, secretes known mediators of inflammation including cytokines, chemokines, and acute phase proteins. These contribute to chronic inflammation and are associated with type-2 diabetes mellitus and cardiovascular disorders through insulin resistance and atherosclerotic lesions (Finley, 2004).

Adipocytes, especially pre-adipocytes during differentiation, synthesize and secrete a number of inflammatory mediators and acute phase reactants such as TNF-α, IL-1, IL-6, IL-8, IL-10, IL-15, MCP-1, prostaglandin $E_2$ ($PGE_2$), hepatocyte growth/scatter factor, and plasminogen activator inhibitor (PAI)-1, among others (Fain et al., 2004; Calabro and Yeh, 2007). The production of these adipokines by adipocytes may be initiated by cross-talk between adipocytes and macrophages (Berg and Scherer, 2005; Tilg and Moschen, 2006) and further stimulate infiltration of macrophages and expression of inflammatory genes in adipose tissue.

Adipocytes secrete MCP-1, which mediates macrophage recruitment to adipose tissue and intensifies macrophage expression of TNF-α. Locally-produced TNF-α affects adipocyte physiology by enhancing expression of other cytokines through activation of the NF-κB signalling pathway, a similar course to that of inflammatory cells (Gonzales and Orlando, 2008). As a result, adipose tissue in obese subjects shows a higher population of macrophages and higher levels of cytokines compared to those in non-obese subjects.

Higher oxidative stress, possibly by free radicals from excess blood sugar and lipid, has been hypothesized to be one cause of the chronic inflammatory state in obesity (Ferroni et al., 2004; Dandona et al., 2005). Macrophages from non-obese subjects have been reported to account for 5 to 10% of cells within adipose tissue; in high caloric diet-induced obesity, macrophage infiltration may account for up to 60% of all cells in adipose tissue (Weisberg et al., 2003). Cytokines produced by infiltrated macrophages in adipose tissue stimulate adipokine expression in adipocytes, establishing a paracrine loop between the two cell types (Suganami et al., 2005). Increased levels of circulating inflammation markers such as IL-6, TNF-α, C-reactive protein (CRP), and haptoglobin have been observed in obese individuals (Das, 2001; Bullo et al., 2003).

Although adipose-infiltrating macrophages serve as major sources of circulating cytokines within adipose tissue (Gonzales and Orlando, 2008), significant amounts of adipokines from adipose tissue reach the overall systemic circulation, as observed in obese individuals (Cottam et al., 2004). Studies have shown a direct correlation between body mass index (BMI) and systemic levels of inflammatory mediators (Berg and Scherer, 2005). During weight gain, adipocytes undergo an enlargement process accompanied by molecular and cellular alterations affecting systemic metabolism (Calabro and Yeh, 2007).

Adipocytes become increasingly hypertropic and produce inflammatory mediators. The elevated level of local or systemic inflammation has been linked to development of obesity-induced complications such as type-2 diabetes (Bullo et al., 2007). Inflammatory cytokines produced in adipose tissue such as TNF-α and IL-6 have been shown to impair insulin signalling by promoting serine phosphorylation of insulin receptor substrate (IRS)-1 and suppressing the expression of insulin-sensitive glucose transporter 4 (Hotamisligil et al., 1993; Feinstein et al., 1993; Hotamisligil, 2003; Guri et al., 2007). Moreover, TNF-α may increase fatty acid-induced systemic insulin resistance by promoting the release of fatty acids from adipose tissue into the bloodstream to act on peripheral tissues such as muscle and liver (Calabro and Yeh, 2007).

Adiponectin is another important regulatory molecule secreted exclusively by adipocytes. This adipokine plays an important role in regulating energy metabolism, insulin resistance, and vascular microenvironment (Kang et al., 2007). Dysregulation of adiponectin production may also contribute to the onset or aggravation of chronic inflammation in obesity and in turn the development of obesity-related pathologies. Adiponectin improves insulin sensitivity in part by upregulating the expression of IRS-1 in skeletal muscles (Kadowaki et al., 2006). It may also inhibit atherosclerosis by downregulating the expression of adhesion molecules in vascular endothelial cells (Kang et al., 2007). However, adiponectin expression in adipocytes is suppressed under inflammatory conditions by pro-inflammatory cytokines such as TNF-$\alpha$ and IL-6, leading to insulin resistance and hyperinsulinemia (Spurlock and Gabler, 2008). Adiponectin levels increase significantly in obese or diabetic subjects by weight loss through lifestyle changes, e.g., exercise and diet (Calabro and Yeh, 2007).

In summary, low-level chronic inflammation in adipose tissue mediated by production and secretion of pro-inflammatory cytokines plays an important role in causing or aggravating the development of obesity-related diseases and health conditions. Therefore, anti-inflammatory strategies are important in reducing the risk of inflammation-mediated complications.

## 13.5 ANTIOXIDANTS AS ANTI-INFLAMMATORY AGENTS

Anti-inflammatory agents exert their therapeutic effects generally by inhibiting the synthesis or actions of pro-inflammatory mediators. Most clinically recognized anti-inflammatory medicines include an arsenal of steroidal or non-steroidal drugs. These drugs exhibit potent activity and are frequently used to treat acute inflammation. However, concerns about their long-term administration have limited their use in the treatment of chronic inflammation.

Natural antioxidant products have attracted growing interest as alternative anti-inflammatory agents. Some Oriental countries have a long history of using certain plants, especially herbs, as traditional medicines for treating inflammation. Certain flavonoids have been identified as anti-inflammatory components within these plants and their extracts (Benavente-Garcia and Castillo, 2008).

Modern epidemiological research has revealed an inverse relationship between consumption of fruits and vegetables rich in antioxidants and the incidence of chronic inflammation and related diseases (Chen et al., 2005; Zhang et al., 2005). This led to identifying individual anti-inflammatory compounds from natural sources. Arguments have been made for the efficacy of natural antioxidants as anti-inflammatory agents, despite their low concentrations and low bioavailability. Sufficient evidence from experimental research and clinical trials supports their use (Singh et al., 2008). Moreover, naturally occurring substances are readily available at lower costs than many anti-inflammatory drugs and produce limited side effects and intolerance (Benavente-Garcia and Castillo, 2008).

Many anti-inflammatory agents have been shown to possess antioxidant activities. These compounds can effectively scavenge ROS and RNS and thus augment antioxidant defenses in a host. Suppression of ROS and RNS by antioxidants such as polyphenols inhibits activation of NF-$\kappa$B and AP-1, the redox-sensitive transcription factors for several inflammatory genes, including those coding for TNF-α, IL-1,-6 and IL-8, VCAM-1, MCP-1, iNOS, COX-1, and other major pro-inflammatory cytokines. Some antioxidants act as direct inhibitors of pro-inflammatory mediators including iNOS and COX-2. Catechins have been reported to exhibit anti-inflammatory activity, possibly due to their ability to scavenge NO, peroxynitrite, and other ROS, to translocate NF-κB and AP-1 and reduce the activity of iNOS and COX-2 (Paquay et al., 2000; Nagai et al., 2002; Tedeschi et al., 2004; Chen et al., 2004; Yu et al., 2005).

Anti-inflammatory activity of catechins has been studied in many cell types, and epigallocatechin gallate (EGCG) has been found to possess the highest potency among all catechins in downregulating cytokine secretion. Kim et al. (2007) demonstrated the ability of EGCG to suppress IL-1 production and COX-2 induction. Moreover, in vivo data show that catechin supplementation significantly reduces the production of ROS including hydrogen peroxide, hypochlorous acid, hydroperoxides, and pro-inflammatory cytokines such as TNF-α and MCP-1 in hemodialysis patients (Hsu et al., 2007).

Other polyphenolic compounds have also been shown to inhibit NF-κB and AP-1 activation or act as direct inhibitors of pro-inflammatory mediators. Quercetin, wogonin, genistein, kaempferol, amentoflavone, resveratrol, curcumin, gingerol, capsaicin, and caffeic acid have been reported to be counteracting agents in the activation of NF-κB and AP-1 (Surh, 2002; Lorenz et al., 2003; Kim et al., 2004; Márquez et al., 2004; Miles et al., 2005; Kang et al., 2007; Gonzales and Orlando, 2008). Documented COX-2 inhibitors include kaempferol, quercetin, luteolin, and galangin (Kim et al., 2004), Anthocyanins have also exhibited inhibitory effects on the release of cytokines MCP-1, IL-1 and TNF-α and suppressive activity on the production of NO in LPS-stimulated macrophages and microglia (Choe et al., 2007; Ho et al., 2007; Lau et al., 2007). Extracts of various Mediterranean dietary plants have been evaluated for their anti-inflammatory effects in mice using the croton oil-induced ear edema model. Phenolic extracts from *Sideritis ozturkii* and *Geranium pratense* subsp. *finitimum* were shown to reduce the inflammatory effects of hind paw-injected carrageenan, PGE2, and TPA (Küpeli et al., 2007a and b). In addition, microbial transformation products of dietary phenolic compounds have also been studied as potential sources of anti-inflammatory agents (Russell et al., 2008).

In summary, obesity is associated with chronic inflammation, especially in adipose tissues. Thus, effective control of inflammation through dietary antioxidant strategies that act to scavenge pro-inflammatory ROS and RNS in these tissues may be an important means for the prevention and treatment of obesity and its related disorders such as metabolic syndrome and diabetes.

## REFERENCES

Adamson GM and Billings RE. Tumor necrosis factor-induced oxidative stress in isolated mouse hepatocytes. *Arch Biochem Biophys* 294, 223–229, 1992.

Akira S and Kishimoto A. NF-IL-6 and NF-κB in cytokine gene regulation. *Adv Immunol* 65, 1–46, 1997.

Amarowicz R, Karamac M, and Shahidi F. Synergistic activity of capelin protein hydrolysates with synthetic antioxidants in a model system. *J Food Lipids* 6, 271–275, 1999.

Anderson RA, Polansky MM. Tea enhances insulin activity. *J Agric Food Chem* 50, 7182–7186, 2002.

Antolovich M et al. Methods for testing antioxidant activity. *Analyst* 127, 183–198, 2002.

Aquilano K et al. Role of nitric oxide synthases in Parkinson's disease: a review on the antioxidant and anti-inflammatory activity of polyphenols. *Neurochem Res* 33, 2416–2426, 2008.

Athukorala Y et al. Potential antioxidant activity of marine red alga *Grateloupia filicina* extracts. *J Food Lipids* 10, 251–265, 2003.

Aw TY. Determinants of intestinal detoxification of lipid hydroperoxides. *Free Rad Res* 28, 637–646, 1998.

Barnes PJ and Karin M. Nuclear factor-κB: a pivotal transcription factor in chronic inflammatory diseases. *New Engl J Med* 336, 1066–1071, 1997.

Beckman KB and Ames BN. The free radical theory of aging matures. *Physiol Rev* 78, 547–581, 1998.

Benavente-Garcia O and Castillo J. Update on uses and properties of citrus flavonoids: new findings in anticancer, cardiovascular, and anti-inflammatory activity. *J Agric Food Chem* 56, 6185–6205, 2008.

Berg AH and Scherer PE. Adipose tissue, inflammation, and cardiovascular disease. *Circ Res* 96, 939–949, 2005.

Berger MM. Can oxidative damage be treated nutritionally? *Clin Nutr* 24, 172–183, 2005.

Brenneisen P et al. Hydrogen peroxide ($H_2O_2$) increases the steady-state mRNA levels of collagenase/MMP-1 in human dermal fibroblasts. *Free Rad Biol Med* 22, 515–524, 1997.

Bullo M et al. Inflammation, obesity and comorbidities: the role of diet. *Pub Health Nutr* 10, 1164–1172, 2007.

Bullo M et al. Systemic inflammation, adipose tissue tumor necrosis factor, and leptin expression. *Obes Res* 11, 525–531, 2003.

Calabro P and Yeh ETH. Obesity, inflammation and vascular disease. In Harris RE, Ed. *Inflammation in the Pathogenesis of Chronic Diseases.* Springer, New York, pp. 63–91, 2007.

Cerutti PA. Prooxidant states and tumor promotion. *Science* 227, 375–381, 1985.

Chen CC et al. Mulberry extract inhibits the development of atherosclerosis in cholesterol-fed rabbits. *Food Chem* 91, 601–607, 2005.

Chen JJ, Bertrand H, and Yu BP. Inhibition of adenine nucleotide translocator by lipid peroxidation products. *Free Rad Biol Med* 19, 583–590, 1995.

Chen WC et al. Catechins prevents substance P-induced hyperactive bladder in rats via the downregulation of ICAM and ROS. *Neurosci Lett* 367, 213–217, 2004.

Choe MR et al. Cyanidin and cyanidin-3-O-β-D-glucoside suppress the inflammatory responses of obese adipose tissue by inhibiting the release of chemokines MCP-1 and MRP-2. *J Food Sci Nutr* 12, 148–153, 2007.

Cohen AB, Chenoweth DE, and Hugli TE. The release of elastase, myeloperoxidase and lysozyme from human alveolar macrophages. *Am Rev Respir Dis* 126, 241–247, 1982.

Conner EM and Grisham MB. Inflammation, free radicals and antioxidants. *Nutrition* 12, 274–277, 1996.

Cottam DR et al. The chronic inflammatory hypothesis for the morbidity associated with morbid obesity: implications and effects of weight loss. *Obes Surg* 14, 589–600, 2004.

Cumby N et al. Antioxidant activity and water-holding capacity of canola protein hydrolysates. *Food Chem* 109, 144–148, 2008.

Curzio M et al. Possible role of aldehydic lipid peroxidation products as chemoattractants. *Int J Tissue React* 9, 295–306, 1987.

Dalton TP, Shertzer HG, and Puga A. Regulation of gene expression by reactive oxygen. *Annu Rev Pharmacol Toxicol* 39, 670–701, 1999.

Dandona P et al. Metabolic syndrome: a comprehensive perspective based on interactions between obesity, diabetes, and inflammation. *Circulation* 111, 1448–1454, 2005.

Das UN. Is obesity an inflammatory condition? *Nutrition* 17, 953–966, 2001.

Davies KJ. Oxidative stress, antioxidant defenses, and damage removal, repair, and replacement systems. *IUBMB Life* 50, 279–289, 2000.

Dean RT et al. Biochemistry and pathology of radical-mediated protein oxidation. *Biochem J.* 324, 1–18, 1997.

Erickson MC. Lipid oxidation of muscle foods. In Akoh CC and Min DB, Eds. *Food Lipids: Chemistry, Nutrition and Biotechnology*, 3rd ed. Marcel Dekker, New York, pp. 365–412, 2002.

Fain JN et al. Comparison of the release of adipokines by adipose tissue, adipose tissue matrix, and adipocytes from visceral and subcutaneous abdominal adipose tissues of obese humans. *Endocrinology* 145, 2273–2282, 2004.

Feinstein R et al. Tumor necrosis factor-α suppresses insulin-indcued tyrosine phosphorylation of insulin receptor and its substrates. *J Biol Chem* 268, 26055–26068, 1993.

Ferroni P et al. Inflammation, insulin resistance, and obesity. *Curr Atheroscler Rep* 6, 424–431, 2004.

Finley JW. Phenolic antioxidants and prevention of chronic inflammation. *Food Tech* 58, 42–46, 2004.

Fiocchi C. Inflammatory bowel disease: etiology and pathogenesis. *Gastroenterology* 115, 182–205, 1998.

Floyd RA and Hensley K. Oxidative stress in brain aging. Implications for therapeutics of neurodegenerative diseases. *Neurobiol Aging* 23, 795–807, 2002.

Foley DJ and White LR. Dietary intake of antioxidants and risk of Alzheimer's disease. *JAMA* 287, 3261–3263, 2002.

Frears ER et al. Inactivation of tissue inhibitor of metalloproteinase-1 by peroxynitrite. *FEBS Lett* 381, 21–24, 1996.

Gonzales AM and Orlando RA. Curcumin and resveratrol inhibit nuclear factor-κB-mediated cytokine expression in adipocytes. *Nutr Metab* 5, 17–29, 2008.

Graham HN. Green tea composition, consumption, and polyphenol chemistry. *Prev Med* 21, 334–350, 1992.

Guri AJ et al. Dietary ascorbic acid ameliorates glucose tolerance and obesity-related inflammation in db/db mice fed high-fat diets. *Clin Nutr* 26, 107–116, 2007.

Gutteridge JM. Lipid peroxidation and antioxidants as biomarkers of tissue damage. *Clin Chem* 41, 1819–1828, 1995.

Halliwell B and Gutteridge JMC. Oxygen radicals and the nervous system. *Trends Neurosci* 8, 22–26, 1985.

Ho SC et al. Suppressive effect of a proanthocyanidin-rich extract from longan [*Dimocarpus longan* Lour.] flowers on nitric oxide production in LPS-stimulated macrophage cells. *J Agric Food Chem* 55, 10664–10670, 2007.

Hopia A and Heinonen M. Comparison of antioxidant activity of flavonoid aglycons and their glycosides in methyl linoleate. *J Am Oil Chem Soc* 76, 139–144, 1999.

Hotamisligil GS. Inflammatory pathways and insulin action. *Int J Obes Relat Metab Disord* 27, S53–S55, 2003.

Hotamisligil GS, Shargill NS, and Spiegelman BM. Adipose expression of tumor necrosis factor-α: direct role in obesity-linked insulin resistance. *Science* 259, 87–91, 1993.

Hsu SP et al. Chronic green tea extract supplementation reduces hemodialysis-enhanced production of hydrogen peroxide and hypochlorous acid, atherosclerotic factors and proinflammaotry cytokines. *Am J Clin Nutr* 86, 1539–1547, 2007.

Ikeda I. Multifunctional effects of green tea catechins on prevention of the metabolic syndrome. *Asia Pac J Clin Nutr* 17, 273–274, 2008.

Ischiropoulos H and Al-Mehdi AB. Peroxynitrite-mediated oxidative protein modifications. *FEBS Lett* 364, 279–282, 1995.

Ischiropoulos H. Biological tyrosine nitration: a pathophysiological function of nitric oxide and reactive oxygen species. *Arch Biochem Biophys* 356, 1–11, 1998.

Ishihara L, Brayne C. A systematic review of nutritional risk factors of Parkinson's disease. *Nutr Res Rev* 18, 259–282, 2005.

Janssen-Heininger YMW et al. Cooperativity between oxidants and tumor necrosis factor in the activation of newclear factor (NF)-κB. *Am J Resp Cell Mol Biol* 20, 942–952, 1999.

Jayatilleke A and Shaw S. Stimulation of monocyte interleukin-8 by lipid peroxidation products: a mechanism for alcohol-induced liver injury. *Alcohol* 16, 119–123, 1998.

Jourd'Heuil D, Vaananen P, and Meddings JB. Lipid peroxidation of the brush-border membrane: membrane physical properties and glucose transport. *Am J Physiol* 264, G1009–G1015, 1993.

Juhel C et al. Green tea extract (AR25®) inhibits lipolysis of triglycerides in gastric and duodenal medium *in vitro*. *J Nutr Biochem* 11, 45–51, 2000.

Kadowaki T et al. Adiponectin and adiponectin receptors in insulin resistance, diabetes, and the metabolic syndrome. *J Clin Invest* 116, 1784–1792, 2006.

Kamata H and Hirata H. Redox regulation of cellular signaling. *Cell Signal* 11, 1–14, 1999.

Kang JH et al. Capsaicin, a spicy component of hot peppers, modulates adipokine gene expression and protein release from obese-mouse adipose tissues and isolated adipocytes, and suppresses the inflammatory responses of adipose tissue macrophages. *FEBS Lett* 581, 4389–4396, 2007.

Kim HP et al. Anti-inflammatory plant flavonoids and cellular action mechanisms. *J Pharmacol Sci* 96, 229–245, 2004.

Kim SJ et al. Epigallocatechin-3-gallate suppresses NF-κB activation and phosphorylation of p38 MAPK and JNK in human astrocytoma U373 MG cells. *J Nutr Biochem* 32, 1720–1725, 2007.

King CC, Jefferson MM, and Thomas EL. Secretion and inactivation of myeloperoxidase by isolated neutrophils. *J Leuk Biol* 61, 293–302, 1997.

Kruidenier L and Verspaget HW. Oxidative stress as a pathogenic factor in inflammatory bowel disease: radicals or ridiculous? *Aliment Pharmacol Ther* 16, 1997–2015, 2002.

Küpeli E et al. Phenolic compounds of *Sideritis ozturkii* and their in vivo anti-inflammatory and antinociceptive activities. *J Ethnopharmacol* 112, 356–360, 2007a.

Küpeli E et al. Estimation of antinociceptive and anti-inflammatory activity of *Geranium pretense* subsp. *finitimum* and its phenolic compounds. *J Ethnopharmacol* 114, 234–240, 2007b.

Lau FC, Shukitt-Hale B, and Joseph JA. Nutritional intervention in brain aging. In: Harris RD, Ed. *Inflammation in the Pathogenesis of Chronic Diseases.* Springer, New York, pp. 299–318, 2007.

Leopoldini M et al. Iron chelation by the powerful antioxidant flavonoid quercetin. *J Agric Food Chem* 54, 6343–6351, 2006.

Lih-Brody L et al. Increased oxidative stress and decreased antioxidant defenses in mucosa of inflammatory bowel disease. *Dig Dis Sci* 41, 2078–2086, 1996.

Liyana-Pathirana C, Dexter J, and Shahidi F. Antioxidant properties of wheat as affected by pearling. *J Agric Food Chem* 54, 6177–6184, 2006.

Lorenz P et al. Oxyresveratrol and resveratrol are potent antioxidants and free radical scavengers. *Nitric Oxide* 9, 64–76, 2003.

Lubec G. The hydroxyl radical: from chemistry to human disease. *J Invest Med* 44, 324–346, 1996.

Márquez N et al. Caffeic acid phenethyl ester inhibits T-cell activation by targeting both nuclear factor of activated T-cells and NF-κB transcription factors. *J Pharmacol Exp Ther* 308, 993–1001, 2004.

Mattson MP and Magnus T. Aging and neuronal vulnerability. *Nat Rev Neurosci* 7, 278–294, 2006.

Miles EA, Zoubouli P, and Calder PC. Differential anti-inflammatory effects of phenolic compounds from extra virgin olive oil identified in human whole blood cultures. *Nutrition* 21, 389–394, 2005.

Min DB and Boff JM. Chemistry and reaction of singlet oxygen in food. *Comp Res Food Sci Food Safety* 1, 56–72, 2002.

Mohamed-Ali V, Pinkney JH, and Coppack SW. Adipose tissue as an endocrine and paracrine organ. *Int J Obes Relat Metab Disord* 22, 1145–1158, 1998.

Morcillo EJ, Estera J, and Cortijo J. Oxidative stress and pulmonary inflammation: pharmacological intervention with antioxidants. *Pharmcol Res* 40, 393–404, 1999.

Mulier B et al. Hydrogen peroxide-induced epithelial injury: the protective role of intracellular nonprotein thiols (NPSH). *Eur Resp J* 11, 384–391, 1998.

Nagai K et al. Epigallocatehin gallate protects against NO stress-induced neuronal damage after ischemia by acting as an antioxidant. *Brain Res* 956, 319–322, 2002.

Nichenametla SN et al. A review of the effects and mechanisms of polyphenolics in cancer. *Crit Rev Food Sci Nutr* 46, 161–183, 2006.

Niki E. Free radicals in chemistry and biochemistry. In Hiramatsu M et al., Eds. *Food and Free Radicals.* Plenum Press, New York, pp. 1–10, 1997.

Paquay JB et al. Protection against nitric oxide toxicity by tea. *J Agric Food Chem* 48, 5768–5772, 2000.

Pokorny J. Are natural antioxidants better-and safer-than synthetic antioxidants? *Eur J Lipid Sci Technol* 109, 629–642, 2007.

Poulsen HE, Prieme H, and Loft S. Role of oxidative DNA damage in cancer initiation and promotion. *Eur J Cancer Prev* 7, 9–16, 1998.

Prescott SM and Fitzpatrick FA. Cyclooxygenase-2 and carcinogenesis. *Biochim Biophys Acta* 1470, M69–M78, 2000.

Rahman I and MacNee W. Oxidative stress and regulation of glutathione in lung inflammation. *Eur Respir J* 16, 534–554, 2000.

Rahman I and MacNee W. Role of oxidants/antioxidants in smoking-induced airway diseases. *Free Rad Biol Med* 21, 669–681, 1996.

Richter C and Schweizer, M. Oxidative stress in mitochondria. In Scandalios JG, Ed. *Oxidative Stress and the Molecular Biology of Antioxidant defenses.* Cold Spring Harbor *Monograph Series 34.* Cold Spring Harbor Laboratory Press, Woodbury, NY, pp. 169–200, 1997.

Ross R. Atherosclerosis: an inflammatory disease. *New Engl J Med* 340, 115–126, 1999.

Russell WR et al. Anti-inflammatory implications of the microbial transformation of dietary phenolic compounds. *Nutr Cancer* 60, 636–642, 2008.

Selley ML. Determination of the lipid peroxidation product 4-hydroxy-2-nonenal in clinical samples by gas chromatography-negative-ion chemical ionization mass spectrometry of the O-pentafluorobenzyl oxime. *J Chromatogr B Biomed Sci Appl* 691, 263–268, 1997.

Shahidi F and Amarowicz R. Antioxidant activity of protein hydrolyzates from aquatic species. *J Am Oil Chem Soc* 73, 1197–1199, 1996.

Shahidi F and Naczk M. *Phenolics in Food and Nutraceuticals.* CRC Press, Boca Raton, FL, 2003.

Shahidi F and Zhong Y. Lipid oxidation: measurement methods. In Shahidi F, Ed. *Bailey's Industrial Oil and Fat Products,* Vol. 1, John Wiley & Sons, Hoboken, NJ, pp. 357–386, 2005a.

Shahidi F and Zhong Y. Antioxidants: regulatory status. In Shahidi F, Ed. *Bailey's Industrial Oil and Fat Products,* Vol. 1, John Wiley & Sons, Hoboken, NJ, pp. 491–512, 2005b.

Shahidi F, Antioxidants in food and food antioxidants, *Nahrung 44,* 158–163, 2000.

Shahidi F, Liyana-Pathirana CM, and Wall DS. Antioxidant activity of white and black sesame seeds and their hull fractions. *Food Chem* 99, 478–483, 2006.

Shahidi F, Alasalvar C, and Liyana-Pathirana C M. Antioxidant phytochemicals in hazelnut kernel [*Corylus avellana* L.] and hazelnut byproducts. *J Agric Food Chem* 55, 1212–1220, 2007.

Singh M et al. Challenges for research on polyphenols from foods in Alzheimer's disease: bioavailability, metabolism, and cellular and molecular mechanisms. *J Agric Food Chem* 56, 4855–4873, 2008.

Spurlock ME and Gabler NK. The development of porcine models of obesity and the metabolic syndrome. *J Nutr* 138, 397–402, 2008.

Sud'ina GF et al. Caffeic acid phenethyl ester as a lipoxygenase inhibitor with antioxidant properties. *FEBS Lett* 329, 21–24, 1993.

Suganami T, Nishida J, and Ogawa Y. A paracrine loop between adipocytes and macrophages aggravates inflammatory changes: role of free fatty acids and tumor necrosis factor α. *Arterioscler Thromb Vasc Biol* 25, 2062–2068, 2005.

Sundell CL et al. AGI-1067: a multifunctional phenolic antioxidant, lipid modulator, anti-inflammatory and antiatherosclerotic agent. *J Pharmacol Exp Ther* 305, 1116–1123, 2003.

Surh YJ. Anti-tumor promoting potential of selected spice ingredients with antioxidative and anti-inflammatory activities: a short review. *Food Chem Toxicol* 40, 1091–1097, 2002.

Surh YJ. Cancer chemoprevention with dietary phytochemicals, *Nat Rev Cancer* 3, 768–780, 2003.

Taylor A. Cataract: relationship between nutrition and oxidation. *J Am Coll Nutr* 12, 138–146, 1993.

Tedeschi E et al. Green tea inhibits human inducible nitric-oxide syntase expression by down-regulating signal transducer and activator of transcription-1α activation. *Mol Pharmacol* 65, 111–120, 2004.

Tilg H and Moschen AR. Adipocytokines: mediators linking adipose tissue, inflammation and immunity. *Nat Rev Immunol* 6, 772–783, 2006.

Tylicki L, Rutkowski B, and Horl WH. Antioxidants: a possible role in kidney protection. *Kidney Blood Press Res* 26, 303–314, 2003.

Unno T et al. Effect of tea catechins on post-prandial plasma lipid responses in human subjects. *Brit J Nutr* 93, 543–547, 2005.

Vercellotti JR, St Angelo AJ, and Spanier AM. In St Angelo AJ, Ed. *Lipid Oxidation in Food,* American Chemical Society, Washington, pp. 1–11, 1992.

Wang X and Tian T. Green tea epigallocatechin gallate: a natural inhibitor of fatty acid synthase. *Biochem Biophys Res Commun* 288, 1200–1206, 2001.

Ward PA. Role of complement, chemokines and regulatory cytokines in acute lung injury. *Ann NY Acad Sci* 796, 104–122, 1996.

Waring P. Redox active calcium ion channels and cell death. *Arch Biochem Biophys* 434, 33–42, 2005.

Weisberg SP et al. Obesity is associated with macrophage accumulation in adipose tissue. *J Clin Invest* 112, 1796–1808, 2003.

Weiss SJ. Tissue destruction by neutrophils. *New Engl J Med* 320, 365–376, 1989.

Wijeratne SSK, Abou-Zaid MM, and Shahidi F. Antioxidant polyphenols in almond and its coproducts. *J Agric Food Chem* 54, 312–318, 2006.

Winrow VR et al. Free radicals in inflammation: second messengers and mediators of tissue destruction. *Brit Med Bull* 49, 506–522, 1993.

Yanishlieva NV and Marinova EM. Stabilisation of edible oils with natural antioxidants. *Eur J Lipid Sci Technol* 103, 752–767, 2001.

Young S and Woodside IS. Antioxidants in health and disease. *J Clin Pathol* 54, 176–186, 2001.

Yu HJ et al. Sympathetic vesicovascular reflex induced by acute urinary retention evokes proinflammatory and proapoptotic injury in rat livers. *Am J Physiol Renal Physiol* 288, F1005–F1014, 2005.

Zhang Y, Vareed SK, and Nair MG. Human tumor cell growth inhibition by non-toxic anthocyanidins, the pigments in fruits and vegetables. *Life Sci* 76, 1465–1472, 2005.

# 14 Role of Exercise and Weight Loss in Reducing Inflammation

*Tongjian You and Michael J. LaMonte*

## CONTENTS



## 14.1 INTRODUCTION

Chronic inflammation is an important risk factor for type 2 diabetes and cardiovascular disease.[1] There is increasing interest in the application of lifestyle interventions such as increased physical activity and alterations in dietary intake, reducing inflammation, and lowering the risk for incident diseases. However, the specific effects of physical fitness and/or exercise training, adiposity, and weight loss on inflammation still need clarification. This chapter discusses the association of physical activity, adiposity and chronic inflammation, and the effects of exercise training and weight loss on inflammation in healthy and diseased individuals.

## 14.2 ROLE OF AEROBIC EXERCISE IN REDUCING INFLAMMATION: OBSERVATIONAL STUDIES

Findings from several studies of the cross-sectional relationship of physical activity (PA) or cardiorespiratory fitness (CRF) and inflammatory biomarkers are summarized in Table 14.1.[2–15] It shows the mean or median values for

**TABLE 14.1**
**Relationship of Physical Activity and Cardiorespiratory Fitness with Measures of Inflammation**

| Study | Participants | PA measure | Covariates | Inflammatory Measure (Mean or Median Values) | | P Value |
|---|---|---|---|---|---|---|
| | | | | Low PA or CRF | High PA or CRF | |
| **Middle Aged Adults** | | | | | | |
| Borodulin et al.[5] | 1,713 men and 2,090 women, 46 ± 12 years | PA (MET-hrs/wk) quartiles from questionnaire on moderate and vigorous PA | Age, WHR, smoking, DM, lipids, medication use, HRT use | CRP (mg/dL) | | |
| | | | | Men: 0.42 | 0.37 | <0.001 |
| | | | | Women: 0.55 | 0.40 | <0.001 |
| Pischon et al.[12] | 405 men (mean age 60 years) and 454 women (mean age 42 years) | PA (MET hours/week) quintiles from questionnaire on moderate and vigorous PA | | CRP (mg/L) | | |
| | | | | Men: 2.0 | 1.5 | NR |
| | | | | Women: 2.8 | 0.9 | NR |
| | | | | IL-6 (pg/mL) | | |
| | | | | Men: 1.4 | 1.3 | NR |
| | | | | Women: 1.6 | 1.5 | NR |
| | | | | TNF-R1 (pg/mL) | | |
| | | | | Men: 955 | 907 | NR |
| | | | | Women: 1035 | 986 | NR |
| | | | | TNF-R2 (pg/mL) | | |
| | | | | Men: 1457 | 1330 | NR |
| | | | | Women: 2145 | 2064 | NR |
| | | | Sex, age, BMI, smoking, NSAID use, diet, TV watching | Per 20 MET hours/week change in PA: | | |
| | | | | CRP: –0.12 mg/L | | 0.06 |
| | | | | IL-6: –0.06 pg/mL | | 0.06 |
| | | | | TNF-R1: –6.27 pg/mL | | 0.19 |
| | | | | TNF-R2: –19.71 pg/mL | | 0.01 |

| | | | | | | |
|---|---|---|---|---|---|---|
| Mora et al.[11] | 27,158 women, 55 ± 7 years | PA (MET hours/week) from questionnaire on moderate and vigorous PA | | CRP (mg/dL) 2.5 | 1.8 | <0.001 |
| | | | | ICAM-1 (ng/ml) 358 | 337 | <0.01 |
| | | | Age, race, BMI, smoking, BP, DM, HRT use | OR (95% CI) | | |
| | | | | CRP >4.2: | | |
| | | | | 1.37 (1.24 to 1.51) | 1.0 | <0.01 |
| | | | | ICAM >411: | | |
| | | | | 1.59 (1.43 to 1.76) | 1.0 | <0.01 |
| Church et al.[6] | 722 men, 50 ± 10 years | CRF (METs) quintiles from maximal ergometry | Age, BMI, smoking, medication use (statins, aspirin), chronic diseases | CRP (mg/dL) 1.6 | 0.8 | <0.001 |
| | | | | OR (95% CI) | | |
| | | | | CRP >2.0: 5.9 (2.7 to 12.5) | 1.0 | 0.001 |
| LaMonte et al.[9] | 135 women, 55 ± 11 years | CRF (METs) tertiles from maximal ergometry | Age, BMI, smoking, DM, HRT or oral contraceptive use | CRP (mg/dL) | | |
| | | | | Black: 4.3 | 4.3 | NS |
| | | | | White 3.0 | 1.7 | <0.05 |
| | | | | Native American: 3.1 | 1.8 | <0.05 |
| Aronson et al.[3] | 829 adults (59% men), 50 ± 9 years | CRF (METs) quartiles from maximal ergometry | Sex, age, BMI, smoking, DM, HTN, TG, HDL-C, medication use (statins, aspirin), HRT use | CRP (mg/dL) 2.4 | 1.7 | <0.001 |
| | | | | OR (95% CI) | | |
| | | | | CRP >3.0: 2.0 (1.25 to 2.5) | 1.0 | NR |
| **Younger Adults** | | | | | | |
| Williams et al.[15] | 400 men, 315 women, 26 years old | CRF (L/minute oxygen uptake) tertiles estimated from submaximal ergometry | | CRP (mg/dL) | | |
| | | | | Men: 2.1 | 1.5 | NR |
| | | | | Women: 4.3 | 1.9 | NR |
| | | | BMI, smoking, BP, oral contraceptive | Per unit change CRF: | | |
| | | | | Men: –0.11 mg/L | | ≤0.01 |
| | | | | Women: –0.18 mg/L | | ≤0.01 |

**TABLE 14.1 (continued)**
**Relationship of Physical Activity and Cardiorespiratory Fitness with Measures of Inflammation**

| Study | Participants | PA measure | Covariates | Inflammatory Measure (Mean or Median Values) | | P Value |
|---|---|---|---|---|---|---|
| | | | | Low PA or CRF | High PA or CRF | |
| Dufaux et al.[7] | 356 male and 103 female athletes; 45 male and 40 female sedentary controls, 17 to 23 years old | Swimming, rowing, distance running | None | CRP (mg/dL) | | |
| | | | | Men | Swimmer: | |
| | | | | Control: 0.50 | 0.10 | 0.26 |
| | | | | | Rower: | |
| | | | | | 0.26 | <0.01 |
| | | | | | Runner: | |
| | | | | | 0.32 | NS |
| | | | | Women | | |
| | | | | Control: 0.39 | Swimmer: | |
| | | | | | 0.11 | <0.001 |
| | | | | | Rower: | |
| | | | | | 0.29 | NS |
| | | | | | Runner: | |
| | | | | | 0.25 | NS |
| **Older Adults** | | | | | | |
| Stauffer et al.[13] | 39 endurance trained (58 ± 7 years) and 75 sedentary (60 ± 8 years) postmenopausal women | PA status by interview | None | CRP (mg/L) | | |
| | | | | HRT user: 1.8 | 0.6 | <0.01 |
| | | | | HRT nonuser: 1.1 | 0.3 | <0.01 |
| | | | | IL-6 (pg/mL) | | |
| | | | | HRT user: 1..5 | 1.2 | NS |
| | | | | HRT nonuser: 1.5 | 0.9 | <0.05 |

| | | | | | | |
|---|---|---|---|---|---|---|
| Geffken et al.[8] | 2,495 men, 3,393 women, ≥65 years | PA (Kcal/week) from questionnaire on moderate and vigorous PA | Sex, age, race, BMI, smoking, DM, HTN, CVD | CRP (mg/L) 2.2 | 1.8 | <0.001 |
| | | | | WBC (×1000/μL) 6.4 | 5.9 | <0.001 |
| | | | | Albumin (g/dL) 4.0 | 4.0 | 0.84 |
| | | | | CRP per 1-SD change PA: –0.21 mg/L | | <0.001 |
| Wannamethee et al.[14] | 3,954 men, 60 to 79 years | PA (ordinal scale) from questionnaire on total daily PA | Age, BMI, smoking, alcohol intake, CVD, exam date | CRP (mg/L) 2.3 | 1.5 | <0.001 |
| | | | | WBC ($10^9$ cells/L) 7.0 | 6.6 | <0.001 |
| | | | | OR (95% CI) | | |
| | | | | CRP ≥4.27: 2.3 (1.7 to 3.1) | 1.0 | <0.001 |
| | | | | WBC ≥8.5 × $10^9$/L: 1.6 (1.2-2.3) | 1.0 | 0.0002 |
| **Individuals with Existing Disease** | | | | | | |
| Wannamethee et al.[14] | 1,245 men with CVD, 60 to 79 years | PA (ordinal scale) from questionnaire on total daily PA | Age, BMI, smoking, alcohol intake, CVD, exam date | CRP (mg/L) 2.9 | 1.8 | <0.001 |
| | | | | WBC ($10^9$ cells/L) 2.1 | 1.9 | <0.001 |
| Albert et al.[2] | 1,153 adults with CVD, mean age ≈63 years | PA (ordinal scale) from questionnaire on frequency of weekly exercise | Age, BMI, smoking, HDL-C, BP, DM, aspirin use | CRP (mg/dL) 3.0 | 1.8 | NR |
| | | | | Multivariate β-coefficient for Ln CRP: 0.07 | –0.20 | 0.02 |
| Aronson et al.[3] | 340 adults, 50 ± 10 years, with NCEP-ATP 3 defined metabolic syndrome | CRF (METs)quartiles from maximal ergometry | Sex, age, BMI, smoking, CVD, statin , aspirin and HRT use | CRP (mg/dL) 4.6 | 2.2 | 0.001 |
| | | | | CRP >3.0 mg/L: 74% | | 35% |
| | | | | CRP per 1-unit change CRF: –0.06 mg/L | | <0.001 |
| McGavock et al.[10] | 28 women, 57 ± 6 years with type 2 diabetes | CRF (mL $O_2$ uptake/kg/minute) from maximal ergometry | Age, BMI, HOMA | CRP (mg/L) 6.3 | 1.9 | <0.05 |
| | | | | CRP per 1-unit change CRF: –0.65 mg/L | | |

BMI = body mass index. BP = blood pressure. CI = confidence interval. CRF = cardiorespiratory fitness. CRP = C-reactive protein. CVD = cardiovascular disease. DM = diabetes mellitus. HDL-C = high density lipoprotein cholesterol. HOMA = homeostasis model of insulin resistance. HRT = hormone replacement therapy. HTN = hypertension. ICAM-1 = intracellular adhesion molecule 1. IL-6 = interleukin 6. Kcal = kilocalories. MET = metabolic equivalent (1 MET = 3.5 mL $O_2$ uptake/kg/min, or 1 kcal/kg/hr). NR = not reported. NS = not significant.

NSAID = nonsteroidal anti-inflammatory drug. OR = odds ratio. PA = physical activity. SD = standard deviation. TG = triglyceride. TNF-R1 = tumor necrosis factor receptor 1. TNF-R2 = tumor necrosis factor receptor 2. WHR = waist-to-hip ratio.

each inflammatory marker among study participants in the lowest and highest PA and CRF categories and, when available, the results from multivariable analyses. C-reactive protein (CRP) was the most frequently reported measure of inflammation, although a small number of studies included other inflammatory biomarkers.[8,11–14]

Generally, significant inverse relationships between PA or CRF and inflammatory measures have been reported in both women and men, in younger and older adults, and among those with existing chronic diseases. Findings are more consistent for CRP than for other biomarkers such as IL-6 or TNF receptor. Absolute mean differences in inflammatory biomarker concentrations between the highest and lowest PA or CRF groups are small but statistically significant.

The association between PA or CRF and inflammatory measures is independent of other factors related to inflammation including adiposity, smoking, diabetes, use of anti-inflammatory medications, and hormone replacement therapy.[2–6,8–12,14,15] In one study, CRP concentrations were 0.21 mg/L lower ($P < 0.001$) with each standard deviation (SD) increment in PA among older adults, even after adjustment for sex, race, adiposity, smoking, prevalent diabetes, and cardiovascular disease (CVD).[8] Low levels of PA and CRF have been strongly associated with the presence of inflammatory markers that exceed clinically relevant concentrations.[3,4 6,11,14] For example, after accounting for differences in sex, age, BMI, smoking, medication use, and CVD risk factors, the odds of having CRP >3.0 mg/L were two-fold greater ($P < 0.05$) in adults in the lowest compared with highest quartile of CRF determined from maximal exercise testing.[4]

Even among adults with clinically manifest diseases such as CVD, diabetes, and metabolic syndrome, higher PA or CRF is related to lower inflammatory biomarkers.[2,3,10,14,16] Among older women with diabetes, each unit increment in CRF (milliliters of $O_2$ uptake per kilogram body weight per minute) was associated with a 0.65 mg/L lower CRP concentration after controlling for age, BMI, and insulin resistance.[10]

To further examine the interrelationships among PA or CRF, adiposity, and inflammation, some investigators cross-tabulated inflammatory biomarkers on joint strata of activity or fitness and adiposity.[4–6,11] Higher levels of PA or CRF are significantly associated with lower inflammatory biomarkers across a broad range of adiposity phenotypes. In a study of 27,158 middle-aged women, multivariable odds ratios for CRP >3.0 mg/L in those who were physically active compared with sedentary, respectively, were 1.0 (referent) and 1.26 ($P < 0.001$) among normal weight (BMI 18.5 to 24.9); 2.68 and 3.11 ($P = 0.003$) among overweight (BMI 25.0 to 29.9); and, 8.25 and 9.86 ($P = 0.01$) among obese participants.[11] In another study, after adjusting for several confounding factors including smoking, prevalent diabetes, and statin use, the values for men with CRP >2.0 mg/L across incremental CRF tertiles were 40, 19, and 15% among those with waist girths <102 cm; and 55, 39, and 18% among those with waist girths ≥102 cm (trend $P < 0.01$ each).[6]

The above findings derive from cross-sectional studies that prohibit causal inferences. Stronger evidence for anti-inflammatory effects of PA or CRF comes from prospective studies relating changes in activity or fitness levels with subsequent measures of inflammation, but such data are sparse. In a study of 2,545 men aged 60 to

79 years who completed two PA questionnaires separated by 20 years, significantly lower concentrations of CRP and WBC were seen in men who became or remained physically active across the two surveys as compared with men who sustained sedentary lifestyles.[14] A few studies prospectively examined the risks of CVD events according to PA and inflammatory biomarkers.[17–19] Among middle-aged women, relative hazards of CVD across incremental PA quartiles were 1.00, 0.73, 0.68, and 0.59 (trend $P < 0.001$) in an age-adjusted model, and only slightly attenuated to 1.00, 0.77, 0.73, 0.64 (trend, $P < 0.001$) after further adjustment for CRP.[19] Additional analyses showed that about 24% of CVD risk associated with PA levels was explained by differences in inflammatory biomarkers.

Not all studies report significant findings for PA or CRF in relation to inflammation.[20–23] In adults 70 to 79 years old, neither grip strength, timed chair raises, or 6-meter walk time were associated with CRP or IL-6 concentrations.[22] Past-year and current PA were not associated with average CRP concentrations across five assessments during a 12-month interval.[21] These discrepancies with the findings summarized in Table 14.1 may result from healthier study samples, greater proportion of sedentary participants, misclassification of PA status due to recall biases in self-reporting, a high proportion of out-of-range inflammatory biomarker concentrations (e.g., CRP >15 mg/L), or simply from chance.

## 14.3 ROLE OF AEROBIC EXERCISE IN REDUCING INFLAMMATION: INTERVENTIONAL STUDIES

Unlike acute exercise that stimulates immune response and increases levels of inflammatory cytokines such as CRP, chronic exercise, especially aerobic exercise, reduces levels of inflammatory markers. For example, circulating CRP levels were lowered by 30% in athletes following 9-month marathon training; however, CRP levels did not change during a non-training period.[24] These findings indicate that even in lean, healthy individuals, exercise training reduces levels of inflammatory markers.

Findings from several studies of the effects of exercise training on circulating levels of inflammatory biomarkers are summarized in Table 14.2.[25–42] The subjects of these studies were healthy lean and obese people or type 2 diabetics and heart disease patients. Typical exercise training included aerobic exercise, resistance exercise, or combined aerobic and resistance exercise. The training term ranged from 4 weeks to 2 years. CRP was measured as an inflammatory marker in most studies. Other markers of chronic inflammation included serum amyloid A (SAA), interleukin-6 (IL-6); interleukin-1 β (IL-1β), interleukin-10 (IL-10), interleukin-18 (IL-18), tumor necrosis factor α (TNF-α), soluble TNF receptor 1 (sTNFR1), soluble TNF receptor 2 (sTNFR2), adiponectin, and plasminogen activator inhibitor 1 (PAI-1).

For most studies, aerobic fitness or other physical variables (i.e., walk distance, maximal workload, and muscle strength) were measured. In addition, body composition changes were recorded in studies involving obese and/or diabetes patients.

Mixed findings were reported on the effects of exercise training on chronic inflammation in healthy individuals.[25–34] Short-term (8 to 12 weeks) studies did not support a consistent exercise training effect in lowering chronic inflammation.

## TABLE 14.2
## Effects of Exercise Training on Measures of Inflammation

| Study | Participants | Exercise Type | Effects on Body Composition and Aerobic Fitness | Effects on inflammation |
|---|---|---|---|---|
| | | **Healthy Individuals** | | |
| Tsukui et al.[25] | 27 lean and obese women | Aerobic exercise, 12 weeks | ↓ body weight by 1.6% $P < 0.01$) ↓ fat mass by 3.4% ($P < 0.01$) | ↓ TNF-α by 83.3% ($P < 0.01$), ↓ sTNFR1 by 17.6% ($P < 0.01$), ↓ sTNFR2 by 20% ($P < 0.01$) |
| Straczkowski et al.[26] | 16 obese women with normal (NGT; n = 8) and impaired glucose tolerance (IGT; n = 8) | Aerobic exercise, 12 weeks | NGT: ↓ body weight by 2.9% ($P < 0.05$), ↓ fat mass by 6.4% ($P < 0.05$), ↓ waist circumference by 1.8% ($P < 0.05$) IGT: ↓ body weight by 2% ($P < 0.05$), ↓ fat mass by 9.2% ($P < 0.05$), ↓ waist circumference by 2.9% ($P < 0.05$) | NGT: ↓ TNF-α by 15.7% ($P < 0.05$), = sTNFR1, ↓ sTNFR2 by 17.3% ($P < 0.05$) IGT: ↓ TNF-α by 21.9% ($P < 0.05$), = sTNFR1, ↓ sTNFR2 by 21.8% ($P < 0.05$) |
| Nassis et al.[27] | 19 overweight and obese girls(13 ± 2 years; BMI = 27 ± 4 kg/m$^2$) | Aerobic exercise, 12 weeks | = body weight, = percent body fat, ↑ $VO_2$max by 18.8% ($P < 0.05$) | = adiponectin, CRP, IL-6 |
| Polak et al.[28] | 25 obese postmenopausal women (BMI = 32 ± 3 kg/m$^2$) | Aerobic exercise, 12 weeks | ↓ body weight by 5.9% $P < 0.001$), ↓ fat mass by 6.4% ($P < 0.001$), ↑ $VO_2$max by 12.8% ($P < 0.05$) | = adiponectin, IL-6, TNF-α, = adipose tissue adiponectin, IL-6, TNF-α mRNA |
| Kondo et al.[29] | 8 lean (BMI = 22 ± 3 kg/m$^2$) and 8 obese (BMI = 30 ± 3 kg/m$^2$) Japanese women | Aerobic exercise, 28 weeks | Lean: = body weight, ↓ fat mass by 20.2% ($P < 0.05$), ↑ $VO_2$max by 17.2% ($P < 0.05$) Obese: ↓ body weight by 11.0% ($P < 0.05$), ↓ fat mass by 20.2% ($P < 0.05$), ↑ $VO_2$max by 12.8% ($P < 0.05$) | Lean: = adiponectin, CRP, TNF-α Obese: ↑ adiponectin by 75% ($P < 0.01$), ↓ CRP by 53.3% ($P < 0.05$), ↓ TNF-α by 36.8% ($P < 0.01$) |

**TABLE 14.2 (continued)**
**Effects of Exercise Training on Measures of Inflammation**

| Study | Participants | Exercise Type | Effects on Body Composition and Aerobic Fitness | Effects on inflammation |
|---|---|---|---|---|
| Stewart et al.[30] | 29 physically active individuals; 31 physically inactive individuals | Aerobic and resistance exercises, 12 weeks | Physically active: = $VO_2$max Physically inactive: ↑ $VO_2$max by 10.4% (P <0.01), ↑ strength by 38.1% (P <0.01) | Physically active: = CRP, IL-6, TNF-α, IL-1β Physically inactive: ↓ CRP by (P <0.01), = IL-6, TNF-α, IL-1β |
| Dekker et al.[31] | 24 lean and obese individuals | Aerobic exercise, 12 weeks | = body weight, = fat mass, ↓ waist circumference by (P <0.05) | = CRP, PAI-1, ↓ IL-6 by (P<0.05) |
| Leick et al.[32] | 13 obese individuals (7 men, BMI = 33 ± 1 kg/m$^2$; 6 women, BMI = 39 ± 4 kg/m$^2$) | Aerobic exercise, 8 weeks | ↑ $VO_2$max by 8% (P <0.05) | = IL-18, ↓ adipose tissue IL-18 mRNA by 20% (P = 0.06) |
| * Olson et al.[33] | 28 overweight women; control: n = 12; exercise: n = 16 | No excercise or resistance exercise, 1 year | Control: = body weight, fat mass, lean mass Exercise: = body weight, fat mass, ↑ lean body mass | Control: = adiponectin, CRP, IL-6 Exercise: ↑ adiponectin by (P <0.01), ↓ CRP by (P <0.01), = IL-6 |
| Devries et al.[34] | 24 lean and obese women | Aerobic exercise, 12 weeks | = body weight, = fat mass | = CRP |
| | | **Individuals with Existing Disease** | | |
| Tisi et al.[35] | 22 patients with intermittent claudication | Active and passive leg exercises, 1 year | Not specified | ↓ CRP by 17% at 3 months (P <0.05) ↓ SAA by 71.4% at 6 months (P <0.01) |
| Larsen et al.[36] | 28 older men with chronic heart failure | Aerobic exercise, 12 weeks | ↑ 6-min walk distance by 8.1% (P <0.001), ↑ maximal workload by 44.7% (P <0.001) | ↓ TNF-α by 12.5% (P = 0.013) =IL-6, IL-8 |

## TABLE 14.2 (continued)
## Effects of Exercise Training on Measures of Inflammation

| Study | Participants | Exercise Type | Effects on Body Composition and Aerobic Fitness | Effects on inflammation |
|---|---|---|---|---|
| Conraads et al.[37] | 23 older patients with chronic heart failure | Combined aerobic and resistance exercise, 16 weeks | ↑ $VO_2$max by 7.5% (P = 0.008) | ↓ sTNFR1 by 15.4% (P = 0.02) = IL-6, TNF-α, sTNFR2 |
| Milani et al.[38] | 235 CHD patients | Phase II cardiac rehabilitation and exercise training 12 weeks | = body weight, ↓ fat mass by 3% (P = 0.0008), ↓ waist circumference by 2% (P = 0.002), ↑ $VO_2$max by 9% (P <0.0001) | ↓ mean CRP by 36% (P <0.0001) ↓ median CRP by 41% (P = 0.002) |
| Zoppini et al.[39] | 16 obese older patients with type 2 diabetes | Aerobic exercise, 24 weeks | = body weight, = waist circumference | = CRP, sTNFRs |
| [a] Brooks et al.[40] | 62 older patients with type 2 diabetes, control: n = 31, exercise: n = 31 | No resistance or other exercise, 16 weeks | Control: = lean mass, ↓ upper and low body muscle strength | Exercise: ↑ lean mass by 2.5%, ↑ upper and low body muscle strength Exercise vs control: ↑ adiponectin (19.6% vs –14.5%, P <0.001) ↓ CRP (–37.1% vs 11.4%, P = 0.05) |
| Oberbach et al.[41] | 40 patients with insulin resistance and type 2 diabetes | Aerobic and resistance exercise, 4 weeks | ↓ fat mass (P <0.05), ↑ $VO_2$max (P <0.05) | ↑ adiponectin (P <0.001) ↓ CRP (P<0.001), = IL-6, IL-10 |
| [a] Walther et al.[42] | 101 male CAD patients, PCI: n = 50, exercise: n = 51 | PCI or exercise, 2 years | PCI: = $VO_2$max Exercise: ↑ $VO_2$max by 10% (P <0.05) | PCI: = CRP, IL-6 Exercise: ↓ CRP by 41% (P <0.05), ↓ IL-6 by 18% (P <0.05) |

*BMI*: body mass index, mean ± standard deviation. CAD = coronary artery disease. CHD = coronary heart disease. CRP = C-reactive protein. IL-10 = interleukin 10. IL-18 = interleukin 18. IL-1β = interleukin 1β. IL-6 = interleukin 6.

*PAI-1* = plasminogen activator inhibitor 1. PCI = percutaneous intervention. SAA = serum amyloid A. sTNFR1 = soluble TNF receptor 1. sTNFR2 = soluble TNF receptor 2. TNF-α = tumor necrosis factor α. $VO_2$max = maximal aerobic capacity.

[a] Randomized controlled trial.

However, longer term (20 weeks to 1 year) studies including a randomized controlled trial indicated that exercise training was very effective in lowering measures of inflammation. These findings indicate that exercise training term is important for an effective inflammation-lowering exercise program. Baseline levels of physical fitness and body composition are also important. For example, 12 weeks of aerobic plus resistance training reduced CRP levels in physically inactive individuals, but not in physically active individuals.[30] Moreover, 28 weeks of aerobic exercise training increased circulating levels of adiponectin and decreased levels of CRP and TNF-$\alpha$ in obese but not lean individuals.[29]

In diseased individuals, exercise training was effective in lowering chronic inflammation.[35–42] Although the intervention terms ranged widely from 4 weeks to 2 years, these studies supported a consistent exercise training effect in lowering inflammation, possibly due to the high baseline levels of inflammatory markers in these patients. Two randomized controlled exercise trials were recently conducted in diseased individuals. One study indicated that resistance training decreased CRP, increased adiponectin levels, and increased insulin sensitivity in patients with type 2 diabetes.[40] The decreased inflammation was related to improved metabolic risk factors. Another study showed exercise training produced significant reductions in CRP and IL-6 in patients with coronary artery disease. Moreover, training-induced reductions in levels of pro-inflammatory cytokines were accompanied by improvement in aerobic fitness ($VO2_{max}$).[42]

Aerobic exercise as a traditional exercise intervention has been widely used to treat chronic inflammation. Interesting[14], two recent randomized trials support that resistance exercise training alone is effective in reducing inflammation in both healthy individuals and type 2 diabetics.[33,40] Future studies, especially randomized controlled studies, are needed to investigate the independent effects of aerobic and resistance training in lowering inflammation, to clarify whether combined aerobic and resistance exercise is more effective than either type of exercise alone, and to further delineate the dose–response characteristics of different exercise modes and specific inflammatory biomarkers.

## 14.4 ROLES OF ADIPOSITY AND WEIGHT LOSS IN REDUCING INFLAMMATION: OBSERVATIONAL STUDIES

Findings from several studies on the cross-sectional relationship between adiposity/weight loss and inflammatory biomarkers are summarized in Table 14.3.[43–50] Shown are mean or median values for each inflammatory marker among study participants in the lowest and highest adiposity category, and when available, also shown are results from multivariable analyses. Most studies report on CRP as a measure of inflammation; a number of studies also included other inflammatory biomarkers.[45–49] Individuals with higher levels of overall adiposity (defined by BMI or body fat mass) and those with higher levels of abdominal obesity (defined by waist circumference or WHR) exhibited significantly higher inflammatory biomarker concentrations compared with their lean peers. Findings were similar for women and men,[44,45,49] for measures of overall and abdominal adiposity,[45,47,49] and in older population samples.[46–50]

**TABLE 14.3**
**Relationship of Adiposity and Measures of Inflammation**

| Study | Participants | Adiposity Measure | Covariates | Inflammatory Measure (Mean or Median Value) | | |
|---|---|---|---|---|---|---|
| | | | | Low Adiposity[a] | High Adiposity[b] | P Value |
| **Younger Adults** | | | | | | |
| Lemieux et al.[43] | 159 men, mean age 43 ± 8 years; mean BMI 30 ± 4 kg/m² | BMI from measured HT and WT | None | CRP (mg/L): | | |
| | | | | 1.9 | 2.9 | <0.05 |
| Ford et al.[44] | 7,325 men and 8,244 women, ≥20 years; 26% obese (BMI ≥30) | BMI from measured HT and WT | | CRP (mg/dL): | | |
| | | | | 2.6 | 4.5 | NR |
| | | | | OR (95% CI) | | |
| | | | | Men, CRP >4.4 mg/L: | | |
| | | | | 1.0 | 4.5 (2.5 to 8.1) | NR |
| | | | | Women, CRP >7.0 mg/L: | | |
| | | | | 1.0 | 6.3 (4.3 to 9.2) | NR |
| Panagiotakos et al.[45] | 1,514 men and 1,528 women, mean age 45 ± 13 years; 20% men and 15% women obese (BMI >30) | BMI from measured HT and WT; waist and hip measures | | Men, BMI: | | |
| | | | | CRP (mg/L) | | |
| | | | | 1.7 | 2.7 | <0.05 |
| | | | | IL-6 (ng/mL) | | |
| | | | | 1.5 | 2.2 | <0.05 |
| | | | | TNF-α (mg/dL) | | |
| | | | | 4.6 | 5.8 | <0.05 |
| | | | | WHR CRP (mg/L) | | |
| | | | | 1.9 | 2.9 | <0.05 |
| | | | | IL-6 (ng/mL) | | |
| | | | | 1.4 | 2.1 | <0.05 |
| | | | | TNF-α (mg/dL) | | |

| | | | |
|---|---|---|---|
| | 4.5 | 6.2 | <0.05 |
| | Women, BMI: CRP (mg/L) | | |
| | 1.3 | 3.9 | <0.05 |
| | IL-6 (ng/mL) | | |
| | 1.4 | 2.1 | <0.05 |
| | TNF-α (mg/dL) | | |
| | 4.2 | 5.4 | <0.05 |
| | WHR CRP (mg/L) | | |
| | 1.6 | 2.8 | <0.05 |
| | IL-6 (ng/mL) | | |
| | 1.3 | 1.9 | <0.05 |
| | TNF-α (mg/dL) | | |
| | 4.9 | 5.9 | <0.05 |
| Age, smoking, PA, BP, lipids, glucose, diet intake | Per 1 SD increment BMI: | | |
| | Men: | | |
| | CRP +0.03 mg/L | | |
| | IL-6 +0.03 ng/mL | | <0.001 |
| | TNF-α +0.02 mg/dL | | 0.01 |
| | Women: | | |
| | CRP +0.03 mg/L | | <0.001 |
| | IL-6 +0.03 ng/mL | | 0.01 |
| | TNF-α +0.02 mg/dL | | 0.02 |
| | Per 1 SD increment waist: | | |
| | Men: | | |
| | CRP +0.04 mg/L | | 0.002 |
| | IL-6 +0.62 ng/mL | | <0.001 |
| | TNF-α +0.04 mg/dL | | 0.009 |
| | Women: | | |
| | CRP +0.63 mg/L | | 0.003 |

**TABLE 14.3 (continued)**
**Relationship of Adiposity and Measures of Inflammation**

| Study | Participants | Adiposity Measure | Covariates | Inflammatory Measure (Mean or Median Value) Low Adiposity[a] | High Adiposity[b] | P Value |
|---|---|---|---|---|---|---|
| | | | | TNF-α +0.04 mg/dL | | 0.008 |
| | | | **Older Adults** | | | |
| Thorand et al.[49] | 641 men and 597 women, 55 to 74 years; mean BMI 28 ± 4 $kg/m^2$ | BMI from measured HT and WT; waist and hip girths; body fat from BIA | Age, smoking, alcohol intake, PA | Percent body fat: | | |
| | | | | Men[c] | | |
| | | | | CRP (mg/L): | | |
| | | | | 1.0 | 1.9 | <0.001 |
| | | | | IL-6 (pg/mL): | | |
| | | | | 0.7 | 2.3 | <0.001 |
| | | | | Women[c] | | |
| | | | | CRP (mg/L): | | |
| | | | | 1.3 | 3.5 | <0.001 |
| | | | | IL-6 (pg/mL): | | |
| | | | | 1.3 | 2.3 | 0.002 |
| | | | | WHR: | | |
| | | | | Men[†] | | |
| | | | | CRP (mg/L): | | |
| | | | | 1.2 | 1.9 | <0.001 |
| | | | | IL-6 (pg/mL): | | |
| | | | | 1.0 | 2.3 | 0.02 |
| | | | | Women[†] | | |
| | | | | CRP (mg/L): | | |
| | | | | 1.5 | 2.7 | <0.001 |
| | | | | IL-6 (pg/mL): | | |

| | | | | | | |
|---|---|---|---|---|---|---|
| | | | | 1.0 | 1.8 | 0.005 |
| Rexrode et al.[47] | 733 women, mean age 56 ± 7 years; mean BMI 25 ± 4 kg/m² | BMI from reported HT and WT; waist and hip girths from self-report | Age, smoking, alcohol intake, PA, HRT use, medication use, DM, HTN | BMI | | |
| | | | | CRP (mg/L): | | |
| | | | | 0.12 | 0.54 | 0.0001 |
| | | | | IL-6 (pg/mL): | | |
| | | | | 1.1 | 1.8 | 0.0001 |
| | | | | Waist girth | | |
| | | | | CRP (mg/L): | | |
| | | | | 0.11 | 0.54 | 0.0001 |
| | | | | IL-6 (pg/mL): | | |
| | | | | 1.1 | 1.8 | 0.0001 |
| | | | | WHR: | | |
| | | | | CRP (mg/L): | | |
| | | | | 0.19 | 0.40 | 0.0001 |
| | | | | IL-6 (pg/mL): | | |
| | | | | 1.24 | 1.6 | 0.0001 |
| | | | | BMI: | | |
| | | | | HRT users: | | |
| | | | | CRP: 0.21 | 0.68 | <0.001 |
| | | | | IL-6: 1.10 | 1.70 | <0.001 |
| | | | | HRT nonusers: | | |
| | | | | CRP: 0.05 | 0.45 | <0.001 |
| | | | | IL-6: 1.10 | 1.81 | <0.001 |
| | | | | OR (95% CI) | | |
| | | | | CRP >0.59 mg/L: | | |
| | | | | BMI: | | |
| | | | | 1.0 | 12.2 (6.4 to 23.0) | <0.001 |
| | | | | Plus waist: | 15.3 (6.9 to 33.8) | <0.001 |
| | | | | Waist | | |

**TABLE 14.3 (continued)**
**Relationship of Adiposity and Measures of Inflammation**

| Study | Participants | Adiposity Measure | Covariates | Inflammatory Measure (Mean or Median Value) | | |
|---|---|---|---|---|---|---|
| | | | | Low Adiposity[a] | High Adiposity[b] | P Value |
| | | | | Plus BMI: | 2.0 (0.9 to 4.6) | NS |
| Barinas-Mitchell et al.[50] | 208 PM women, mean age 59 ± 2 years; mean BMI 27 ± 5 kg/m$^2$ | VAT volume (cm$^3$) from CT scan | None | CRP (mg/L): | | |
| | | | | HRT users | | |
| | | | | 2.2 | 4.3 | <0.01 |
| | | | | HRT nonusers | | |
| | | | | 0.9 | 3.2 | <0.01 |
| Pou et al.[46] | 600 men and 650 women, mean age 60 ± 9 years; mean BMI 28 ± 5 kg/m$^2$ | VAT and SAT from CT scan | Sex, age, smoking, aspirin use, HRT use, PA, alcohol intake | Per 1 SD increment VAT: | | |
| | | | | CRP | | |
| | | | | +1.8 mg/L | | <0.001 |
| | | | | +0.6 mg/L controlling for BMI | | <0.01 |
| | | | | IL-6 | | |
| | | | | +0.5 pg/mL | | <0.001 |
| | | | | +0.2 pg/mL controlling for BMI | | 0.01 |
| | | | | TNF-α | | |
| | | | | +0.04 pg/mL | | 0.06 |
| | | | | 0.00 pg/mL controlling for BMI | | 0.92 |
| | | | | Per 1 SD increment SAT: | | |
| | | | | CRP | | |
| | | | | +1.7 mg/L | | <0.001 |
| | | | | +0.6 mg/L controlling for BMI | | 0.51 |
| | | | | IL-6 | | |
| | | | | +0.5 pg/mL | | <0.001 |
| | | | | +0.1 pg/mL controlling for BMI | | 0.17 |

| | | | | | | |
|---|---|---|---|---|---|---|
| | | | | TNF-α | | |
| | | | | +0.03 pg/mL | | 0.08 |
| | | | | – 0.03 pg/mL controlling for BMI | | 0.31 |
| Schrager et al.[48] | 378 men and 493 women, ≥65 yrs; 25% obese (BMI >30) | BMI from measured HT and WT; waist measure | Sex, age, smoking, PA, DM, HTN, CVD | Sarcopenic obesity[d]: | | |
| | | | | No | Yes | |
| | | | | CRP (mg/L): | | |
| | | | | 2.0 | 4.0 | 0.07 |
| | | | | IL-6 (pg/mL): | | |
| | | | | 1.4 | 1.8 | 0.99 |
| | | | | TNF-α (pg/mL): | | |
| | | | | 2.3 | 2.7 | 0.29 |
| **Individuals with Existing Disease** | | | | | | |
| Ford et al.[44] | 1,945 adults with prevalent DM or IFG; mean age 58 years | BMI from measured HT and WT | Sex, age, race | OR (95% CI) | | |
| | | | | CRP >4.4 in men, >7.0 in women | | |
| | | | | Prevalent IFG | | |
| | | | | 1.0 | 6.0 (1.6 to 22.8) | NR |
| | | | | Prevalent DM | | |
| | | | | 1.0 | 5.5 (2.4 to 12.5) | NR |
| Thorand et al.[49] | 441 adults with prevalent DM or IGT; mean age 64 ± 5 years | Waist and hip measures | Age, smoking, alcohol intake, PA | WHR | | |
| | | | | Men: | | |
| | | | | CRP 1.6 | 2.4 | 0.14 |
| | | | | IL-6 1.2 | 1.9 | 0.09 |
| | | | | Women: | | |
| | | | | CRP 1.2 | 4.1 | <0.001 |
| | | | | IL-6 1.7 | 3.2 | 0.26 |

**TABLE 14.3 (continued)**
**Relationship of Adiposity and Measures of Inflammation**

| Study | Participants | Adiposity Measure | Covariates | Inflammatory Measure (Mean or Median Value) Low Adiposity[a] | High Adiposity[b] | P Value |
|---|---|---|---|---|---|---|
| Aronson et al.[53] | 297 adults with metabolic syndrome; mean age 50 ± 10 years | BMI from measured HT and WT | Sex, age, smoking, HRT use, albumin, WBC, PA | CRP (mg/L)<br>2.0 | 3.3 | <0.001 |

*BIA* = bioelectrical impedance analysis. BMI = body mass index. BP = blood pressure. CI – confidence interval. CRP = C-reactive protein. CVD = cardiovascular disease. DM = diabetes mellitus. HRT = hormone replacement therapy. IFG = impaired fasting glucose. IL = interleukin.

*NR* = not reported. NS = not significant. OR = odds ratio. PA = physical activity. PM = postmenopausal. SAT = subcutaneous adipose tissue. SD = standard deviation. TNF = tumor necrosis factor. VAT = visceral adipose tissue.

*WBC* = white blood cells. WHR = waist-to-hip ratio.

[a] Low adiposity defined, when possible, as normal weight (BMI 18.5 to 29.9) or normal waist circumference ( men <102 cm; women <88 cm) or as lowest study-specific category (e.g., lowest quartile).

[b] High adiposity defined as obesity (BMI ≥30) or abdominal obesity (men ≥102 cm; women ≥88 cm waist circumference) or as highest study specific category.

[c] Subset of participants (n = 797) with normal glucose tolerance.

[d] Sarcopenic obesity defined as the presence of obesity and low muscular strength; reported P values represent interaction of obesity and strength of inflammatory biomarker concentration.

Absolute mean differences in inflammatory biomarker concentrations between the lowest and highest adiposity groups are small but statistically significant. The significant positive association between adiposity and inflammation is seen after controlling for potential confounding factors and after stratifying pro-inflammatory factors such as smoking or hormone replacement therapy (HRT) use.[47,50,51] Adjusted mean CRP concentrations (milligrams per liter) were 1.96 among non-obese never-smoker adults and 3.93 ($P = 0.002$) among obese (BMI ≥30 kg/m$^2$) current smokers.[51] Mean CRP concentrations (milligrams per liter) were 2.2 and 4.3 in postmenopausal HRT users in the lowest and highest quartiles, respectively, of visceral abdominal adipose tissue (VAT) measured by CT scans, and were 0.9 and 3.2 in HRT nonusers in the same VAT categories ($P < 0.01$, each).[50] A study of adults 65 years and older showed that higher concentrations of inflammatory biomarkers are associated with sarcopenic obesity that may be an important link of obesity, inflammation, and functional decline with advanced age.[48]

Some investigators tried to quantify the influence of abdominal adiposity on inflammatory biomarkers over and above the influence of total adiposity.[43,46,47,52] In a study of older women, the significant multivariable association between BMI and CRP was materially unchanged when further adjusted for waist girth; the association between waist and CRP was eliminated after adjustment for BMI.[47] In another study, associations between VAT (measured by CT scan) and inflammation remained highly significant when adjusted for differences in BMI.[46] Moreover, significant positive associations with CRP were seen across incremental VAT tertiles nested within tertiles of subcutaneous adipose tissue (measured by CT scan).[46] Similarly, the presence of abdominal obesity based on elevated waist girth or VAT was associated with higher CRP concentrations when nested within a stratum of total body fat measured by hydrodensitometry.[43] Available data suggest that the location of body fat may be a critical factor affecting chronic inflammation.

Adiposity also is associated with inflammation in adults with existing chronic disease.[44,49,53] After adjusting for sex, age, HRT, and pro-inflammatory factors, mean CRP concentrations among adults with metabolic syndrome were 2.0 and 3.3 mg/L ($P<0.001$) in non-obese and obese individuals, respectively.[53] The adjusted relative odds of having elevated CRP (>4.4 in men, >7.0 in women) was five-fold higher ($P < 0.05$) in obese compared to normal weight adults with diagnosed diabetes.[44] Other investigators examined the joint effects of adiposity and inflammation in predicting new onset disease.[54,55] Two studies of older women revealed a stepwise increase in the multivariable risk of diabetes[54] and CHD[55] across incremental tertiles of CRP and IL-6 among both obese and non-obese participants. Obese women in the highest inflammatory tertiles had a two-fold or greater risk ($P < 0.05$) of developing disease than their non-obese counterparts in the lowest inflammation groups.

The above summarized study findings clearly demonstrate that greater adiposity is associated with higher concentrations of inflammatory biomarkers, even after accounting for differences in other important pro-inflammatory factors. Greater adiposity also adds to chronic disease risk over and above the risk conferred by chronic inflammation.

## 14.5 ROLES OF ADIPOSITY AND WEIGHT LOSS IN REDUCING INFLAMMATION: INTERVENTIONAL STUDIES

Findings from several studies of the effects of dietary weight loss alone on circulating levels of inflammatory biomarkers are summarized in Table 14.4.[56–67] Subjects of these studies were healthy overweight and obese individuals. The diet interventions included very low calorie diet (VLCD) and low calorie diet (LCD). Most were short- to medium-term (3 to 24 weeks) studies. Body weight, fat mass, and waist circumferences were the common physical measures utilized. Circulating levels of CRP, SAA, IL-6, IL-1β, IL-1RA, IL-8, TNF-α, and soluble TNF receptors were the measured inflammation markers.

These weight loss studies reported that LCD or VLCD resulted in weight losses of 5 to 20%. Most studies showed that weight loss reduced levels of CRP, pro-inflammatory cytokines and their soluble receptors and increased levels of adiponectin. A few studies, including a randomized controlled trial, reported that weight loss reduced some, but not other measures of inflammation. This may be due to the relatively short intervention term (3 to 10 weeks) and small weight losses (<10%) in these studies. Indeed, a recent study showed that at least 10% weight loss is needed to achieve a significant reduction in chronic inflammation in obese individuals.[67]

Several intervention studies examined the effects of hypocaloric diet plus other interventions, mainly aerobic exercise, on measures of inflammation in obese individuals. Findings from these studies are summarized in Table 14.5.[68–79] Collectively, the results show that these interventions reduce CRP, IL-6, IL-8, IL-18, TNF-α, TNF soluble receptors, monocyte chemotactic protein 1 (MCP-1), macrophage migration inhibitory factor (MIF), and matrix metallopeptidase 9 (MMP-9), but increase adiponectin in obese healthy and diabetic individuals. Hypocaloric diet plus other interventions seem to have more consistent inflammation-lowering effects than diet alone. However, several studies reported no changes in some measures of inflammation, possibly due to a short intervention term or small amount of weight loss.

Two recent randomized controlled studies reported effects of hypocaloric diet and exercise training on inflammation and metabolic risk factors. Compared to the control group, a 2-year weight loss program by diet and exercise reduced circulating levels of CRP, IL-6, and IL-18, and increased adiponectin in obese women.[73] In addition, the intervention also reduced fasting lipids, glucose, and insulin in these women. Changes in IL-6 and adiponectin were independently related to improvement in insulin sensitivity. Similarly, compared to a control group, a 6-month diet and exercise program decreased body weight, CRP and IL-6 in obese older adults.[77] Waist circumference, lipids, glucose and blood pressure were also decreased in the intervention, but not control group. The number of subjects with metabolic syndrome decreased by 59% in the intervention group, but did not change in the control group.

The independent effect of exercise training on inflammation during dietary weight loss was investigated by a previous study of our group.[80] Six months of hypocaloric diet and diet plus aerobic exercise resulted in similar reductions in body weight, fat mass, waist circumference, and abdominal visceral and subcutaneous fat areas. Diet alone did not change aerobic fitness, while diet plus exercise increased aerobic

## TABLE 14.4
## Effects of Dietary Weight Loss Alone on Measures of Inflammation

| Study | Participants | Intervention | Effects on Body Composition | Effects on Inflammation |
|---|---|---|---|---|
| Bastard et al.[56] | 17 obese women (BMI = 40 ± 7 kg/m$^2$) | VLCD (941 kcal/day) 3 weeks | ↓ BMI by 5.5%, ↓ fat mass by 5.1% | ↓ sTNFR1 by 7.5%, = sTNFR2 |
| Bastard et al.[57] | 14 obese women (BMI = 40 ± 4 kg/m$^2$) | VLCD (941 kcal/day) 3 weeks | ↓ BMI by 5.3% ($P < 0.005$), ↓ fat mass by 8.5% ($P < 0.005$) | ↓ IL-6 by 16.5% ($P = 0.05$), = CRP, TNF-α |
| Heilbronn et al.[58] | 83 obese women (BMI = 34 ± 4 kg/m$^2$) | LCD (1360 kcal/day) 12 weeks | ↓ weight by 7.9 kg | ↓ CRP by 25.9% ($P < 0.001$) |
| Bruun et al.[59] | 34 obese individuals (BMI = 39 ± 3 kg/m$^2$) | VLCD (800 kcal/day, 8 weeks), then weight stabilization (12 weeks) | ↓ weight by 13.1% ($P < 0.001$), ↓ fat mass by 19% ($P < 0.001$) | ↑ IL-8 by 37.5% ($P < 0.05$), ↓ TNF-α by 40% ($P < 0.001$) |
| Bruun et al.[60] | 19 obese men (BMI = 39 ± 3 kg/m$^2$) | LCD (1000 to 1480 kcal/day, 16 weeks), then weight stabilization (8 weeks) | ↓ weight by 14.7% ($P < 0.001$) | ↑ IL-6 by 24.4% ($P < 0.001$), ↑ IL-8 by 29.5% ($P < 0.001$), ↓ TNF-α by 29.2% ($P < 0.001$) |
| Xydakis et al.[61] | 40 obese individuals with metabolic syndrome (BMI = 39 + 6 kg/m$^2$) | VLCD (600 to 800 kcal/day, 4 to 6 weeks) | ↓ weight by 7% ($P < 0.001$) | ↓ CRP by 14% ($P = 0.02$), = TNF-α, adiponectin |
| Arvidsson et al.[62] | 40 obese women (BMI = 37 ± 4 kg/m$^2$) | LCD (600 kcal/day deficit, 10 wks) | ↓ weight by 7.4% ($P < 0.0001$), ↓ fat mass by 12.6% ($P < 0.001$) | ↓ IL-6 by 6.2% ($P < 0.05$), = TNF-α, adiponectin |
| Salas-Salvado et al.[63] | 19 morbidly obese individuals (BMI = 48 ± 9 kg/m$^2$) | LCD ( 800 kcal/day, 6 weeks) | ↓ weight by 9% ($P < 0.001$), ↓ fat mass by 3.5% ($P < 0.05$) | ↓ CRP by 35.5% ($P < 0.05$), ↓ SAA by 56.2% ($P < 0.001$), ↓ IL-6 by 32.8% ($P < 0.05$), ↑ sTNFR2 by 21.5% ($P < 0.05$), = TNF-α, sTNFR1 |
| Shin et al.[64] | 129 overweight and obese women (MAO: n = 106, BMI = 28 ± 3 kg/m$^2$; MHO: n = 23, BMI = 27 ± 2 kg/m$^2$) | LCD (300 kcal/day deficit, 12 weeks) | MAO: ↓ weight by 3.2%<br>MHO: ↓ weight by 2.8% | MAO: ↓ CRP by 21.1% ($P < 0.05$), = IL-6<br>MHO: = CRP, = IL-6 |

## TABLE 14.4 (continued)
## Effects of Dietary Weight Loss Alone on Measures of Inflammation

| Study | Participants | Intervention | Effects on Body Composition | Effects on Inflammation |
|---|---|---|---|---|
| Bougoulia et al.[65] | 36 PM obese women (BMI = 39 ± 7 kg/m²) | LCD ( 600 kcal/day, 24 weeks) | ↓ weight by 19.2% (P <0/.001), ↓ fat mass by 18.4% (P <0.001) | ↓ CRP by 22.4% (P <0.01), ↓ IL-6 by 85.9% (P <0.001) |
| [a] de Mello et al.[66] | 34 obese subjects (10 controls: BMI = 32 ± 2 kg/m²; 24 weight reduction: BMI = 33 ± 3 kg/m²) | Control: no change in diet<br>Weight reduction: LCD (120 kcal/day deficit, 12 weeks), then weight maintenance (21 weeks) | Control: = weight and fat mass<br>Weight reduction: ↓ weight by 4.9% (P <0.05), ↓ fat mass by 8.8% (P <0.05) | Control: ↑ TNF-α by18.4% (P <0.05), = adiponectin, CRP, IL-6, IL-1β, IL-1RA<br>Weight reduction: ↑ adiponectin by 19.6% (P <0.001), ↓ CRP by 41.5% (P <0.001), ↓ IL-6 by 7.1% (P <0.001), ↑ TNF-α by 37.6% (P <0.001), = IL-1β, IL-1RA |
| Madsen et al.[67] | 93 obese subjects (mean BMI = 38 kg/m²) | VLCD (800 kcal/day, 8 weeks) | ↓ weight by 13.1% (P <0.01), ↓ waist circumference by 9.7% (P <0.01) | ↓ adiponectin by 22.3% (P <0.01), = CRP |

*BMI*: body mass index, mean ± standard deviation. CRP = C-reactive protein. IL-1RA = IL-1 receptor antagonist.

*IL-1*β = interleukin 1β. IL-6 = interleukin 6. IL-8 = interleukin 8. LCD = low calorie diet. MAO = metabolically abnormal obese. MHO = metabolically healthy obese. PM = premenopausal. SAA = serum amyloid A. sTNFR1 = soluble TNF receptor 1. sTNFR2 = soluble TNF receptor 2. TNF-α = tumor necrosis factor α. VLCD = very low calorie diet.

[a] Randomized controlled trial.

fitness. Diet alone did not change any measures of inflammation, but diet plus exercise reduced levels of CRP, IL-6, IL-6sR, and sTNFR1. In addition, reductions in IL-6sR and sTNFR1 were related to improvement in aerobic fitness. Therefore, adding exercise training to dietary weight loss is more effective than diet alone in reducing measures of inflammation. These findings further support the importance of exercise training as a component that should be included in a weight loss program.

## 14.6 MECHANISMS FOR EFFECTS OF EXERCISE AND WEIGHT LOSS ON INFLAMMATION

Two important sources are responsible for the elevated inflammatory state in the circulation. In obese individuals, adipose tissue is a major organ for releasing inflammatory cytokines.[81] The production and secretion of cytokines by adipose tissue are

**TABLE 14.5**
**Effect of Dietary Weight Loss and Other Interventions on Measures of Inflammation**

| Study | Participants | Intervention | Effects on Body Composition | Effects on Inflammation |
|---|---|---|---|---|
| Dandona et al.[68] | 38 obese women (BMI = 36 ± 6 kg/m$^2$) | LCD (925 to 1150 kcal/day) and aerobic exercise, 1 to 2 years | ↓ weight by 12.3% (P <0.0001) | ↓ TNF-α by 23.8% (P <0.02) |
| Gallistl et al.[69] | 49 obese children and adolescents (BMI = 27 ± 5 kg/m$^2$) | LCD (908 to 1194 kcal/day) and physical activities, 3 weeks | ↓ BMI by 5.2% (P <0.005), ↓ fat mass by 12.5% (P <0.005) | ↓ IL-6 by 48.7% (P <0.05) |
| Ziccardi et al.[70] | 56 obese PM women (BMI = 37 ± 2 kg/m$^2$) | LCD (1300 kcal/day), aerobic exercise, behavioral and nutritional counseling, liposuction, 1 year | ↓ BMI by 12.6% (P <0.001) | ↓ IL-6 by 46.5% (P <0.01), ↓ TNF-α by 31% (P <0.01) |
| Esposito et al.[71] | 40 obese women (BMI = 36 ± 3 kg/m$^2$) | LCD (1300 kcal/day), behavioral and nutritional counseling 1 year | ↓ BMI by 12.4% (P <0.01) | ↓ IL-18 by 40.5% (P <0.01) |
| Monzillo et al.[72] | 24 obese healthy and diabetic individuals (BMI = 37 ± 4 kg/m$^2$) | LCD (500 kcal/day deficit) and aerobic exercise, 24 weeks | ↓ weight by 7.0% (P <0.001) | ↓ IL-6 by 14.8% (P<0.05), ↓ TNF-α (P = 0.059), = CRP, sTNFR1, adiponectin |
| [a] Esposito et al.[73] | 120 obese women (60 control: BMI = 35 ± 2 kg/m$^2$; 60 intervention: BMI = 35 ± 2 kg/m$^2$) | Control: diet and exercise education Intervention: LCD (1300 to 1500 kcal/day) and aerobic exercise, 2 years | Control: ↓ weight by 3.2% (P = 0.01), ↓ WHR by 2.3% (P = 0.03) Intervention: ↓ weight by 14.7% (P <0.001), ↓ WHR by 9.3% (P <0.001) | Control: = adiponectin, CRP, IL-6, IL-18 Intervention: ↑ adiponectin by 48.2% (P = 0.02), ↓ CRP by 34.4% (P = 0.01), ↓ IL-6 by 32.6% (P = 0.01), ↓ IL-18 by 30.2% (P = 0.02) |

## TABLE 14.5
## Effect of Dietary Weight Loss and Other Interventions on Measures of Inflammation

| Study | Participants | Intervention | Effects on Body Composition | Effects on Inflammation |
|---|---|---|---|---|
| Marfella et al.[74] | 67 obese PM women (BMI = 37 ± 5 kg/m$^2$) | LCD (1300 kcal/day), aerobic exercise, behavioral and nutritional counseling, 1 year | ↓ BMI by 13.6% (P <0.001) | ↓ CRP by 44.1% (P <0.02), ↓ IL-6 by 61.9% (P <0.01), ↓ IL-18 by 29.9% (P <0.01), ↓ TNF-α by 31% (P <0.01) |
| Dvorakova-Lorenzova et al.[75] | 40 obese women (BMI = 32 ± 4 kg/m$^2$) | Dietary education and aerobic exercise, 9 weeks | ↓ weight by 7.7% (P <0.001), ↓ waist circumference by 8.3% (P <0.001) | ↓ CRP by 30% (P <0.001), = adiponectin, IL-6 |
| Bruun et al.[76] | 27 obese individuals (BMI = 46 ± 9 kg/m$^2$) | Hypocaloric diet and aerobic exercise, 15 weeks | ↓ weight by 13% (P <0.001), ↓ fat mass by 10% (P <0.001), ↓ waist circumference by 7.3% (P <0.001) | ↑ adiponectin by 32.7% (P <0.001), ↓ CRP by 28.6% (P <0.05), ↓ IL-6 by 26.1% (P <0.01), ↓ IL-8 by 14.3% (P <0.05), ↓ MCP-1 by 13.6% (P <0.01),= TNF-α |
| *Villareal et al.[77] | 27 obese older individuals (10 control: BMI = 39 ± 5 kg/m$^2$; 17 intervention: BMI = 39 ± 5 kg/m$^2$) | Control: no changes in diet and exercise Intervention: LCD (750 kcal/day deficit), aerobic, resistance, balance, flexibility exercises, 6 months | Control: = weight and waist Intervention: ↓ weight by 8.4% (P <0.001), ↓ waist by 8.7% (P <0.05) | Control: = CRP, IL-6 Intervention: ↓ CRP by 42.6% (P <0.05), ↓ IL-6 by 50% (P = 0.05) |

**TABLE 14.5**
**Effect of Dietary Weight Loss and Other Interventions on Measures of Inflammation**

| Study | Participants | Intervention | Effects on Body Composition | Effects on Inflammation |
|---|---|---|---|---|
| Bruun et al.[78] | 23 obese individuals (BMI = 46 ± 9 kg/m²) | Hypocaloric diet and aerobic exercise, 15 weeks | ↓ weight by 12.7% (P <0.001), ↓ fat mass by 10% (P <0.001), | ↓ waist circumference by 7.3% (P <0.001) ↓ IL-18 by 22% (P <0.001) |
| Sheu et al.[79] | 23 obese women (BMI = 33 ± 6 kg/m²) | LCD (500 to 1000 kcal/day deficit) and light exercise, 12 weeks | ↓ weight by 5.1% (P <0.05), ↓ waist circumference by 5.5% (P <0.05) | ↑ adiponectin by 33.9% (P <0.05), ↓ CRP by 48.9% (P <0.05), ↓ MIF by 66.6% (P <0.05), = IL-6, TNF-α, MMP-9 |

*BMI*: body mass index, mean ± standard deviation. CRP = C-reactive protein. IL-18 = interleukin 18. IL-6 = interleukin 6. IL-8 = interleukin 8. LCD = low calorie diet. MCP-1 = monocyte chemotactic protein-1. MIF = macrophage migration inhibitory factor. MMP-9 = matrix metallopeptidase 9.PM = premenopausal. sTNFR1 = soluble TNF receptor 1. TNF-α = tumor necrosis factor α.

[a] Randomized controlled trial.

elevated with adiposity.[82] Peripheral blood mononuclear cells release large amounts of inflammatory cytokines, and cytokine expression in these cells is elevated with obesity.[83]

Exercise training alone does not influence adipose tissue cytokine expression or release.[28,32] However, hypocaloric diet plus exercise is effective in reducing adipose tissue cytokine gene expression.[76,79] These observations suggest that weight loss may be needed for reductions in adipose tissue cytokine levels. Exercise training alone reduces pro-inflammatory cytokines and increases anti-inflammatory cytokines released by peripheral mononuclear cells in individuals with elevated risk for CVD, including obese individuals.[84] Mechanisms through which exercise training modifies mononuclear cell cytokine production are still unknown. However, muscle-derived cytokines, such as IL-6, may play a role.[85] When exercise is performed regularly, chronic adaptations in the immune system cause lower cytokine release from immune cells such as mononuclear cells.

Dietary weight loss alters adipose tissue and mononuclear cell cytokine production in obese individuals. Weight loss reduced pro-inflammatory cytokine and increased anti-inflammatory cytokine production in adipose tissue, and changes of these measures in adipose tissue were similar to changes in the circulation.[62] A recent study reported that diet plus exercise lowered pro-inflammatory cytokines and increased anti-inflammatory cytokine expression in peripheral mononuclear cells.[79] Intranuclear binding of NF-κB, a transcription factor that regulates pro-inflammatory genes, was reduced by the intervention. However, whether weight loss alone changes cytokine production by peripheral mononuclear cells is unclear. More studies are needed to clarify the independent effect of weight loss and the underlying mechanism through which it lowers inflammation.

## 14.7 SUMMARY

Chronic inflammation is a novel risk factor for several clinical diseases. Increasing evidence indicates that inflammation levels are related to levels of physical activity and adiposity. Treatment of chronic inflammation by exercise training and dietary weight loss has shown encouraging results. Exercise term is critical for training to produce positive effects on inflammation. However, mixed results were reported regarding the effects of exercise intensity, duration, frequency, and volume on inflammation. Current evidence confirms that a hypocaloric diet or diet plus exercise training may reduce chronic inflammation. While the amount of weight loss is important, exercise may have additional inflammation-reducing effects. Future studies are needed to study the dose–response effects of exercise and hypocaloric diet on inflammation, and further investigate the cellular mechanisms by which exercise and weight loss reduce chronic, systemic inflammation.

## ACKNOWLEDGMENTS

The authors are supported by faculty start-up funds from the School of Public Health and Health Professions, State University of New York at Buffalo.

## REFERENCES

1. Pearson TA et al. Markers of inflammation and cardiovascular disease: application to clinical and public health practice. *Circulation* 2003; 107: 499–511.
2. Albert MA, Glynn RJ, and Ridker PM. Effect of physical activity on serum C-reactive protein. *Am J Cardiol* 2004; 93: 221–225.
3. Aronson D et al. The association between cardiorespiratory fitness and C-reactive protein in subjects with the metabolic syndrome. *J Am Coll Cardiol* 2004; 44: 2003–2007.
4. Aronson D et al. C-Reactive protein is inversely related to physical fitness in middle-aged subjects. *Atherosclerosis* 2004; 176:173–179.
5. Borodulin K et al. Associations of leisure time physical activity, self-rated physical fitness, and estimated aerobic fitness with serum C-reactive protein among 3,803 adults. *Atherosclerosis* 2006; 185: 381–387.
6. Church TS et al. Associations between cardiorespiratory fitness and C-reactive protein in men. *Arterioscler Thromb Vasc Biol* 2002; 22: 1869–1876.
7. Dufaux B et al. C-reactive protein serum concentrations in well-trained athletes. *Int J Sports Med* 1984; 5:102–106.
8. Geffken DF et al. Association between physical activity and markers of inflammation in a healthy elderly population. *Am J Epidemiol* 2001; 153: 242–250.
9. LaMonte MJ et al. Cardiorespiratory fitness and C-reactive protein among a tri-ethnic sample of women. *Circulation* 2002; 106: 403–406.

10. McGavock JM et al. Low cardiorespiratory fitness is associated with elevated C-reactive protein levels in women with type 2 diabetes. *Diabetes Care* 2004; 27: 320–325.
11. Mora S et al. Association of physical activity and body mass index with novel and traditional cardiovascular biomarkers in women. *JAMA* 2006; 295: 1412–1419.
12. Pischon T et al. Leisure-time physical activity and reduced plasma levels of obesity-related inflammatory markers. *Obes Res* 2003; 11: 1055–1064.
13. Stauffer BL et al. Plasma C-reactive protein is not elevated in physically active post-menopausal women taking hormone replacement therapy. *J Appl Physiol* 2004; 96: 143–148.
14. Wannamethee SG et al. Physical activity and hemostatic and inflammatory variables in elderly men. *Circulation* 2002; 105: 1785–1790.
15. Williams MJ et al. C-reactive protein and cardiorespiratory fitness in young adults. *Eur J Cardiovasc Prev Rehabil* 2005; 12: 216–220.
16. LaMonte MJ, Ainsworth BE, and Durstine JL. Influence of cardiorespiratory fitness on the association between C-reactive protein and metabolic syndrome prevalence in racially diverse women. *J Womens Health* 2005; 14: 233–239.
17. Boekholdt SM et al. Physical activity, C-reactive protein levels and the risk of future coronary artery disease in apparently healthy men and women: the EPIC-Norfolk prospective population study. *Eur J Cardiovasc Prev Rehabil* 2006; 13: 970–976.
18. Rothenbacher D et al. Physical activity, coronary heart disease, and inflammatory response. *Arch Intern Med* 2003; 163: 1200–1205.
19. Mora S et al. Physical activity and reduced risk of cardiovascular events: potential mediating mechanisms. *Circulation* 2007; 116: 2110–2118.
20. Manns PJ et al. Physical activity, body fat, and serum C-reactive protein in postmenopausal women with and without hormone replacement. *Am J Hum Biol* 2003; 15: 91–100.
21. Rawson ES et al.Body mass index, but not physical activity, is associated with C-reactive protein. *Med Sci Sports Exerc* 2003; 35: 1160–1166.
22. Taaffe DR et al. Bone mineral density in older non-Hispanic, Caucasian, and Mexican-American women: relationship to lean and fat mass. *Ann Hum Biol* 2000; 27: 331–344.
23. Verdaet D et al. Association between leisure time physical activity and markers of chronic inflammation related to coronary heart disease. *Atherosclerosis* 2004; 176: 303–310.
24. Mattusch F et al. Reduction of the plasma concentration of C-reactive protein following nine months of endurance training. *Int J Sports Med* 2000; 21: 21–24.
25. Tsukui S et al. Moderate-intensity regular exercise decreases serum tumor necrosis factor-α and HbA1c levels in healthy women. *Int J Obes Relat Metab Disord* 2000; 24: 1207–1211.
26. Straczkowski M et al. Changes in tumor necrosis factor-α system and insulin sensitivity during an exercise training program in obese women with normal and impaired glucose tolerance. *Eur J Endocrinol* 2001; 145: 273–280.
27. Nassis GP et al. Aerobic exercise training improves insulin sensitivity without changes in body weight, body fat, adiponectin, and inflammatory markers in overweight and obese girls. *Metabolism* 2005; 54: 1472–1479.
28. Polak J, Klimcakova E, Moro C, et al. Effect of aerobic training on plasma levels and subcutaneous abdominal adipose tissue gene expression of adiponectin, leptin, interleukin 6, and tumor necrosis factor α in obese women. *Metabolism* 2006; 55: 1375–1381.
29. Kondo T, Kobayashi I, and Murakami M. Effect of exercise on circulating adipokine levels in obese young women. *Endocr J* 2006; 53: 189–195.
30. Stewart LK et al. The influence of exercise training on inflammatory cytokines and C-reactive protein. *Med Sci Sports Exerc* 2007; 39: 1714–1719.

31. Dekker MJ et al. An exercise intervention without weight loss decreases circulating interleukin-6 in lean and obese men with and without type 2 diabetes mellitus. *Metabolism* 2007; 56: 332–338.
32. Leick L et al. Adipose tissue interleukin-18 mRNA and plasma interleukin-18: effect of obesity and exercise. *Obesity* 2007; 15: 356–363.
33. Olson TP et al. Changes in inflammatory biomarkers following one-year of moderate resistance training in overweight women. *Int J Obes* 2007; 31: 996–1003.
34. Devries MC et al. Endurance training without weight loss lowers systemic, but not muscle, oxidative stress with no effect on inflammation in lean and obese women. *Free Radic Biol Med* 2008; 45: 503–511.
35. Tisi PV et al. Exercise training for intermittent claudication: does it adversely affect biochemical markers of the exercise-induced inflammatory response? *Eur J Vasc Endovasc Surg* 1997; 14: 344–350.
36. Larsen AI et al. Effect of aerobic exercise training on plasma levels of tumor necrosis factor α in patients with heart failure. *Am J Cardiol* 2001; 88: 805–808.
37. Conraads VM et al. Combined endurance/resistance training reduces plasma TNF-α receptor levels in patients with chronic heart failure and coronary artery disease. *Eur Heart J* 2002; 23: 1854–1860.
38. Milani RV, Lavie CJ, and Mehra MR. Reduction in C-reactive protein through cardiac rehabilitation and exercise training. *J Am Coll Cardiol* 2004; 43: 1056–1061.
39. Zoppini G et al. Effects of moderate-intensity exercise training on plasma biomarkers of inflammation and endothelial dysfunction in older patients with type 2 diabetes. *Nutr Metab Cardiovasc Dis* 2006; 16: 543–549.
40. Brooks N et al. Strength training improves muscle quality and insulin sensitivity in Hispanic older adults with type 2 diabetes. *Int J Med Sci* 2006; 4: 19–27.
41. Oberbach A et al. Effect of a 4-week physical training program on plasma concentrations of inflammatory markers in patients with abnormal glucose tolerance. *Eur J Endocrinol* 2006; 154: 577–585.
42. Walther C et al. Regular exercise training compared with percutaneous intervention leads to a reduction of inflammatory markers and cardiovascular events in patients with coronary artery disease. *Eur J Cardiovasc Prev Rehabil* 2008; 15: 107–112.
43. Lemieux I et al. Elevated C-reactive protein: another component of the atherothrombotic profile of abdominal obesity. *Arterioscler Thromb Vasc Biol* 2001; 21: 961–967.
44. Ford ES. Body mass index, diabetes, and C-reactive protein among U.S. adults. *Diabetes Care* 1999; 22: 1971–1977.
45. Panagiotakos DB et al. The implication of obesity and central fat on markers of chronic inflammation: the ATTICA study. *Atherosclerosis* 2005; 183: 308–315.
46. Pou KM et al. Visceral and subcutaneous adipose tissue volumes are cross-sectionally related to markers of inflammation and oxidative stress: the Framingham Heart Study. *Circulation* 2007; 116: 1234–1241.
47. Rexrode KM et al. Relationship of total and abdominal adiposity with CRP and IL-6 in women. *Ann Epidemiol* 2003; 13: 674–682.
48. Schrager MA et al. Sarcopenic obesity and inflammation in the InCHIANTI study. *J Appl Physiol* 2007; 102: 919–925.
49. Thorand B et al. Sex differences in the relation of body composition to markers of inflammation. *Atherosclerosis* 2006; 184: 216–224.
50. Barinas-Mitchell E et al. Serum levels of C-reactive protein are associated with obesity, weight gain, and hormone replacement therapy in healthy postmenopausal women. *Am J Epidemiol* 2001; 153: 1094–1101.
51. Levitzky YS et al. Relation of smoking status to a panel of inflammatory markers: the Framingham offspring. *Atherosclerosis* 2008.

52. Hak AE et al. Association of C-reactive protein with measures of obesity, insulin resistance, and subclinical atherosclerosis in healthy, middle-aged women. *Arterioscler Thromb Vasc Biol* 1999; 19: 1986–1991.
53. Aronson D et al. Obesity is the major determinant of elevated C-reactive protein in subjects with the metabolic syndrome. *Int J Obes Relat Metab Disord* 2004; 28: 674–679.
54. Pradhan AD et al. C-reactive protein, interleukin 6, and risk of developing type 2 diabetes mellitus. *JAMA* 2001; 286: 327–334.
55. Pradhan AD et al. Inflammatory biomarkers, hormone replacement therapy, and incident coronary heart disease: prospective analysis from the Women's Health Initiative observational study. *JAMA* 2002; 288: 980–987.
56. Bastard JP et al. Variations in plasma soluble tumour necrosis factor receptors after diet-induced weight loss in obesity. *Diabetes Obes Metab* 2000; 2: 323–325.
57. Bastard JP et al. Elevated levels of interleukin 6 are reduced in serum and subcutaneous adipose tissue of obese women after weight loss. *J Clin Endocrinol Metab* 2000; 85: 3338–3342.
58. Heilbronn LK, Noakes M, and Clifton PM. Energy restriction and weight loss on very-low-fat diets reduce C-reactive protein concentrations in obese, healthy women. *Arterioscler Thromb Vasc Biol* 2001; 21: 968–970.
59. Bruun JM et al. Opposite regulation of interleukin-8 and tumor necrosis factor-α by weight loss. *Obes Res* 2002; 10: 499–506.
60. Bruun JM et al. Association between measures of insulin sensitivity and circulating levels of interleukin-8, interleukin-6 and tumor necrosis factor-alpha. Effect of weight loss in obese men. *Eur J Endocrinol* 2003; 148: 535–542.
61. Xydakis AM et al. Adiponectin, inflammation, and the expression of the metabolic syndrome in obese individuals: the impact of rapid weight loss through caloric restriction. *J Clin Endocrinol Metab* 2004; 89: 2697–2703.
62. Arvidsson E, Viguerie N, Andersson I, et al. Effects of different hypocaloric diets on protein secretion from adipose tissue of obese women. *Diabetes* 2004; 53: 1966–1971.
63. Salas-Salvadó J et al. Subcutaneous adipose tissue cytokine production is not responsible for the restoration of systemic inflammation markers during weight loss. *Int J Obes* 2006; 30: 1714–1720.
64. Shin MJ et al. Weight loss effect on inflammation and LDL oxidation in metabolically healthy but obese (MHO) individuals: low inflammation and LDL oxidation in MHO women. *Int J Obes* 2006; 30: 1529–1534.
65. Bougoulia M, Triantos A, and Koliakos G. Plasma interleukin-6 levels, glutathione peroxidase and isoprostane in obese women before and after weight loss: association with cardiovascular risk factors. *Hormones* 2006; 5: 192–199.
66. de Mello VD et al. Effect of weight loss on cytokine messenger RNA expression in peripheral blood mononuclear cells of obese subjects with the metabolic syndrome. *Metabolism* 2008; 57: 192–199.
67. Madsen EL et al. Weight loss larger than 10% is needed for general improvement of levels of circulating adiponectin and markers of inflammation in obese subjects: a 3-year weight loss study. *Eur J Endocrinol* 2008; 158: 179–187.
68. Dandona P et al. Tumor necrosis factor-α in sera of obese patients: fall with weight loss. *J Clin Endocrinol Metab* 1998; 83: 2907–2910.
69. Gallistl S et al. Changes in serum interleukin-6 concentrations in obese children and adolescents during a weight reduction program. *Int J Obes Relat Metab Disord* 2001; 25: 1640–1643.
70. Ziccardi P et al. Reduction of inflammatory cytokine concentrations and improvement of endothelial functions in obese women after weight loss over one year. *Circulation* 2002; 105: 804–809.

71. Esposito K et al. Weight loss reduces interleukin-18 levels in obese women. *J Clin Endocrinol Metab* 2002; 87: 3864–3866.
72. Monzillo LU et al. Effect of lifestyle modification on adipokine levels in obese subjects with insulin resistance. *Obes Res* 2003; 11: 1048–1054.
73. Esposito K et al. Effect of weight loss and lifestyle changes on vascular inflammatory markers in obese women: a randomized trial. *JAMA* 2003; 289: 1799–1804.
74. Marfella R et al. Effect of weight loss on cardiac synchronization and pro-inflammatory cytokines in premenopausal obese women. *Diabetes Care* 2004; 27: 47–52.
75. Dvoráková-Lorenzová A et al. The decrease in C-reactive protein concentration after diet and physical activity induced weight reduction is associated with changes in plasma lipids, but not interleukin-6 or adiponectin. *Metabolism* 2006; 55: 359–365.
76. Bruun JM et al. Diet and exercise reduce low-grade inflammation and macrophage infiltration in adipose tissue but not in skeletal muscle in severely obese subjects. *Am J Physiol Endocrinol Metab* 2006; 290: E961–E967.
77. Villareal DT et al. Effect of lifestyle intervention on metabolic coronary heart disease risk factors in obese older adults. *Am J Clin Nutr* 2006; 84: 1317–1323.
78. Bruun JM et al. Interleukin-18 in plasma and adipose tissue: effects of obesity, insulin resistance, and weight loss. *Eur J Endocrinol* 2007; 157: 465–471.
79. Sheu WH et al. Effect of weight loss on proinflammatory state of mononuclear cells in obese women. *Obesity* 2008; 16: 1033–1038.
80. You T et al. Effects of hypocaloric diet and exercise training on inflammation and adipocyte lipolysis in obese postmenopausal women. *J Clin Endocrinol Metab* 2004; 89: 1739–1746.
81. Kershaw EE and Flier JS. Adipose tissue as an endocrine organ. *J Clin Endocrinol Metab* 2004; 89: 2548–2556.
82. Mohamed-Ali V et al. Production of soluble tumor necrosis factor receptors by human subcutaneous adipose tissue in vivo. *Am J Physiol* 1999; 277: E971–E975.
83. Ghanim H et al. Circulating mononuclear cells in the obese are in a pro-inflammatory state. *Circulation* 2004; 110: 1564–1571.
84. Smith JK et al. Long-term exercise and atherogenic activity of blood mononuclear cells in persons at risk of developing ischemic heart disease. *JAMA* 1999; 281: 1722–1727.
85. Petersen AM and Pedersen BK. The anti-inflammatory effect of exercise. *J Appl Physiol* 2005; 98: 1154–1162.

# 15 Adipose Tissue and Anti-Inflammatory Pharmacotherapy

*Peter G. Bradford and Atif B. Awad*

## CONTENTS



## 15.1 INTRODUCTION

The hypothesis that low grade chronic systemic inflammation involving adipocyte-derived factors links obesity to clinical metabolic disorders is attracting growing support. Observational and experimental models of obesity have revealed that significant amounts of pro-inflammatory mediators are synthesized and secreted by adipose tissue. It is hypothesized that the release of pro-inflammatory adipokines and the cessation of synthesis of normally anti-inflammatory mediators may be caused by adipocyte hypertrophy or hyperplasia that promotes local hypoxia and cellular stress (de Luca and Olefsky, 2008).

Part of this process may involve macrophage recruitment, conversion of macrophages from tissue remodeling M2 types to classically activated tissue-destructive M1 types, and recruitment of inflammation-primed monocytes into adipose tissue (de Luca and Olefsky, 2008; Shah et al., 2008). Thus, it is important to identify therapeutics, both drugs and pharmaconutrients that may suppress adipose tissue inflammation and ultimately serve to alleviate clinical metabolic distress.

This chapter reviews marketed drugs, drugs in development, and pharmaconutrients that affect the synthesis and secretion of adipokines and the actions of these adipokines on systemic inflammation. Anti-inflammatory actions include the suppression of adipose tissue differentiation. The differentiation processes potentially affected by drugs and nutrients include the cessation of growth of proliferating preadipocytes. Drugs and pharmaconutrients covered in this chapter include leptin, adiponectin, peroxisome proliferator-activator receptor (PPAR) therapeutics, tumor necrosis factor (TNF)-α blockers, resistin, vitamin D, vitamin A, and polyunsaturated fatty acids (PUFAs). See Table 15.1.

## 15.2 LEPTIN: THERAPEUTIC AGENT

Leptin is one of the principal hormones secreted by adipose tissue. Leptin is the product of the *ob* gene and it encodes a 16-kDa protein identified in serum that acts on leptin receptors in the hypothalamus to reduce food intake and increase energy expenditure. Besides these actions, leptin exerts pleiotropic immune modulating

**TABLE 15.1**
**Adipose Tissue-Targeted Anti-inflammatory Therapeutics**

| Adipose Target | Therapeutic Class | Status | Application |
|---|---|---|---|
| Leptin | Leptin receptor chimeras | Investigational | Multiple sclerosis |
| | Leptin receptor antagonists | Investigational | Autoimmune diseases |
| | Anti-leptin antibodies | Investigational | Multiple sclerosis, type 1 diabetes |
| | Leptin peptide antagonists | Investigational | Angiogenesis inhibitor |
| | Recombinant leptin | FDA Phase 2B | Obesity |
| Adiponectin | Globular adiponectin domain | Investigational | Anti-inflammatory |
| | Osmotin analogs | Investigational | Insulin resistance, metabolic syndrome |
| PPARα | Fibrate drugs | FDA-approved | Hyperlipidemia |
| PPARγ | Thiazolidinediones | FDA-approved | Antidiabetic |
| PPARβ/δ | PPARβ and δ agonists | Investigational | Anti-inflammatory, metabolic syndrome |
| TNF-α | TNF-α blockers | FDA-approved | Anti-inflammatory |
| Resistin | Anti-resistin antibodies | Investigational | Anti-inflammatory |
| Vitamins A, D, and PUFAs | Essential nutrients | Commercial | Cardioprotective |

*Note:* The adipose tissue-associated targets of current and potential pharmacotherapy and nutritional modification listed above have application in chronic inflammatory diseases.

activities (Matarese et al., 2005, 2008; Härle and Straub, 2006; Fantuzzi, 2006; Li et al., 2006). Obesity promotes high leptin levels and this is associated with adipose tissue infiltration by pro-inflammatory macrophages that secrete cytokines including TNF-α, IL-6, and IL-1β. This hyperleptinemia is also associated with increased levels of pro-inflammatory T helper cell type 1 (Th1) cytokines (IFN-γ, IL-2, TNF-α) that themselves are elevated in CNS inflammatory lesions of experimental autoimmune encephalomyelitis (EAE), an animal model for MS, and other autoimmune phenotypes. Leptin-deficient mice are resistant to induction of EAE and this protection is reversed by leptin administration (Matarese et al., 2008).

Leptin antagonists are under investigation as immune modulators, especially in autoimmune diseases (Gertler, 2006; Materese et al., 2008). Chimeric glycoproteins containing the extracellular binding domain of the leptin receptor and the Fc portion of immunoglobulin have been shown to neutralize leptin in mice with EAE (DeRosa et al., 2006). The clinical course and progression of disease were ameliorated with improved clinical scores, reduced disease relapses, and a switch to a Th2 regulatory cytokine profile.

Other pharmacologic approaches to antagonizing endogenous leptin action are in development and include the generation of leptin point mutations that act as leptin antagonists with no agonistic activity (Salomon et al., 2006), the use of small leptin peptide antagonists (Gonzalez et al., 2006), and the preparation of anti-leptin receptor neutralizing antibodies (Fazeli et al., 2006). Each approach has shown promise in various in vitro and in vivo bioassays; but none has yet been submitted to the U.S. Food & Drug Administration as an investigational new drug or via a new drug application.

Recombinant leptin is currently in trials for the treatment of lipodystrophy (Oral et al., 2002; Ebihara et al. 2007). Physiologic replacement dosing with twice-daily injections improved glucose and triglyceride levels and reduced insulin resistance in a study demonstrating its efficacy and safety (Ebihara et al. 2007). In November 2007, Amylin Pharmaceuticals announced positive results of a 24-week proof-of-concept phase 2B clinical study including approximately 200 overweight and obese subjects to evaluate dosing combinations of pramlintide, an analog of the natural amylin hormone amylin and recombinant human leptin in the treatment of obesity (www.amylin.com/pipeline/pramlintide.cfm, 2007). The pramlintide and metreleptin treatment significantly reduced body weight on average by 12.7%.

## 15.3 ADIPONECTIN: POTENTIAL THERAPEUTIC AGENT

Adiponectin has attracted attention as a therapeutic target because of its anti-diabetic and anti-atherogenic activities. Adiponectin is an adipose tissue-derived factor associated with the maintenance of insulin sensitivity, the control of endothelial activation, and the limitation of other aspects of inflammation (Kadowaki and Yamauchi, 2005; Teoh et al. 2008; Guerre-Millo, 2008). Adiponectin is a 244-amino acid polypeptide that circulates in the blood at levels relatively high compared to many hormones. Full-length and globular adiponectins bind to cell surface receptors AdipoR1 and AdipoR2 that are expressed in muscle, liver, and other tissues. When activated the receptors stimulate downstream AMP-activated protein kinases, PPARα, and p38

MAP kinase. Knockout studies confirm that AdipoR1 and AdipoR2 play important roles in the regulation of insulin sensitivity in vivo (Kadowaki et al. 2008).

Adiponectin acts to suppress inflammation and counter the metabolic disorders associated with type 2 diabetes, atherosclerosis, and obesity. Polymorphisms of the adiponectin gene that decrease adiponectin levels are associated with insulin resistance and susceptibility to diabetes (Kadowaki and Yamauchi, 2005). Also, reductions of plasma adiponectin levels are observed in states of insulin resistance, visceral adiposity, and the associated metabolic syndrome. In adiponectin null mice, neointimal endovascular proliferation can be attenuated by restoring adiponectin levels to normal. Thus, normal circulating levels of adiponectin may serve to maintain insulin sensitivity of tissues and a basal non-inflammatory state in the vascular system.

Potential adiponectin therapeutic agents include the globular adiponectin domain, the naturally occurring 145-amino acid C-terminus of the protein. The globular adiponectin domain may prove easier to synthesize and administer. In experimental systems, subcutaneous injections of recombinant globular adiponectin domain to adiponectin-deficient mice reduced leukocyte–endothelial interactions and, when administered to wild-type mice, protected against TNF-α-induced leukocyte endothelial interaction (Ouedraogo et al., 2007). In in vitro systems, globular adiponectin domain suppressed superoxide generation and enhanced eNOS activity in endothelial cells treated with oxidized LDL (Motoshima et al., 2004).

Osmotin, a plant anti-fungal protein, has a structure similar to the globular adiponectin domain and has been shown to be a ligand for the yeast homologue of AdipoR (Narasimhan et al., 2005). The yeast system may provide a background for the efficient screening of chemical, naturally occurring, and mutationally generated potential adiponectin receptor agonists.

## 15.4 PPAR THERAPEUTIC AGENTS

PPAR α, β/ δ, and γ are either current or proposed targets for adipose tissue-centered anti-inflammatory pharmacotherapies. PPARs constitute a family of nuclear receptors that heterodimerize with 9-cis-retinoic acid retinoid X receptors to regulate transcription of target genes. The focus of determining PPAR activity in adipose tissue is to identify specific targeted therapies.

### 15.4.1 PPARα Therapeutics: Fibrate Drugs

PPARα is the receptor for fibrate drugs as well as the endogenous fatty acid regulators docosahexaenoic acid, arachidonic acid, and linoleic acid (Moraes et al., 2006). Fibrate drugs are hypolipidemic agents that lower serum triglycerides predominantly by decreasing very low density lipoproteins; however, these drugs also affect adipocyte differentiation and secretion of inflammatory cytokines from adipocytes. In adipocytes, PPARα is an intracellular receptor that modulates carbohydrate and fat metabolism as well as cellular differentiation. The most widely used fibrate drugs include fenofibrate (Tricor), gemfibrazol (Lopid), and clofibrate (Atromid-S). Fenofibrate has been shown to reduce significantly the release of the inflammatory mediator TNF-α from adipocytes in hypercholesterolemic animal studies and this activity correlated with weight loss (Zhao and Wu, 2004). Fibrates as a class also enhance adiponectin levels, partly through activation of adipose PPARα (Hiugi et al., 2007).

Large clinical studies including the Bezafibrate Infarction Prevention (BIP) study and the Fenofibrate Intervention and Event Lowering in Diabetes (FIELD) study supported the positive effects of fibrates in significantly reducing cardiovascular events, myocardial infarctions, and reducing new onset diabetes (BIP Study Group, 2000; FIELD Study Investigators, 2005). Subanalysis of the BIP data showed significantly increased adiponectin levels (10%) in the fibrate group over a 2-year follow-up and significantly reduced development of new diabetes cases in the highest tercile of adiponectin during the 6-year follow-up (Hiugi et al., 2007). In cultured 3T3-L1 adipocytes, benzafibrate and fenofibrate treatment increased adiponectin promoter activity and levels of adiponectin mRNA and protein. Plasma adiponectin levels were increased by bezafibrate treatment in wild-type but not PPARα-knockout mice. Thus, fibrate drugs may reduce inflammatory cytokine release from adipocytes while enhancing adiponectin secretion through PPARα-mediated effects in adipose tissue.

### 15.4.2 PPARγ Therapeutics: Thiazolidinedione Drugs

PPARγ is highly expressed in adipose tissue and it is a target for thiazolidinedione drugs that include pioglitazone (Actos) and rosiglitazone (Avandia). Both Actos and Avandia are used widely as antidiabetic drugs (Yki-Järvinen, 2004; Nissen and Wolski, 2007). PPARγ was first identified as a mediator of adipocyte differentiation and nutrient metabolism (Spiegelman et al., 1997). However, the role of adipose PPARγ in mediating the actions of thiazoloidinedione drugs is not clear. For example, the targeted deletion of PPARγ in adipose tissue of mice does not induce insulin resistance in muscle, whereas muscle-specific PPARγ deletion does cause such resistance (Yki-Järvinen, 2004).

On the other hand, loss of subcutaneous fatty tissue accompanies PPARγ deletion and the rosiglitazone antidiabetic can reverse hypertriglyceridemia, hyperglycemia, and hyperinsulinemia in normal mice, although it is ineffective in lipoatrophic mice. PPARγ ligands favor adipocyte uptake of circulating fatty acids and alter the secretion of adipocyte-derived factors. As part of the overall anti-inflammatory activity, PPARγ agonists repress pro-inflammatory genes like iNOS, TNF-α, and MMP in macrophages (Lehrke and Lazar, 2005). In addition, PPARγ ligands appear to be particularly beneficial in inflammatory bowel disease, reducing mucosal damage, ulceration, TNF-α secretion and an overall downregulation of the inflammatory response in animal models of colitis (Rogler, 2006). In addition, the forced overexpression of PPARγ in mucosal epithelial cells was associated with reduced experimental inflammation.

### 15.4.3 PPARβ and δ Therapeutics: Experimental Agents

PPARδ is widely expressed and studies of receptor knockout models revealed roles for PPARδ in developing adipose tissue mass and modulating skin inflammatory responses (Barish et al., 2006). Based on ligand screening and crystallographic analyses, PPARδ has been shown to bind long chain polyunsaturated fatty acids (docosahexaenoic, arachidonic, and linoleic acids). These studies suggest that PPARδ may be a key regulator and therapeutic target for the metabolic syndrome. Administration of the GW501616 experimental PPARδ-selective agonist to genetically obese (*db/db*) mice reduced triglyceride accumulation (Wang et al., 2003). In addition, experimental overexpression of PPARδ in transgenic mice indicated that PPARδ promotes reduced body weight, deceased fat accumulation in adipocytes, and reduced circulating free fatty acids (Wang et al., 2003).

Studies of PPARβ null mice showed a requirement for PPARβ for maximal adipocyte differentiation and that PPARβ potentiated the actions of PPARδ in this effect (Matsusue et al., 2004). In 3T3-L1 adipocytes, the GW501616 PPARβ/δ-selective agonist blocked production of the IL-6 inflammatory cytokine. Furthermore, PPARβ/δ expression was reduced in Zucker diabetic fatty rats while expression of the inflammatory IL-6 mediator was enhanced (Rodriguez-Calvo et al., 2008). These observations are consistent with the hypothesis that PPARδ protects against inflammation and consequent obesity; and that PPARδ pharmacologic agonists such as GW501616 may provide anti-inflammatory therapeutic outcomes. Preclinical studies with selective PPARβ and δ agonists have indicated that these agents possess anti-inflammatory activities and that a principal application will be modulating the inflammatory response (Kilgore and Billin, 2008).

## 15.5 TNF-α BLOCKERS

Tumor necrosis factor-α (TNF-α) is a pro-inflammatory cytokine produced by adipose tissue and evidence indicates that it plays a significant role in the metabolic dysfunctions associated with obesity (Cawthorn and Sethi, 2008). In obesity-related type 2 diabetes, the membrane and soluble forms of TNF-α in adipose tissue are increased which, in turn, affect local adipose function and systemic energy metabolism (Xu et al., 2002). The source of this TNF-α is not clear but certainly includes adipose tissue macrophages, whose density increases in obesity, as well as non-macrophage adipose cells (Weisberg et al., 2003; De Taeye et al., 2007).

Locally derived TNF-α affects the functioning of resident adipocytes in mediation of glucose and fatty acid uptake and the secretion of insulin-sensitizing adipokines (Cawthorn and Sethi, 2008). Thus, TNF-α is a potential therapeutic target in reducing the chronic, low grade systemic inflammation associated with insulin resistance, type 2 diabetes, and dyslipidemia.

Several anti-TNF-α therapeutic agents are currently FDA-approved for the treatment of various inflammatory diseases including rheumatoid arthritis, Crohn's disease, juvenile idiopathic arthritis, psoriatic arthritis, plaque psoriasis, and ankylosing spondylitis. These drugs are classified as TNF-α blockers and include etanercept (Enbrel), infliximab (Remicade), adalimumab (Humira), and certolizumab pegol (Cimzia). All exhibit specific high affinity binding activity for TNF-α and effectively neutralize circulating TNF-α.

Etanercept (Enbrel) is a fusion protein consisting of the extracellular ligand-binding portion of the human p75 TNF-α receptor linked to the Fc portion of human IgG1. Infliximab (Remicade) is a chimeric IgG1κ monoclonal antibody composed of the human Fc portion of IgG1 and the murine variable region that high specific binding affinity for human TNF-α. Adalimumab (Humira) is a fully humanized recombinant IgG1 monoclonal antibody with human-derived heavy and light chain variable regions specific for human TNF-α which is fused to human IgG1κ-constant regions. Certolizumab pegol (Cimzia) is a recombinant humanized Fab' fragment specific for TNF-α which is conjugated to a large polyethylene glycol polymer.

Is there a potential therapeutic role for TNF-α blockers in obesity-associated inflammation and insulin resistance? Studies in mice showed that targeted disruption of the TNF-α gene results in significantly improved insulin sensitivity in both diet-induced obesity and obesity associated with the *ob/ob* genetic model (Ventre et al., 1997; Uysal et al., 1997). However, initial studies with morbidly obese human subjects failed to show any effect of infliximab therapy on levels of TNF-α or GLUT4 mRNA in adipose tissue, perhaps attributable to the low drug distribution of the monoclonal antibody into adipose tissue (Di Rocco et al., 2004).

In a 3-month study, etanercept treatment of rheumatoid arthritis patients showed increased adiponectin levels (Lewicki et al., 2008a). Significantly, long-term therapy in patients with autoimmune disorders allowed closer study of the effects of TNF-α blockers on lipid profiles and insulin sensitivity (Huvers et al., 2007; Popa et al., 2007; Garcês et al., 2008; Gonzalez-Gay et al., 2008; Lewicki et al., 2008b). Surprisingly, long-term therapy with infliximab in patients with rheumatoid arthritis led to a pro-atherogenic pattern of plasma lipids (Popa et al., 2007; Lewicki et al., 2008). However, other studies that compared 1-year therapy with infliximab and etanercept showed no significant changes in lipid levels with etanercept therapy, whereas triglyceride levels were significantly elevated after 1-year infliximab therapy (Garcês et al., 2008; Lewicki et al., 2008). Studies will continue to determine whether suppression of inflammation with specific TNF-α blockers contributes to reduction in systemic risks.

## 15.6 RESISTIN

Resistin is a small cysteine-rich secretory protein of the FIZZ (found-in-inflammatory-zone) protein family. Resistin is expressed at high levels by murine adipocytes and acquired its name because mice treated with this adipose-specific secretory factor developed insulin resistance (Steppan et al., 2001). In mice, resistin regulates the expression of pro-inflammatory cytokines including IL-1, TNF-α, and IL-6. However, human adipocytes express only low levels of resistin, although resistin transcripts are higher in primary cultures of human undifferentiated pre-adipocytes. Human resistin may be derived mainly from macrophage and synoviocytes and its levels have been associated with inflammation and atherosclerosis (Ahima and Lazar, 2008).

Potential therapeutics applications related to resistin include resistin antagonists and anti-resistin antibodies for treatment of autoimmune inflammatory disorders such as rheumatoid arthritis, osteoarthritis, and other fibrotic diseases (Baumeister, 2008). Concentrations of resistin in rheumatoid arthritis have been observed to be 10-fold higher than in osteoarthritis and positive correlations have been found between resistin levels and systemic inflammatory markers in both osteoarthritis and rheumatoid arthritis patients (Schäffler et al., 2003). The source of this resistin may be activated macrophage-like cells in synovium (Baumeister, 2008). Thus, avenues of potential therapy may be found in the development of neutralizing antibodies reactive with resistin or direct antagonists to the resistin receptor.

## 15.7 NUTRIENTS: VITAMIN D, VITAMIN A, AND PUFAS

Dietary factors that regulate adipocyte differentiation may play a significant role in the expansion of adipose tissue, inflammation and the development of obesity. Dietary components including the fat-soluble vitamins and their metabolites have been studied as regulators of differentiation in adipose tissue.

1,25-Dihydroxyvitamin $D_3$ inhibits adipocyte proliferation in porcine systems and this is associated with increased cellular differentiation and apoptosis (Zhuang et al., 2007). Vitamin D treatment increased the expression of markers of terminal adipocyte differentiations including lipoprotein lipase, stearoyl-CoA desaturase, phosphoenolpyruvate carboxykinase, glycerol-3-phosphate dehydrogenase, and glucose transporter- 4. Thus, this action of vitamin D may decrease inflammation that may be associated with adipose tissue expansion.

Vitamin A has also been shown to have an effect on adipocyte differentiation. In mice, retinoic acid, the oxidized product of retinol (vitamin A), has been shown to decrease the content of white adipose tissue and increase that of brown adipose tissue, resulting in an overall decrease in body weight. Adipocytes were reduced in size and the number of multilocular adipocytes was increased in adipose tissues of treated mice.

In 2004, the US Food and Drug Administration gave "qualified health claim" status to the ω-n-3 polyunsaturated fatty acids (PUFAs) known as eicosapentaenoic acid and docosohexaenoic acid, stating that research shows that consumption of these fatty acids may reduce the risk of coronary heart disease. In experimental animal systems, diets rich in PUFAs, in particular ω-n-3 PUFAs, decrease adipose tissue mass and suppress development of obesity in animal models (Madsen et al., 2005). PUFAs induce adipocyte differentiation and activate anti-inflammatory PPARγ in pre-adipocytes. Thus, these actions may underlie the protection that diets rich in PUFAs lend against cardiovascular disease (Sacks and Campos, 2006).

## 15.8 CONCLUSIONS

Adipocytes are active endocrine organs that secrete a variety of adipokines that affect inflammation and regulate the susceptibility and progression of clinically important metabolic diseases (Figure 15.1). The study of adipose tissue function as it relates to local and systemic inflammation will hopefully identify lifestyle changes along with

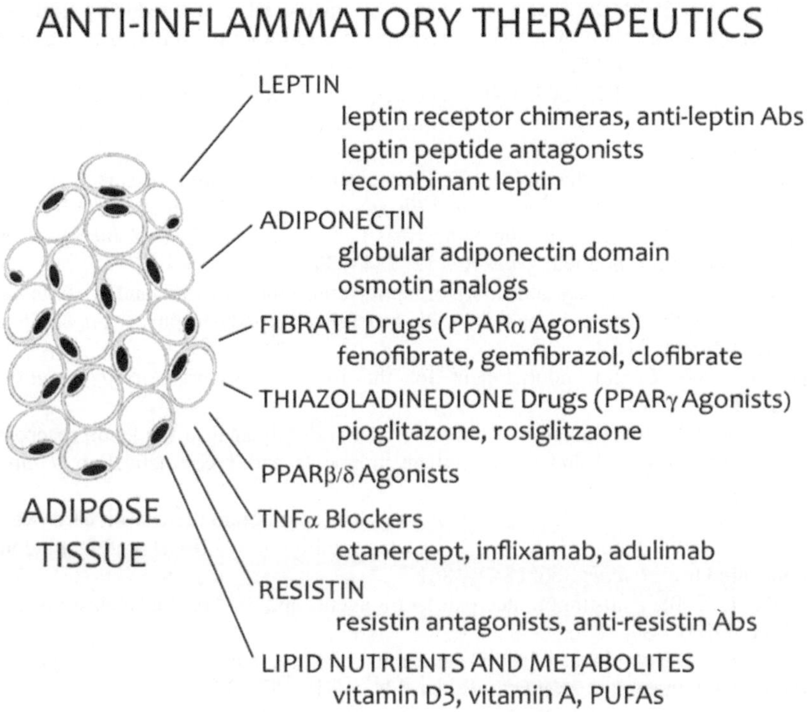


**FIGURE 15.1** Adipose tissue anti-inflammatory therapeutic agents including FDA-approved drugs, investigational drugs, and lipid nutrients have the potential to inhibit inflammation associated with obesity. Leptin inhibitors and adiponectin analogs inhibit the secretion of inflammatory cytokines and are under investigation for use in diabetes and metabolic syndrome, disorders associated with obesity, and chronic low-grade inflammation. PPAR-activating drugs have been used clinically since 1999 for diabetes therapy to improve insulin sensitivity while also activating anti-inflammatory targets. TNF-α blockers have been FDA-approved for over 10 years and are directly anti-inflammatory. Blockers of resistin action have the potential to sensitize tissues to insulin and reduce levels of pro-inflammatory cytokines. Dietary PUFAs and lipid-soluble vitamins $D_3$ and A activate anti-inflammatory signals, suppress obesity, and may reduce the incidence of coronary heart disease.

potential therapeutic targets that will combat inflammatory diseases such as obesity, cardiovascular disease, and related disorders.

## REFERENCES

Ahima RS and Lazar MA. 2008. Adipokines and the peripheral and neural control of energy balance. *Mol Endocrinol* 22: 1023–1031.

Barish GD, Narkar VA, and Evans RM. 2006. PPARγ: a dagger in the heart of the metabolic syndrome. *J Clin Invest* 116: 590–597.

Baumeister K and Johnson & Johnson. 2008. Resistin antagonists and their use. World Intellectual Property Organization, WO/2008/052111.

BIP Study Group. 2000. Secondary prevention by raising HDL and reducing triglycerides in patients with coronary artery disease: the benzafibrate infarction prevention study. *Circulation* 102: 21–27.

Cawthorn WP and Sethi JK. 2008. TNF-α and adipocyte biology. *FEBS Lett* 582: 117–131.

de Luca C and Olefsky JM. 2008. Inflammation and insulin resistance. *FEBS Lett* 582: 97–105.

De Taeye BM et al. 2007. Macrophage TNF-α contributes to insulin resistance and hepatic steatosis in diet-induced obesity. *Am J Physiol Endocrinol Metab* 293: E713–E725.

Di Rocco P et al. 2004. Lowered tumor necrosis factor receptors, but not increased insulin sensitivity, with infliximab. *Obes Res* 12: 734–739.

Ebihara K et al., 2007. Efficacy and safety of leptin-replacement therapy and possible mechanisms of leptin actions in patients with generalized lipodystrophy. *J Clin Endocrinol Metab* 92: 532–541.

Fantuzzi G. 2006. Leptin: nourishment for the immune system. *Eur J Immunol* 36: 3101–3104.

Fazeli M et al. 2006. Identification of a monoclonal antibody against the leptin receptor that acts as an antagonist and blocks human monocyte and T cell activation. *J Immunol Methods* 312: 190–200.

FIELD Study Investigators. 2005. Effects of long-term fenofibrate therapy on cardiovascular events in 9795 people with type 2 diabetes mellitus (the FIELD study): randomised controlled trial. *Lancet* 366: 1849–1861.

Garcês SP et al. 2008. Anti-tumour necrosis factor agents and lipid profile: a class effect? *Ann Rheum Dis* 67: 895–896.

Gertler A. 2006. Development of leptin antagonists and their potential use in experimental biology and medicine. *Trends Endocrinol Metab* 17: 372–378.

Gonzalez-Gay MA et al. 2008. High-grade inflammation, circulating adiponectin concentrations and cardiovascular risk factors in severe rheumatoid arthritis. *Clin Exp Rheumatol* 26: 596–603.

Guerre-Millo M. 2008. Adiponectin: and update. *Diabetes Metab* 34: 12–18.

Härle P and Straub RH. 2006. Leptin is a link between adipose tissue and inflammation. *Ann NY Acad Sci* 1069: 454–462.

Hiugi A et al. 2007. Effects of peroxisome proliferator-activated receptor ligands, bezafibrate, on adiponectin levels. *Arteriosclerosis, Thromb Vasc Biol* 27: 635–642.

Huvers F et al. 2007. Improved insulin sensitivity by ant-TNF-α antibody treatment in patients with rheumatoid diseases. *Ann Rheum Dis* 66: 558–559.

Javor ED et al. 2005. Long-term efficacy of leptin replacement in patients with generalized lipodystrophy. *Diabetes* 54: 1994–2005.

Kadowaki T and Yamauchi T. 2005. Adiponectin and adiponectin receptors. *Endocrine Rev* 26: 439–451.

Kadowaki T, Yamauchi T, and Kubota N. 2008. The physiological and pathophysiological role of adiponectin and adiponectin receptors in the peripheral tissues and CNS. *FEBS Lett* 582: 74–80.

Kilgore KS and Billin AN. 2008. PPARβ/δa ligands as modulators of the inflammatory response. *Curr Opin Investig Drugs* 9: 463–469.

Lehrke M and Lazar MA. 2005. The many faces of PPARγ. *Cell* 123: 993–999.

Lewicki M, Kotyla P, and Kucharz E. 2008a. Etanercept increases adiponectin level in women with rheumatoid arthritis. *Clin Rheumatol* 27: 1337–1338.

Lewicki M et al. 2008b. The lipids after one year therapy with infliximab and etanercept in woman with rheumatoid arthritis. *Ann Rheum Dis* Online 4 Dec.

Li J, Li F, and Zhao A. 2006. Inflammation and leptin. *Drug Disc Today* 3: 387–393.

Madsen L, Petersen RK, and Kristiansen K. 2005. Regulation of adipocyte differentiation and function by polyunsaturated fatty acids. *Biochim Biophys Acta* 1740: 266–286.

Matarese G, Moschos S, and Mantzoros CS. 2005. Leptin in immunology. *J Immunol* 173: 3137–3142.

Matarese G, Procaccine C, and De Rosa V. 2008. The intricate interface between immune and metabolic regulation: a role for leptin in the pathogenesis of multiple sclerosis? *J Leuk Biol* 84: 893–899.

Matsusue K, Peters JM, and Gonzalez FJ. 2004. PPARβ/δ potentiates PPARγ-stimulated adipocyte differentiation. *FASEB J* 18: 1477–1479.

Moraes LA, Piqueras L, and Bishop-Bailey D. 2006. Peroxisome proliferator-activated receptors and inflammation. *Pharmacol Ther* 110: 371–385.

Motoshima H, Wu X, Mahadev K, and Goldstein BJ. 2004. Adiponectin suppresses proliferation and superoxide generation and enhances eNOS activity in endothelial cells treated with oxidized LDL. *Biochem Biophys Res Commun* 315: 264–271.

Nissen SE and Wolski K. 2007. Effect of rosiglitazone on the risk of myocardial infarction and death from cardiovascular causes. *New Engl J Med* 356: 2457–2471.

Ouedraogo R et al. 2007. Adiponectin deficiency increases leukocyte–endothelium interactions via upregulation of endothelial cell adhesion molecules. *J Clin Invest* 117 :1718–1726.

Popa C et al. 2007. Modulation of lipoprotein plasma concentrations during long-term anti-TNF- therapy in patients with active rheumatoid arthritis. *Ann Rheum Dis* 66: 1503–1507.

Rodriguez-Calvo R et al. 2008. Activation of peroxisome proliferator-activated receptor β/γ inhibits lipopolysaccharide-induced cytokine production in adipocytes. *Diabetes* 57: 2149–2157.

Rogler G. 2006. Significance of anti-inflammatory effects of PPARγ agonists? *Gut* 55: 1067–1069.

Sacks FM and Campos H. 2006. Editorial: Polyunsaturated fatty acids, inflammation, and cardiovascular disease: time to widen our view of the mechanisms. *J Clin Endocrinol Metab* 91: 398–400.

Schäffler A et al. 2003. Adipokines in synovial fluids. *JAMA* 290: 1709–1710.

Shah A, Mehta N, and Reilly MP. 2008. Adipose inflammation, insulin resistance, and cardiovascular disease. *J Parent Ent Nutr* 32: 638–644.

Spiegelman BM et al. 1997. PPARγ and the control of adipogenesis. *Biochemie* 79: 111–112.

Steppan CM et al. 2001. The hormone resistin links obesity to diabetes. *Nature* 409: 307–312.

Tenenbaum A et al. 2004. Peroxisome proliferator-activated receptor ligand bezafibrate for prevention of type 2 diabetes mellitus in patients with coronary artery disease. *Circulation* 109: 2197–2202.

Tenenbaum A et al. 2005. Effect of bezafibrate in incidence of type 2 diabetes mellitus in obese patients. *Eur Heart J* 26: 2032–2038.

Teoh H et al. 2008. Adiponectin deficiency promotes endothelial activation and profoundly exacerbates sepsis-related mortality. *Am J Physiol Endocrinol Metab* 295: E658–E664.

Uysal KT et al. 1997. Protection from obesity-induced insulin resistance in mice lacking TNF-α function. *Nature* 389: 610–614.

Ventre J et al. 1997. Targeted disruption of the tumor necrosis-α gene: metabolic consequences in obese and nonobese mice. *Diabetes* 46: 1526–1531.

Wang Y-X et al. 2003. Peroxisome proliferator-activated receptor δ activates fat metabolism to prevent obesity. *Cell* 113: 159–170.

Weisberg S et al. 2003. Obesity is associated with macrophage accumulation in adipose tissue. *J Clin Invest* 112: 1796–1808.

www.amylin.com/pipeline/pramlintide.cfm. 2007. Pramlintide and leptin combination treatment demonstrates 12.7% weight loss in 24-week clinical obesity study.

Xu H et al. 2002. Altered tumor necrosis factor-α processing in adipocytes and increased expression of transmembrane TNF-α in obesity. *Diabetes* 51: 1876–1883.

Yki-Järvinen H. 2004. Thiazolidinediones. *New Engl J Med* 351:1106-1118.

Zhao S-P and Wu J. 2004. Fenfibrate reduces tumor necrosis factor-α serum concentration and adipocyte secretion of hypercholesterolemic rabbits. *Clin Chim Acta* 347: 145–150.
Zhuang H, Lin Y, and Yang G.2007. Effects of 1,25-dihydroxyvitamin $D_3$ on proliferation and differentiation of porcine pre-adipocytes in vitro. *Chem Biol Interact* 170: 114–123.

# 16 Conclusions and Future Directions

*Peter G. Bradford and Atif B. Awad*

## CONTENTS



## 16.1 CONCLUSIONS

Perhaps the principal conclusion to be drawn from *Adipose Tissue and Inflammation* is that adipose tissue functions as an essential endocrine organ that produces and secretes an identifiable set of specific adipokines and that this adipokine production and secretion arise from varying physiologic and pathologic states. In states of obesity, the production profile of these adipokines is altered to one that promotes low-grade systemic inflammation. Recent research implicates overnutrition and obesity as inducers of inflammatory responses in adipose and peripheral tissues, causing or contributing to those metabolic defects that underlie type 2 diabetes, hypertension, and cardiovascular diseases.

The most productive research over the last 10 years has relied upon the concept of the "adipose organ" as a discrete anatomic and endocrine entity that produces specific adipokines that regulate essential functions such as glucose and lipid metabolism, blood pressure, steroid hormone modulation, and inflammation. The complexity of this adipose organ stems in part from the fact that in addition to adipocytes, adipose tissue contains non-adipocyte cell types (endothelial, smooth muscle, and fibroblastic cells) and stromal components (monocytes, macrophages, and preadipocytes). The concept of adipokines as biologically active substances secreted by adipose tissue must be broadened to include substances secreted by all cell types within the adipose tissue. It is clear from the material presented here in *Adipose Tissue and Inflammation* that these adipokines exert local autocrine and paracrine actions and that these actions affect adiposity, adipocyte metabolism, and inflammatory responses within adipose tissue but also have important roles in the regulation of overall energy metabolism and inflammation through systemic actions in the brain, liver, muscle, and other sites.

Specific cytokines, adipokines, and transcription factors have been intimately and causally linked to adipose tissue inflammation and the systemic inflammation associated with obesity. In 1993 Hotamisligil and colleagues first described the increased expression of TNF-α in adipose tissue and showed that this secreted TNF-α interfered with the action of insulin.

Now a host of adipose tissue hormones and mediators have been shown to regulate local and systemic tissue inflammation and this regulation has been discovered to be mediated by the actions of classic inflammatory signal transducers, kinases, and transcription factors, specifically c-Jun N-terminal kinase (JNK), the inhibitor of κB kinase-β (IKKβ), and nuclear factor NF-κB. Overnutrition induces these inflammatory mediators in adipose tissue and in central nervous system target sites that control the set points of nutritional balance and feeding behavior. These pro-inflammatory adipokines include TNF-α as well as leptin, resistin, C-reactive protein, IL-1, IL-6, and others. All these adipokines are increased in obesity and evidence suggests their causative involvement in obesity-linked metabolic and cardiovascular diseases.

On the other hand, the adipose-specific adiponectin cytokine exhibits anti-inflammatory activity and is substantially reduced in obesity, suggesting it may play a protective role by apposing the development of obesity-related diseases. These are simplistic summaries of the activities of adipose tissue and secreted adipokines. Fortunately, we have in-depth contributions from worldwide experts who detail the normal physiology and complexities of action of hormones and adipokines that affect adipose tissue and the roles that these agents play in obesity as they promote low-grade local and systemic inflammation.

In *Adipose Tissue and Inflammation,* we see how local and systemic inflammation is affected by hormonal factors including insulin, growth hormone, glucocorticoids, and prostaglandins; by dietary factors including fatty acids, polyphenols, phytosterols, phytoestrogens, and antioxidants; by lifestyle changes involving diet, exercise, and weight loss; and by new and investigative advances in pharmacotherapy. Thus, the second principal conclusion that is apparent from these analyses is that the inflammation associated with obesity and present within adipose tissue may be regulated by a host of controllable factors such as hormone levels, dietary antioxidant polyphenols and phytosterols, caloric content of diet, exercise, and targeted pharmacotherapy.

## 16.2 FUTURE DIRECTIONS

Many truly exciting scientific areas seem to represent the forefronts for future research and development in the area of adipose tissue and inflammation. These include: (1) genetic studies to identify obesity-associated and obesity–inflammation linkage genes, (2) investigations into whether dietary restrictions will improve measures of systemic inflammation and reduce the incidence of associated diseases, (3) development of anti-inflammatory therapeutics as means for reducing the morbidity and mortality associated with chronic diseases such as type 2 diabetes, hypertension, and cardiovascular disease, and (4) investigations into controlling oxidative stress as a potential means to affect obesity-associated maladies.

### 16.2.1 Genetic Studies to Identify Obesity-Associated Genes

In addition to body mass index (BMI), lifestyle, and family history, recent linkage analyses have identified 11 to 18 genes that may predict the likelihood of developing type 2 diabetes. Despite these advances in genome-wide association studies, it is likely that each of these genes contributes very modestly to the overall variance of diabetes risk. The case for identifying genetic variants affecting adiposity is even more complex given the interactions of nature and nurture in the outcome. Despite this, twin studies show that 40 to 70% of BMI variation is inherited and recent studies have shown that systems in the brain that control satiety show variances linked to obesity. To add to the significance of these findings, three large cohort studies involving 25,000 to 30,000 subjects used obesity-specific phenotyping and genome-wide analyses to identify 15 additional loci linked to BMI. Questions remain but as genome-wide association studies grow larger, the power of association will increase and more definite answers will ensue.

### 16.2.2 Dietary Restriction Studies

Calorie-restricted diets have been shown to reduce pancreatic inflammation and the incidence of cancer. As adipose tissue is a major source of inflammatory cytokines, it is not unexpected that calorie restriction influences the levels of systemic and local inflammatory mediators derived from adipose tissue. For example, high-fat diets and diets rich in saturated fatty acids such as palmitate induce sustained JNK1 activation, one of the essential signal transducers associated with chronic low-grade inflammation as seen in obesity. Efforts to understand the complex interactions among specific dietary components, obesity, and inflammation should provide immediate health benefit in terms of readily translatable diet and lifestyle changes.

### 16.2.3 Anti-Inflammatory Therapeutics for Chronic Systemic Diseases

Anti-inflammatory therapeutics represents an emerging area for treatment of chronic diseases such as type 2 diabetes, hypertension, and cardiovascular disease. The link established now between adipose tissue and inflammation suggests that therapeutics targeted to reduce adipose inflammation will provide relief for those diseases. The science outlined in this book highlights several areas for therapeutic development. Drugs and pharmaconutrients that affect the synthesis and actions of adipokines are in development or currently used to treat systemic inflammation. TNF-α blockers are already widely prescribed for their anti-inflammatory activities, fibrate and thiazolidinedione-based drugs are marketed for their lipid lowering and anti-diabetic activities. However, both also exert potent anti-inflammatory effects through direct effects on adipose tissue to enhance adiponectin secretion and to repress the pro-inflammatory actions of adipose tissue macrophages. Adiponectin-based drugs are under investigation both for their ability to suppress inflammation and to counter the metabolic disorders associated with type 2 diabetes, atherosclerosis, and obesity. IL-1 receptor antagonist drugs are FDA-approved anti-inflammatory agents that clearly decrease systemic inflammation in autoimmune diseases like rheumatoid arthritis. Advanced clinical trials have now shown similar anti-inflammatory activities along with improved glycemia in type 2 diabetic patients. Since adipose tissue is an essential source of inflammatory cytokines, it will be beneficial to assess effects of these broadly anti-inflammatory agents on adipose function in obesity.

### 16.2.4 Role of Oxidative Stress in Obesity-Associated Morbidities

Oxidative stress is a contributory pathogenic factor in obesity-related disorders. Obesity is associated with increased free radical production, depletion of cellular antioxidants, and an overall increase in oxidative stress. An earlier book in the CRC Press–Taylor & Francis series edited by Lester Packer and Helmut Sies was dedicated entirely to this topic and titled *Oxidative Stress and Inflammatory Mechanisms in Obesity, Diabetes, and the Metabolic Syndrome.* The realization that adipose tissue is a central source of local and systemic inflammatory mediators has made it the focus of intense investigation in recent years. Reactive oxygen and nitrogen species are produced in adipose tissue under conditions of oxidative and metabolic stress. The role of these reactive species in contributing to the production and secretion of pro-inflammatory adipokines and how this may affect obesity-associated morbidities must be investigated.

*Adipose Tissue and Inflammation* focuses on the contributions of adipose tissue to local and systemic inflammation. Adipose tissue produces a multitude of metabolic and hormonal signals that affect systemic endocrine equilibrium. Through the investigations reported here, we now appreciate that in states of obesity, cells within adipose tissue exhibit an inflammatory capacity and secrete a distinct profile of adipokines that contribute to the pathogenesis of diabetes, hypertension, and cardiovascular disease.

# Index

## S



## T



## U



## V



## W